# Thermal Engineering in Power Systems

# International Series on Developments in Heat Transfer

## Objectives

The Developments in Heat Transfer book Series publishes state-of-the-art books and provides valuable contributions to the literature in the field of heat transfer. The overall aim of the Series is to bring to the attention of the international community recent advances in heat transfer by authors in academic research and the engineering industry.

Research and development in heat transfer is of significant importance to many branches of technology, not least in energy technology. Developments include new, efficient heat exchangers, novel heat transfer equipment as well as the introduction of systems of heat exchangers in industrial processes. Application areas include heat recovery in the chemical and process industries, and buildings and dwelling houses where heat transfer plays a major role. Heat exchange combined with heat storage is also a methodology for improving the energy efficiency in industry, while cooling in gas turbine systems and combustion engines is another important area of heat transfer research.

To progress developments within the field both basic and applied research is needed. Advances in numerical solution methods of partial differential equations, high-speed, efficient and cheap computers, advanced experimental methods using LDV (laser-doppler-velocimetry), PIV (particle-image-velocimetry) and image processing of thermal pictures of liquid crystals, have all led to dramatic advances during recent years in the solution and investigation of complex problems within the field.

The aims of the Series are achieved by contributions to the volumes from invited authors only. This is backed by an internationally recognised Editorial Board for the Series who represent much of the active research worldwide. Volumes planned for the series include the following topics: Compact Heat Exchangers, Engineering Heat Transfer Phenomena, Fins and Fin Systems, Condensation, Materials Processing, Gas Turbine Cooling, Electronics Cooling, Combustion-Related Heat Transfer, Heat Transfer in Gas-Solid Flows, Thermal Radiation, the Boundary Element Method in Heat Transfer, Phase Change Problems, Heat Transfer in Micro-Devices, Plate-and-Frame Heat Exchangers, Turbulent Convective Heat Transfer in Ducts, Enhancement of Heat Transfer and other selected topics.

# Thermal Engineering in Power Systems

Edited by:

**R.S. Amano**
*University of Wisconsin-Milwaukee, USA*

**&**

**B. Sundén**
*Lund University, Sweden*

**WIT***PRESS* **Southampton, Boston**

**R.S. Amano**
*University of Wisconsin-Milwaukee, USA*

**B. Sundén**
*Lund University, Sweden*

Published by

**WIT Press**
Ashurst Lodge, Ashurst, Southampton, SO40 7AA, UK
Tel: 44 (0) 238 029 3223; Fax: 44 (0) 238 029 2853
E-Mail: witpress@witpress.com
http://www.witpress.com

For USA, Canada and Mexico

**WIT Press**
25 Bridge Street, Billerica, MA 01821, USA
Tel: 978 667 5841; Fax: 978 667 7582
E-Mail: infousa@witpress.com
http://www.witpress.com

British Library Cataloguing-in-Publication Data

A Catalogue record for this book is available
from the British Library

ISBN: 978-1-84564-062-0
ISSN: 1369-7331

Library of Congress Catalog Card Number: 2008921657

*The texts of the papers in this volume were set individually by the authors or under their supervision.*



Printed in Great Britain by Athenaeum Press Ltd.

# Contents

**Chapter 6**
**Modern CFD application on aerothermal engineering aspects of natural draft cooling towers** ........ **171**
*D. Bohn & K. Kusterer*



**Chapter 7**
**Innovative gas turbine cooling techniques** ........ **199**
*R.S. Bunker*



**Chapter 8**
**Hot gas path heat transfer characteristics/active cooling of turbine components** ........ **231**
*T. Simon & J. Piggush*

**Chapter 11**

# Preface

This book focuses on thermal engineering topics for power systems. The topic ranges over all power systems that include fossil power plants, associated environmental cleaning methods, safety of nuclear power plants, development of virtual engineering design environments, cooling technologies and associated problems related to high-temperature gas turbine blades, highly efficient compressors, steam turbines, and hybrid cars. The objective is to present the current state-of-the-art technologies in various power systems such as power plants, aero-jet engines, utilities, thermal management, and safety issues in nuclear power plants.

The first chapter presents the relevance of heat transfer and heat exchangers for development of sustainable energy systems. The authors use numerous research examples and demonstrate why heat transfer and heat exchangers are important in the development of sustainable energy systems. The key factors necessary to develop sustainable energy systems are discussed.

The second chapter deals with advanced technologies for clean and efficient energy conversion in power systems. The main focus is an efficient fossil energy use with low pollution in all kinds of power plants, in all industrial sectors. The use of energy by many countries is growing steadily worldwide due to the desire for higher standards of living and increased productivity. Efficient energy use is favorable for better productivity, product quality, costs, and quality of human life, but energy use adversely impacts our environment.

The third chapter describes virtual engineering and the design of power systems. This chapter examines how these virtual engineering design environments will be created and utilized to build a new generation of power plants. Examples of how virtual engineering is currently being used in the power industry are given. In addition, the US Department of Energy's use of virtual engineering in the development of near-zero-emission power plants is discussed.

The fourth chapter focuses mainly on steam turbine power systems. The emphasis is on efficient fossil energy utilization in power generation together with low pollution issues in conventional thermal power plants. This chapter briefly highlights the fouling problem in power plant water walls and proposes a monitoring, inspection and maintenance schedule. The information provides a quick guide to the commonly faced operation problems and methods to enhance energy conversion efficiency.

The fifth chapter presents enhancement of nuclear power plant safety by condensation-driven passive heat removal systems. This chapter discusses the numerous ways in which passive condensation heat transfer enhances nuclear power plant safety in current and future nuclear power plants. The discussion extends to US-design light water reactors and US reactor safety codes, although there are many commonalities to reactors of other designs. The physical phenomena are described and the state-of-the art in analysis methods are presented. Challenges for improved analysis are summarized.

The sixth chapter presents modern CFD application on aero-thermal engineering aspects of natural draft cooling towers. The chapter demonstrates various designs of cooling towers using computer simulation technology.

The seventh and eighth chapters present more on the detailed technology development of gas turbine blade cooling systems. Both these chapters give better understanding of the physics of engine heat transfer by developing from simple ideas to complex heat transfer phenomena. The authors address how the engineer applies heat transfer tools available in the literature to support designs which will advance engine life and enhance machine efficiencies. If successful, the newcomer will establish a foothold in the technology and the more experienced engineer will be reminded of some basic concepts.

The ninth and tenth chapters cover more detailed studies on compressors for both industrial usage and aircraft engines. Whereas Chapter seven presents an overview and details of blade designs for both axial and centrifugal compressors and Chapter 8 focus mainly on centrifugal compressors, these two chapters discuss the developments for better understanding of the flow in an impeller and the contributions of this knowledge towards better and advanced impeller designs.

The final chapter presents the current development of hybrid cars. This chapter summarizes modern hybrid car systems and their thermal management issues. These thermal issues are important since commercial hybrid cars employ gasoline/diesel-electric systems, which are the major factors for improving tank to wheel efficiencies.

All of the chapters follow a unified outline and presentation to aid accessibility and the book provides invaluable information to both graduate researchers and industrial research engineers/scientists.

We are grateful to the authors and reviewers for their excellent contributions. We also thank the ceaseless help provided by the staff members of WIT Press, in particular Mr Brian Privett and Mrs Elizabeth Cherry, for their strong encouragement in the production of this book. Finally our appreciation goes to Professor Carlos Brebbia who gave us support and encouragement to complete this project.

**Ryoichi S. Amano and Bengt Sundén**
2008

# CHAPTER 1

# Relevance of heat transfer and heat exchangers for the development of sustainable energy systems

B. Sundén[1] & L. Wang[2]
[1]*Division of Heat Transfer, Department of Energy Sciences, Lund University, Lund, Sweden.*
[2]*Siemens Industrial Turbines, Finspong, Sweden.*

## Abstract

There are many reasons why heat transfer and heat exchangers play a key role in the development of sustainable energy systems as well as in the reduction of emissions and pollutants. In general, all attempts to achieve processes and thermodynamic cycles with high efficiency, low emissions and low costs include heat transfer and heat exchangers to a large extent. It is known that sustainable energy development can be achieved by three parallel approaches: by reducing final energy consumption, by improving overall conversion efficiency and by making use of renewable energy sources. In these three areas, it is important to apply advanced heat transfer and heat exchanger technologies, which are explained extensively in this chapter. In addition, heat transfer and heat exchangers are important in protecting the environment by reducing emissions and pollutants. To illustrate this, several research examples from our group are used to demonstrate why heat transfer and heat exchangers are important in the development of sustainable energy systems. It can be concluded that the attempt to provide efficient, compact and cheap heat transfer methods and heat exchangers is a real challenge for research. To achieve this, both theoretical and experimental investigations must be conducted, and modern techniques must be adopted.

## 1 Introduction

The concept of sustainable development dates back to several decades, and it was brought to the international agenda by the World Commission on Environment and

Development in 1987 [1]. At the same time, it also provided the henceforth most commonly used definition of sustainable development, describing it as development which meets the needs of the present without comprising the ability of future generations to meet their own needs. This concept has indeed expressed people's concern about the growing fragility of the earth's life support systems, i.e. the use of the available resources on our planet. Among the aspects concerned, energy is certainly a very important part, and sustainable energy systems have become the worldwide concern among scientific and political communities as well as among ordinary people.

Today, the production of electricity and heat is mainly based on finite primary energy sources. Fossil fuels are combusted in such large amounts that flue gas emissions have affected the environment, e.g. green house effect and toxic pollutants. A general approach to improve the degree of sustainability of the energy supply lies in the following three aspects: reducing final energy consumption, improving overall conversion efficiency and making use of renewable sources [2]. To reduce final energy consumption is an obvious approach, which requires more energy efficient process components and systems. The energy source requirement for the same energy output can be brought down by improving overall conversion efficiency. To use renewable energy sources other than fossil fuels, such as hydropower, biomass, wind and solar energy, is an attractive approach because they are sustainable in nature.

In all three aspects, it was found that heat transfer and heat exchangers play an important role. For instance, increasing the efficiency in thermal processes for heat and power generation requires increasing the highest temperature in the process and it has to be increased further in the future. To enable the materials of the equipment, e.g. in gas turbine units, to withstand such high temperatures, cooling is needed. In this chapter, several examples will be illustrated to stress the importance of the relevant heat transfer and heat exchangers in the development of sustainable energy systems. Examples will also be given to illustrate that heat transfer and heat exchanger technologies can bring down the emissions of green house gases and other pollutants. It can be concluded that the attempt to provide efficient, compact and cheap heat transfer methods and heat exchangers is a real challenge for research, and that both theoretical and experimental investigations must be carried out and modern scientific techniques must be adopted to develop sustainable energy systems.

## 2 Reduction of energy consumption

The process industry remains one of the biggest sectors in consuming energy. A typical process, shown in Fig. 1, consists of three parts: chemical plant, utility plant and heat recovery network [3]. The purpose of the chemical plant is to produce products from raw materials with the supply of energy from both the utility plant and the heat recovery network. The utility plant produces power, hot utility and cold utility. The heat recovery network, which consists of many

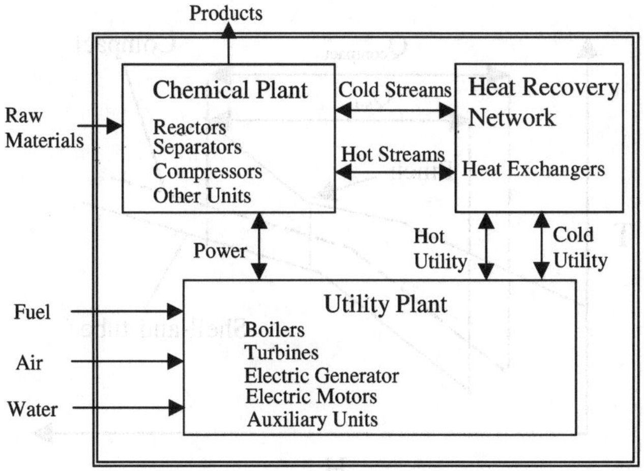


Figure 1: Total processing system.

heat exchangers, aims to recover heat from hot streams to heat cold streams. Maximizing heat recovery in the heat recovery network can bring down both energy consumption and consequently flue gas emission from the utility plant. Therefore, reduction in energy consumption requires the optimization of the heat recovery network, i.e. heat exchanger networks. Advanced heat exchanger technologies can improve the efficiency of heat exchanger networks. Such technologies include compact heat exchangers, multi-stream heat exchangers, heat transfer enhancement, mini- and micro-heat exchangers, etc. [4]. Using these technologies, current processes can be improved and the final energy demands can be reduced.

Conventional heat exchangers in process industries are shell-and-tube heat exchangers. There are several disadvantages in using such units, e.g. low ratio of surface to volume, tendency of severe fouling, use of multi-pass design, low efficiency due to a relatively high pressure drop per unit of heat transfer in the shell side, etc. Most of these disadvantages are due to the relatively large hydraulic diameter. To overcome these disadvantages, compact heat exchangers have been developed. A compact heat exchanger is one which incorporates a heat transfer surface with area density (or compactness) of above 700 $m^2/m^3$ on at least one of the fluid sides [5]. The common types include plate heat exchangers (PHEs), plate-fin heat exchangers, tube-fin heat exchangers, etc. In the process industries using compact heat exchangers, energy consumption can be reduced in addition to the reduced capital cost and complexity of the plant.

Compact heat exchangers usually have a small hydraulic diameter, which results in high heat transfer coefficients. This will reduce the unit size and weight, hence the unit capital cost. In addition, the high heat transfer coefficients permit compact heat exchangers to operate under conditions with small temperature differences. This is significant in the optimization of heat exchanger networks. In the pinch

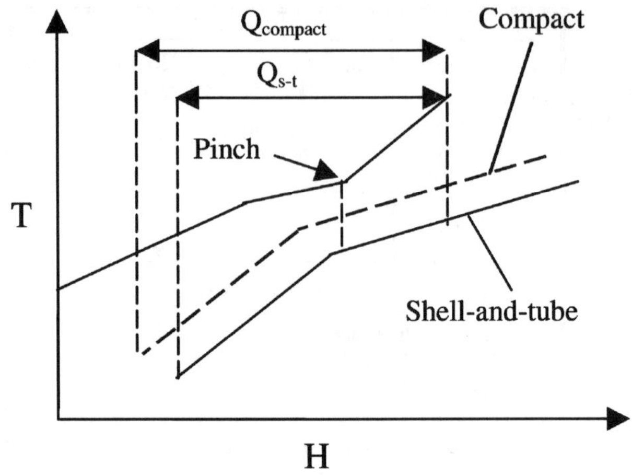


Figure 2: Composite curves.

analysis method for the design of heat exchanger networks [6], the minimum temperature difference is the decisive parameter to construct the so-called composite curves, which are shown schematically in Fig. 2. By using compact heat exchangers, the minimum temperature difference can be reduced significantly compared to shell-and-tube heat exchangers. This makes the two lines in the composite curves approach very close to each other, which means that the heat recovery is enlarged, and at the same time, the external utility requirements are reduced. Therefore, the utility consumption in the entire plant is reduced. Due to the high heat transfer coefficients and low unit capital costs, the total capital cost for the heat recovery system can still be lower than that using shell-and-tube heat exchangers.

A multi-stream heat exchanger is a good option, when too many heat exchanger units are required. In the optimization of heat exchanger networks using the pinch technology, a large number of exchangers are often required when the network is designed in terms of two-stream exchangers. This not only increases the capital cost but also increases the complexity of the network. Therefore, it may challenge the optimal solution, and relaxation has to be made. Using multi-stream heat exchangers might be a good way to circumvent this problem, and it offers a number of potential benefits including large savings in capital and installation costs, reduction in physical weight and space, better integration of the process, etc. However, the streams connected to them should not be too far away in physical space to save piping costs. Common multi-stream heat exchangers include multi-stream plate-fin heat exchangers, multi-stream PHEs, etc. [7].

Heat transfer enhancement for shell-and-tube heat exchangers should be also considered in the optimization of heat exchanger networks. It reduces the capital cost because of the small size needed for a given duty. It also reduces the temperature driving force, which reduces the entropy generation and increases the second law efficiency. In addition, heat transfer enhancement enables heat exchangers to operate at a smaller velocity but still achieve the same or even higher heat transfer coefficient. This means that a reduction in pressure drop, corresponding to less power utilization, may be achieved. All these advantages have made heat transfer enhancement technology attractive in heat exchanger applications. For the tube side, different geometries (e.g. low-finned tubes, twisted tubes, grooved tubes) and tube inserts (e.g. twist taped inserts, wire coil inserts, extended surface inserts) have been developed [8]. For the shell side, improvements have been also made, e.g. helical baffles and twisted tube heat exchangers [4].

More heat transfer and heat exchanger technologies are available to improve the process, and consequently to reduce the final energy consumption. These may include micro- and mini-heat exchangers, integrated chemical reactor heat exchangers, etc. Due to the space constraints in this chapter, these technologies are not explored in detail. However, the possibilities of their application in process industries should not be underestimated.

## 3 Improved efficiency of energy conversion

There are many ways to improve the efficiency of thermal power plants, but heat transfer and heat exchangers play a significant role in all means. This can be highlighted by considering as an example a power plant that uses gas turbines. The original Brayton cycle for the power plant only needs a compressor, a combustion chamber and a power turbine; this concept can be found in any textbook on thermodynamics, e.g. Cengel and Boles [9]. However, the thermal efficiency is usually very low in such systems, and improvements can be made by employing the concept of intercooling, recuperation (regeneration) and reheating. Such a flow sheet is illustrated in Fig. 3, and the corresponding thermodynamic cycle is shown in Fig. 4. Two stages of gas compression are provided to reduce the power consumption for compression due to the lower inlet temperature of the gas in the second compression stage by using an intercooler. Because the compression power required is reduced, the net power output is certainly increased. The concept of recuperation is the utilization of energy in the turbine exhaust gases to heat the air entering the combustion chamber, thus saving a certain amount of fuel in the combustion process. This will certainly increase the overall thermal efficiency as well. In addition, the turbine output can be increased by dividing the expansion into two or more stages, and reheating the gas to the maximum permissible temperature between the stages. Although the power output is improved, the cost of additional fuel will be heavy unless a heat exchanger is also employed. These concepts can be also seen in the thermodynamic cycle in Fig. 4. The cycle 1-2-3-4-1 corresponds to the simple Brayton cycle. The cycle 9-11-12-2 represents the intercooling and the cycle 15-14-13-4 represents the reheating. The cycles 4-7-12-5 and 4-6-2-5 represent recuperation in the case of intercooling and no intercooling, respectively. This concept has already been incorporated in some real gas turbines, e.g. LMS100 from GE makes use of an intercooler, Mercury 50 from Solar Turbine makes use of a recuperator and GT24/26 from ASLTOM uses sequential combustion. These features significantly increase the efficiency of gas turbines, and a great deal of work has been done for the design of reliable heat exchangers that are operated at higher temperatures.

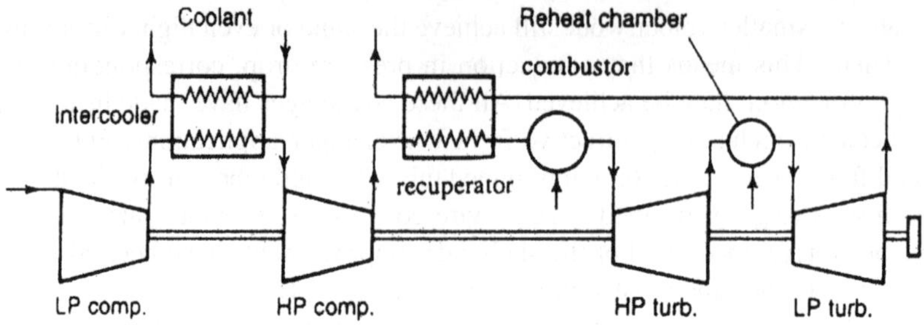


Figure 3: Power plant with intercooler, recuperator and reheater.

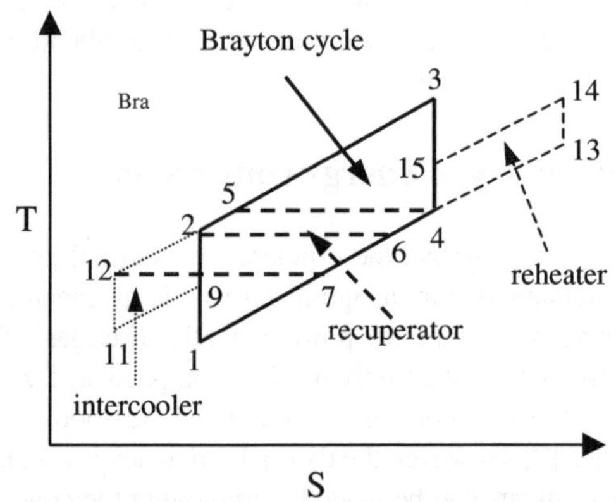


Figure 4: Thermodynamic cycle of a gas turbine power plant with intercooler, recuperator and reheater.

The thermal efficiency and power output of gas turbines will increase with increasing turbine rotor inlet temperature, which corresponds to the temperature at point 3 in Fig. 4. This is the reason why modern advanced gas turbine engines operate at high temperatures (ISO turbine inlet temperature in the range of 1200–1400°C), and the trend is to operate at even higher temperatures. To enable this, in addition to material innovation cooling technologies must be developed for the combustion chamber, turbine blade, guide vane, etc. Over the years, film cooling, convection cooling and impingement cooling have been developed for both combustion chamber (see Fig. 5) and turbine blade (see Fig. 6), and the technique of transpiration cooling is still under development due to engineering difficulties. In addition, more advanced high temperature materials such as superalloys of single crystals and ceramic coating significantly contribute to the high turbine inlet temperature operation. With these advanced cooling technologies, reliable and high-efficiency power plants can be sustained.

The above blade cooling technologies are for air-cooled gas turbines, but a new turbine cooling concept is available, i.e. steam cooled gas turbines. Steam provides several benefits over air as a cooling medium. First, steam provides higher heat

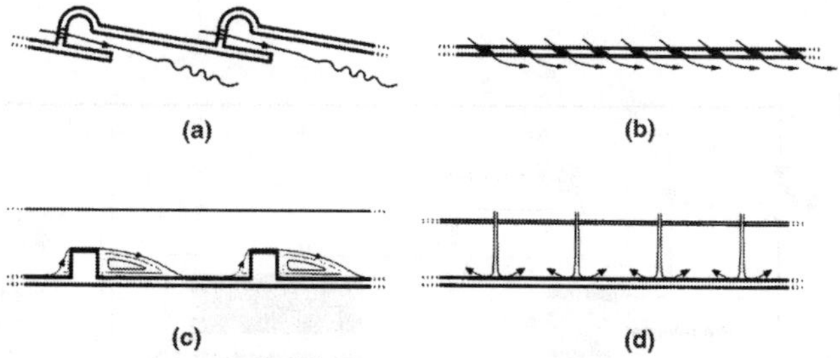


Figure 5: Cooling concepts of combustion chamber: (a) Film cooling; (b) Transpiration cooling; (c) Enhanced convective cooling; (d) Impingement cooling.

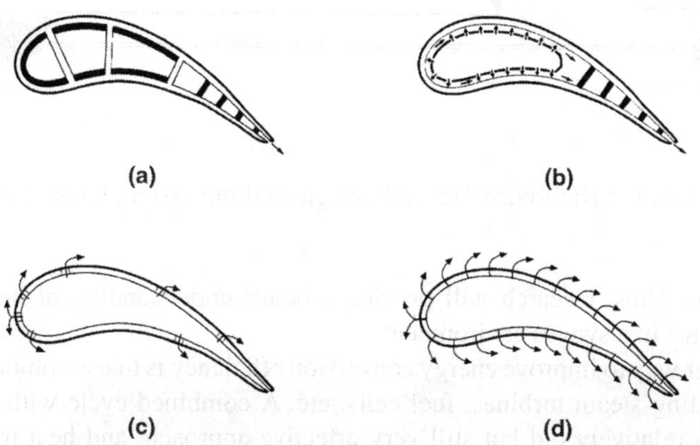


Figure 6: Cooling concepts of gas turbine blade: (a) convection cooling; (b) impingement cooling; (c) film cooling; (d) transpiration cooling.

transfer characteristics because its heat capacity is higher than that of air. Second, the use of steam as a cooling medium reduces the use of cooling air, which means that more cooling air is available for the combustion process, which contributes to improving emissions. Third, reduction in cooling air results in less temperature dilution of the hot gas caused, while mixing with the cooling air. This increases the turbine inlet temperature, which results in more power availability. Finally, no ejection of cooling air to the main gas flow means aerodynamic loss is minimized. With this technology, the efficiency of the gas turbine is greatly enhanced; the best example of this is GE's H class gas turbine, which is the first gas turbine to achieve 60% efficiency in the combined cycle power plants. However, to design such turbines, the heat transfer characteristics of steam as a cooling medium must be thoroughly understood, which requires extensive research. For the gas side, the use of different fuels can lead to a significant change in properties for the gas in the turbine part. The Integrated Gasification Combined Cycle application operates on hydrogen, and consequently the syngas will increase the amount of heat transferred to rotating and stationary airfoils due to increased moisture content and

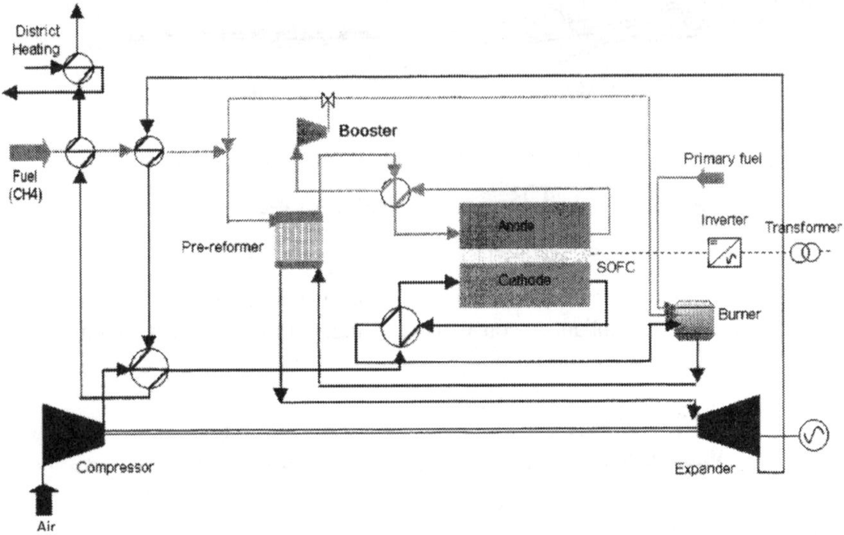


Figure 7: Reference fuel cell and gas turbine system layout [11].

mass flow. Thus, research will provide a better understanding of heat transfer mechanisms in a syngas environment.

Another way to improve energy conversion efficiency is to use combined cycles incorporating steam turbines, fuel cells, etc. A combined cycle with steam turbines is a relatively old but still very effective approach, and heat transfer and heat exchangers play a significant role in this approach without any doubt. Here, a brief discussion is given for the heat transfer issues associated with fuel cells. Figure 7 shows a typical configuration for a combined cycle using both a gas turbine and a fuel cell. As is well known, fuel cells can convert the chemical energy stored in the fuel into electrical and thermal energy through electrochemical processes. Because these processes are not subject to the Carnot cycle limitation, high electrical efficiencies can be obtained. Typical fuel cell types include phosphoric acid fuel cells, proton exchange membrane, solid oxide fuel cell and molten carbonate fuel cell, etc. [10].

The operation principle indicates that heat and mass transfer play an important role in fuel cells [10]. One typical fuel cell construction is the flat plate design for solid oxide fuel cells, shown in Fig. 8. As can be seen, the fuel in fuel ducts has both heat and mass transfer on the top wall with the anode, and the air in air ducts has both heat and mass transfer on the bottom wall with the cathode. In addition, two-phase flows exist in fuel ducts after a part of the fuel is consumed. Therefore, the conditions of fluid flow and heat transfer in air and fuel ducts have great effects on the performance of fuel cells and consequently the entire power cycle. Most of the current designs are based on constant values of Nusselt number and friction factor. Such rough estimations cannot meet future developments, and considerable research efforts must be given to this complex heat and mass transfer problem.

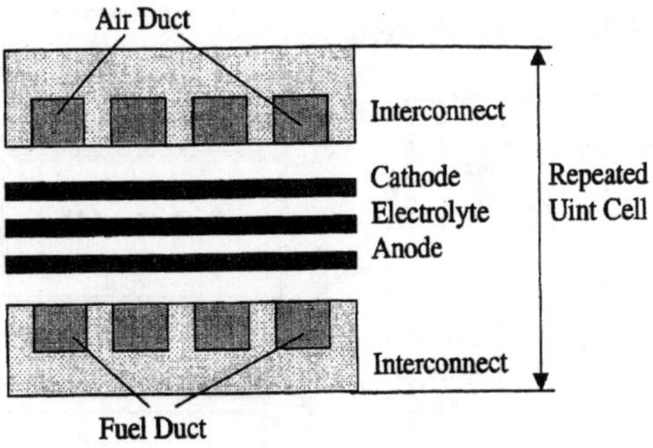


Figure 8: Structure of a unit of a fuel cell.

The above analysis demonstrates that high efficiency of power conversion can be reached with the help of relevant heat transfer and heat exchanger technologies. Therefore, attempts to provide compact, efficient heat transfer methods and heat exchangers and at the same time allowing a cheap and relatively simple manufacturing technique are real challenges for research.

## 4 Use of renewable energy

Hydropower, biomass, wind and solar energy are regarded the most important renewable and sustainable energy sources. Hydropower is, of course, dependent on the earth's contour, and it is not substantial for those countries with flat earth surface. Biomass appears to be an attractive option for many countries, and technologies for the conversion of biomass into electricity and heat are highly similar to the technologies for other solid fossil fuels. Wind and solar energies are strongly fluctuating sources, but they are very clean, with no pollutant emissions and have received great attention. In these renewable energy systems, heat transfer and heat exchangers play an important role as in those systems described earlier.

Consider now a simple solar energy system as an example. Figure 9 is a schematic view of a typical domestic hot water heating system designed for residential applications. When there is sun, the photovoltaic (PV) module produces power, which runs a small circulating pump. Antifreeze is pumped through the solar collectors and is heated. The fluid then returns to a reservoir in the heat exchanger module. Water coils in the reservoir absorb the heat from the solar fluid. The domestic water flows through these heat exchanger coils by natural thermosiphon action. As the water is heated, it rises and returns to the top of the tank, drawing cold water from the bottom of the tank into the heat exchanger.

It should be pointed out that no external heat exchanger as shown in Fig. 9 was used historically. Instead, the heat exchangers were coils of copper pipes located at the bottom of the solar storage tank. The current design shown in Fig. 9 has a number of advantages. First, the system performance is enhanced. External heat

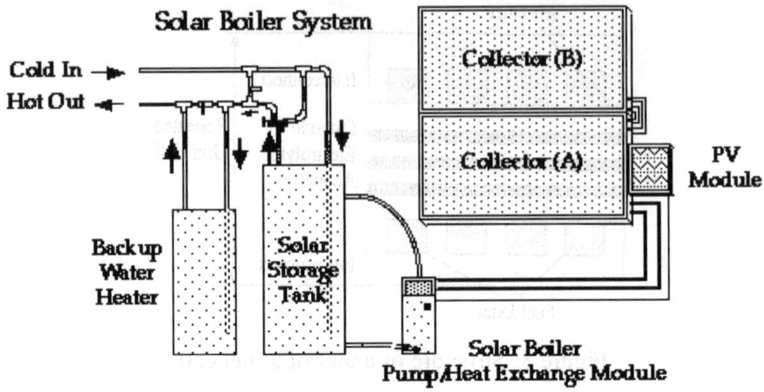


Figure 9: A solar energy system (from Solar-Works Inc.).

exchangers can be configured so that the potable water circulates by natural convection (i.e. it thermosiphons), which means that excellent temperature stratification can be achieved in the storage tank. With the hot water remaining at the top of the tank, usable hot water is available more rapidly with an external heat exchanger. Second, the thermodynamic efficiency is improved with the external heat exchanger configuration. The rate of heat transfer is directly proportional to the difference in temperature between the water being heated and the antifreeze from the solar collectors. With the external heat exchanger configuration, the heat exchanger coil is always surrounded by the very cold water, which means that thermal efficiency is greatly improved. Third, low cost can be achieved due to the long lifetime of the external heat exchanger compared to the solar tank. The external heat exchanger can be saved when the solar tank develops a leak, and thus cost saving is achieved. However, the heat transfer mechanism involved in the external heat exchanger is highly complex. Both forced convection and natural convection have important impacts. Shell-and-tube heat exchangers may serve well in this condition, but compact heat exchangers (such as PHEs) also claim superior operating condition. Because this practical application is still in its infancy, more research is expected in the future.

In addition, the solar collector using the PV module is a special heat exchanger. On one side of the surface, solar energy (radiant energy) is absorbed. This energy is transferred to the second side of the coolant. This is a quite complex heat transfer problem, not only because it involves both the radiant and the convection heat transfer but also because it is a time-dependent issue. The solar energy varies with time and location, and this must be taken into account in the use of this renewable energy.

The importance of heat transfer and heat exchangers has been illustrated for the solar energy system. Similar conclusions can be reached when dealing with the other types of renewable energy systems. However, they are not fully exploited here due to space constraints.

## 5 Reduction of emission and pollutant

Heat transfer and heat exchangers are also important in reducing emissions and pollutants. As illustrated earlier, they play an important role in the development of sustainable energy systems. The reduction of final energy consumption means less prime energy (e.g. fossil fuels) consumption, which results in overall reduction in emissions and pollutants. Improved efficiency of power plants certainly also reduces the primary energy consumption as well as the consequent emissions. Alternative fuels like biofuels (including biomass and waste utilization) are said to be neutral in terms of $CO_2$. The other renewable energy sources – solar, hydropower and wind – simply are clean enough and no emissions exist at all. In addition, by considering the pressure drop and associated pressure losses (work loss) in the heat transfer processes and attempting to reduce it, the consumption of electricity will be decreased, which is also beneficial. Therefore, heat transfer and heat exchangers are important for the protection of the environment, with regard to their role in the development of sustainable energy system.

The above influences on emissions and pollutants are obviously the indirect effect. However, heat transfer and heat exchangers can also have a direct effect on reducing emissions and pollutants in many situations. One example is their presence in internal combustion engines. In diesel combustion engines, exhaust gas recycling (EGR) was used for a while because this has been found to be an efficient method to reduce $NO_x$. However, particle emissions are increased and the engine performance is reduced. It has been recognized that if the exhaust gas is cooled in a heat exchanger, the above-mentioned problems can be overcome or at least partially avoided. In addition, the $NO_x$ emission will be further reduced as shown in Fig. 10. In this situation, several factors must be considered. First, due to the limited space in automobiles, an EGR cooler must be compact and lightweight. Second, because the cooling water is taken from the total engine cooling water, the amount of cooling water for the EGR cooler is limited and must be kept as small as possible. This means that the EGR cooler must have high thermal efficiency. Third, the EGR cooler is always subject to unsteady or oscillatory operation and is also severely affected by fouling, which means that the operating reliability and lifetime are extremely important in selecting the heat exchanger type. Therefore, a compact heat exchanger (e.g. a brazed plate heat exchanger) may be a better option, although shell-and-tube heat exchangers are currently often used in automobiles. To design an EGR cooler giving very reliable performance and durability, further research must be carried out.

Another example is the combustion chamber in gas turbine systems. It is well known that the production of $NO_x$ is related to high flame temperatures. One way to reduce the flame temperature is to use high air to fuel ratios [13]. This means that much more compressor air is needed for combustion and consequently less air is available for the cooling of combustion chambers and turbine blades. However, low temperature zones lead to unburned hydrocarbons. Thus, the emission control and the cooling system are coupled and need careful attention. This is another evidence that heat transfer design has a direct effect on reducing emissions and pollutants.

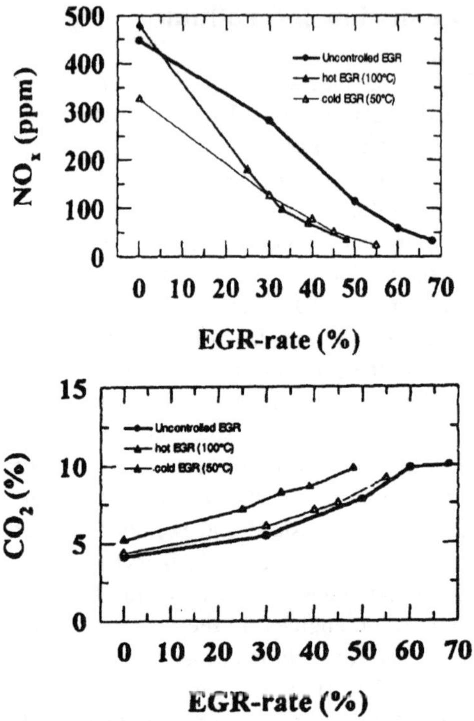


Figure 10: Effect of cooled EGR on $NO_x$ and $CO_2$ emissions [12].

## 6 Some examples of recent research

### 6.1 Case study of a heat exchanger network design using the pinch technology

A heat recovery system at a Swedish pulp mill has been investigated. At the mill, there is a big amount of hot water and thin liquor coming from the washing and bleaching process. These hot streams exchange heat with some cold streams, which will be used in the digesting plant. Since the hot streams labelled 2 and 3 contain a small content of fibres and some other substances, fouling may occur quite easily. Therefore, the process is very appropriate for PHEs because of the characteristic of easy cleaning. Specially designed PHEs, called wide gap PHEs, are used for the streams 2 and 3.

The network investigated contains three hot streams and five cold streams. The existing network is presented in grid form in Fig. 11. All the existing heat exchangers are PHEs and the total heat transfer area is 1436.5 m$^2$. The heat capacity flow rate, supply and target temperatures, physical properties and allowable pressure drop of each stream are given in Table 1. It should be pointed out that the allowable pressure drops are treated as the pressure drops to promote

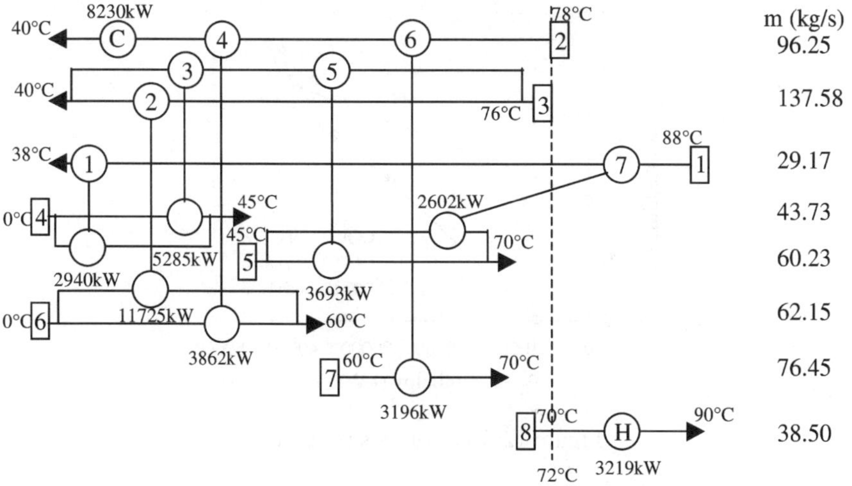


Figure 11: Grid structure of the existing plant–DTMIN = 6°C.

Table 1: Base data of streams.

| Stream no. | Supply temp. (°C) | Target temp. (°C) | $\dot{m}$ (kg/s) | Pr | $C_p$ (J/(kg·K)) | $\rho$ (kg/m$^3$) | $\mu$ (kg/(m·s)) | $\Delta P$ (kPa) |
|---|---|---|---|---|---|---|---|---|
| 1 | 88 | 38 | 29.17 | 4.0 | 3800 | 1050 | 0.0004 | 25 |
| 2 | 78 | 40 | 96.25 | 3.0 | 4180 | 990 | 0.0005 | 70 |
| 3 | 76 | 40 | 137.58 | 3.0 | 4180 | 990 | 0.0005 | 80 |
| 4 | 0 | 45 | 43.73 | 7.0 | 4180 | 990 | 0.001 | 40 |
| 5 | 45 | 70 | 60.23 | 3.0 | 4180 | 990 | 0.0005 | 50 |
| 6 | 0 | 60 | 62.15 | 5.0 | 4180 | 990 | 0.0008 | 60 |
| 7 | 60 | 70 | 76.45 | 3.0 | 4180 | 990 | 0.0004 | 60 |
| 8 | 70 | 90 | 38.50 | 2.0 | 4180 | 990 | 0.0004 | 30 |

heat transfer in the channels, which means that the pressure drops in the connecting pipes and some other additional ones are already excluded.

### 6.1.1 Grassroots design

The composite curves are plotted in Fig. 12 for DTMIN 6°C.

The optimal hot and cold utilities as well as the estimated total heat transfer area are calculated. The optimal hot and cold utility requirements are 1788 and 6800 kW, respectively. By comparing these figures with those in Fig. 11, it is obvious that the hot and cold utility consumption in the existing network could be reduced by 44.5% and 17.4%, respectively.

For the total heat transfer area, the estimation is carried out based on the proposed method. The total heat transfer area is a function of both DTMIN and the

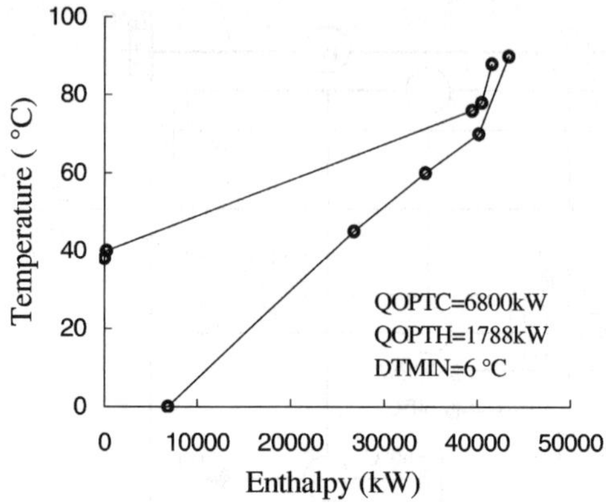


Figure 12: The composite curves.

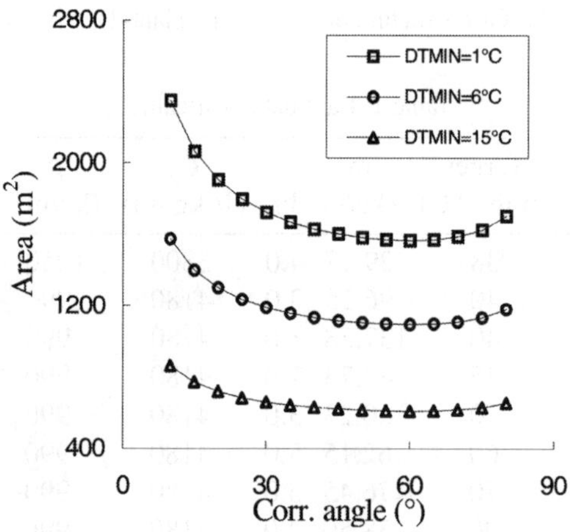


Figure 13: Area estimation.

corrugation angle, which is shown in Fig. 13. It is obvious that there is an optimal corrugation angle corresponding to the minimum area after DTMIN is specified. In addition, it is easy to understand that the heat transfer area becomes larger for lower DTMIN for the same corrugation angle. For DTMIN = 6°C, the minimum total heat transfer area is 1095 $m^2$, and the optimal corrugation angle is about 60°. This value is much lower than the existing exchanger area. This is probably due to the fact that the existing exchangers do not fully use the allowable pressure drop. In addition, a small part of the exchangers have lower corrugation angles, which also increases the total heat transfer area. In the calculation, the fouling resistances for stream 1, streams 2 and 3, and the rest are taken as 0.0001, 0.0003 and 0.00008 $m^2 \cdot K/W$, respectively. The hydraulic diameters for the wide gap PHEs and the normal PHEs are 0.022 and 0.008 m, respectively.

The variation in DTMIN causes variations in utility consumption, heat transfer area and most likely in the structure of the network. The variation in the corrugation angle causes variation in heat transfer area. Therefore, the optimal DTMIN and corrugation angle should be determined before any network generation. The annual costs for energy and exchanger area can be estimated by the following relationship:

$$\text{Capital cost} = 2700 \cdot \text{Area}^{0.85}$$

$$\text{Energy cost} = 1400 \cdot \text{Hot utility} + 400 \cdot \text{Cold utility}$$

The estimation for the capital cost is based on some experience of a PHE manufacturer and the estimation for the energy cost is provided by the staff at the mill. The hot utility is a live steam, and the cold utility is the normal cold water. The units for cost, area and utility are Swedish Crown (SEK), $m^2$ and kW. Now, it is possible to plot a graph of the total annualized cost versus DTMIN and the corrugation angle. The plot is given in Fig. 14, and the optimal DTMIN and corrugation angle are close to 1°C and 62°, respectively. The optimal DTMIN is quite small because the energy cost is the dominant part in the total cost.

The Pinch design suggested by Linnhoff *et al.* [6] is employed to design the network. The minimum DTMIN is taken as 6°C after considering the minimum temperature difference for PHEs. The optimal corrugation angle for this value of DTMIN is close to 60°. The final design is shown in Fig. 15. After the detailed calculation is carried out, the total heat transfer area is 1247 $m^2$. The deviation between predicted and calculated values is about 12%. Considering the fact that the vertical alignment is assumed in the prediction while it is actually not in the network synthesis, this deviation is acceptable for the pre-optimization. Hence, it demonstrates that the suggested method is suitable for the optimization of heat exchanger networks using PHEs.

As for the potential use of multi-stream PHEs, the heat exchangers 2 and 4, 3 and 5, and 6 and 7 are likely to be combined as three-stream PHEs. By doing so,

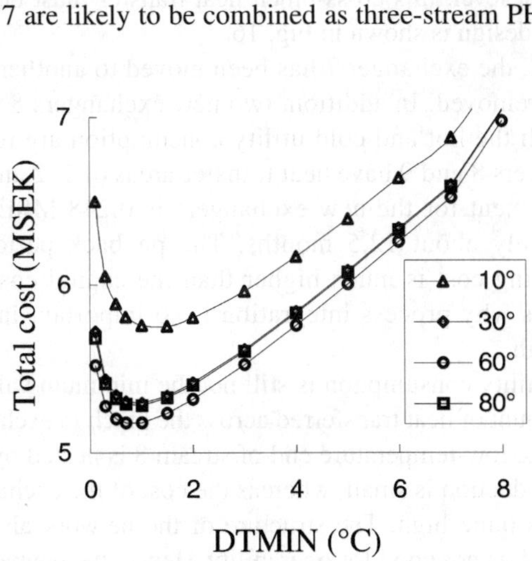


Figure 14: Indication of optimum DTMIN.

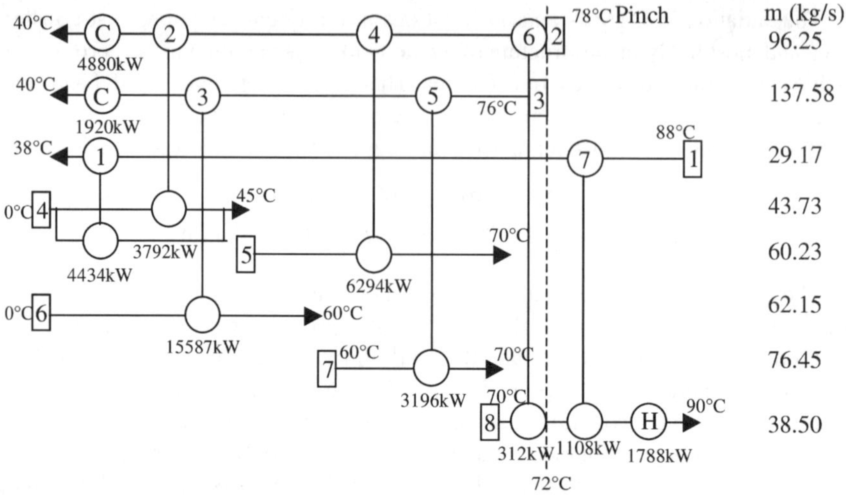


Figure 15: Grassroots design–DTMIN = 6°C.

the capital cost and installation cost are greatly reduced. The process is also made more integrated. The operability concern can be solved by several ways suggested above. Although the detailed calculation is not carried out here, the potential for use of multi-stream PHEs is obvious.

### 6.1.2 Retrofit design

The reason for the excessive consumption of both hot and cold utility is that there is heat transfer (exchange 7 in Fig. 11) across the pinch point. To reduce the utility cost to the optimal level, this cross-pinch heat transfer must be eliminated. The suggested retrofit design is shown in Fig. 16.

As can be seen, the exchanger 7 has been moved to another place and some of the plates are removed. In addition, two new exchangers 8 and 9 are added. By doing so, both the hot and cold utility consumption are reduced. The two new heat exchangers 8 and 9 have heat transfer areas of 122 and 82 $m^2$, respectively. The investment for the new exchangers is 0.248 MSEK, and the payback period is only about 12.5 months. The payback period is very short because the running cost is much higher than the capital cost in this case. It also demonstrates why process integration is so important in industry where energy cost is high.

However, the utility consumption is still not the minimum value because there is still a small amount of heat transferred across the pinch in exchanger 9. This can be eliminated if the low-temperature end of stream 8 is heated by stream 2. However, the energy reduction is small, whereas the cost of the exchanger, connecting pipe and others is quite high. The structure of the network also becomes more complicated, which is not good for operability. Hence, the payback period will be rather long.

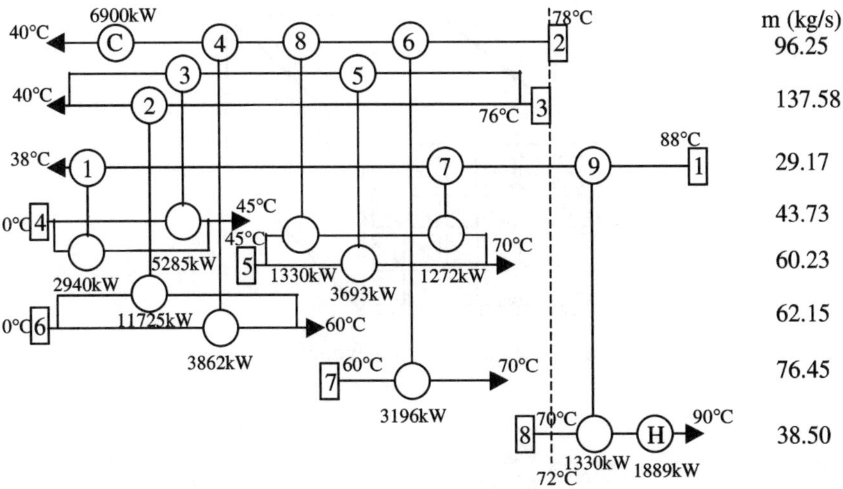


Figure 16: Grid structure of the retrofit design.

### 6.2 High temperature heat exchangers

High temperature heat exchanger technology has become important for improving the performance of power generation. There is a need to develop various type of high temperature heat exchangers in different applications such as hydrogen production, reforming process of solid oxide fuel cells, generation of high temperature gas, low emission power plants. In this section, monolithic heat exchangers are considered and some specific problems are addressed.

#### 6.2.1 Monolithic exchangers

**6.2.1.1 Ceramic monolith structures** Ceramic monolith structures are used in the industry today and they are produced in large numbers by using the extrusion technique. They are unibody structures composed of interconnected repeating cells or channels (Figs. 17 and 18). They are increasingly under development and evaluation for many new reactor applications [14, 15], e.g. chemical process and refining industries, catalytic combustion, low emission power plants. However, monoliths are mainly used where only one fluid flows through all the channels. An example is the monolithic exhaust structure in automotive applications. In endothermic and slow reactions such as steam reforming of hydrocarbons, large amounts of heat are needed to maintain reaction rates. If the catalysts were deposited on tubes, usage of monoliths would be more efficient, leading to greater reaction rates and a smaller reactor [16]. Additionally, there would be a great improvement in mechanical integrity. Especially, it would be advantageous if two fluids in monolithic channels can exchange heat and/or mass. The reason why monoliths are not widely used in these applications is because of complex technique for feeding and distributing the two fluids in and out of the channels.

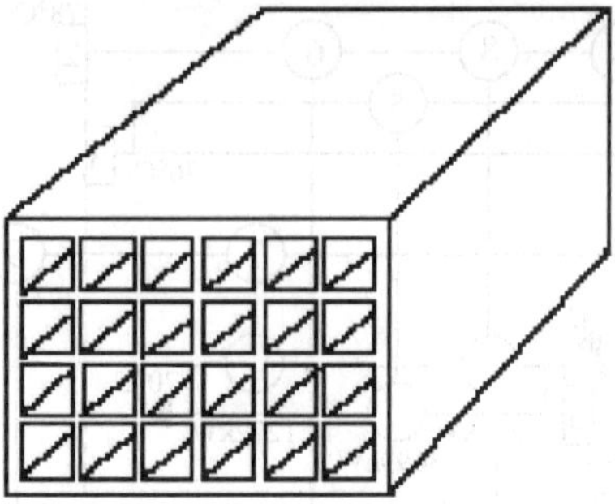

Figure 17: Monolithic 'honeycomb' structure.

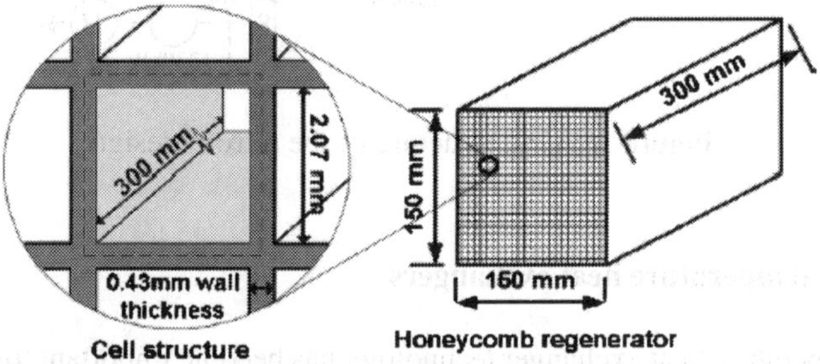


Figure 18: Honeycomb regenerator and dimensions of solid material and flow path. The dashed line binds one symmetrical cell.

Selimovic and Sunden [17] focused on the compact ceramic heat exchanger where two fluids are fed and distributed into individual channels in a multi-channels structure. Their study shows three different approaches of modelling: analytical, experimental and numerical modelling. The exchanger is of monolithic shape where heat and mass is transferred in rectangular channels. Usually, for the pressure drop calculations of standard channel shapes, different available correlations can be applied. However, when these channels are attached to a manifolded and connected to other components, complex geometries are involved and then modelling with correlation parameters may be unsuccessful. Similar to PHEs, the pressure drop, as well as thermal performance, depends on distribution of fluid. Therefore, it is important to investigate how good the flow distribution is from the main port pipe into the channels. The analytical investigation made here includes both U- and Z-type configurations.

Monolithic 'honeycomb' structure has been manifolded by two stage manifolds where either U-type or Z-type manifolds can be used to distribute the flow rate uniformly through each branch. This stage manifold can be compared to the manifolding of PHEs. The main difference compared to PHEs is that each branch will further divide the flow to the monolithic structure with specified channel arrangement. This stage manifolding is called I-type manifold here. More detailed picture of I-type manifold can be observed in Fig. 19. Concerning the monolithic

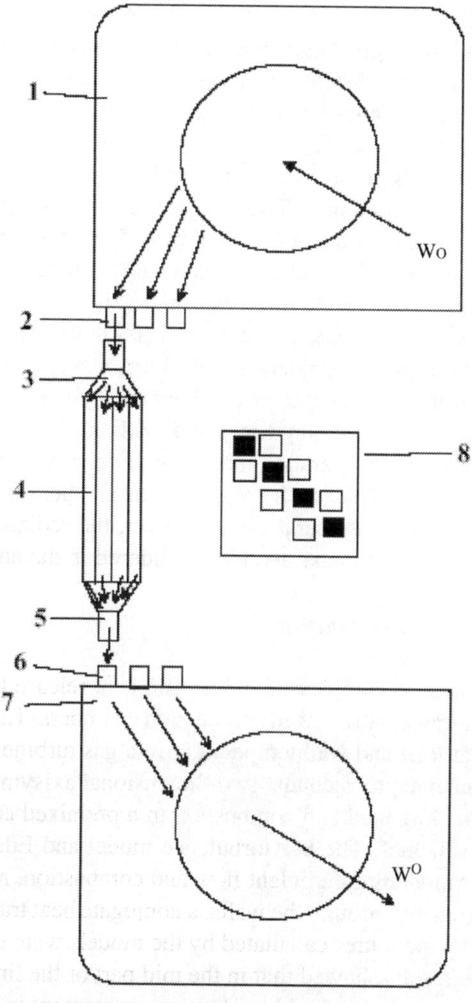


Figure 19: I-type manifold assembly: 1 and 2 – manifold top, 3 – dividing plates, 4 – monolithic channels, 5 – collecting plate, 6 and 7 – manifold bottom, 8 – checkerboard channel arrangement.

channels, two different gas distributions (channel arrangements) are investigated: the checkerboard and linear (Figs. 18 and 19). The important physical characteristics are then the size of the channel through which the gaseous reactants and products traverse wall thickness, and the total monolith's compactness.

Rafidi and Blasiak [18] developed a two-dimensional simulation model to find out the temperature distribution of the solid storing material and flowing gases and other thermal and flow parameters for this regenerator and compared the computed results with experiments. Because of geometric symmetry of the honeycomb structure, mathematical analysis was made on one honeycomb cell, or matrix, that formed a small part of the regenerator cross-section along the flow path. The regenerator is composed of two different materials along the heat exchangers, one is 0.2 m long alumina and the other is 0.1 m long cordierite.

Figure 19 shows the dimensions of a heat regenerator used in a twin-type 100 kW HiTAC (high temperature air combustion) regenerative burner. The regenerator dimensions are 150 × 150 × 300 $mm^3$. The cell size is attributed to 100 cells/$in^2$ and hence, the specific heat transfer area is 4200 $m^2/m^3$. All flue gases generated by combustion are sucked again by the burners and pass through the regenerators. The honeycomb compact heat regenerator has relatively high effectiveness of about 88% and recovers 72% of energy contained in combustion flue gases at nominal operating conditions. Consequently, the energy storage and the pressure drop are calculated and the thermal performance of the honeycomb heat regenerator is evaluated at different switching times and loading. The model takes into account the thermal conductivity parallel and perpendicular to flow direction of solid and flowing gases. It considers the variation in all thermal properties of solid materials and gases with temperature. Moreover, the radiation from combustion flue gases to the storage materials was considered in the analysis.

### 6.3 Heat load prediction in combustors

Different phenomena such as complex flow field and heat release by combustion are involved in the heat transfer process in combustion chambers. This section concerns prediction of heat load and wall temperature in a gas turbine combustor by taking different phenomena into account. Two-dimensional axisymmetric models were used to model the flow field and combustion in a premixed combustor with two different cooling schemes. The $k$–$\varepsilon$ turbulence model and Eddy Dissipation Concept were used for modelling turbulent flow and combustion, respectively. In the modelling of heat transfer through the walls, a conjugate heat transfer formulation was applied. The temperatures calculated by the models were compared with experimental data. The results showed that in the mid part of the liner, the prediction of the wall temperature is good, although worse agreement is found in other parts. In addition, radiative heat transfer has been studied. The results showed that radiative heat transfer in simple and ribbed duct cooling schemes can increase the average inner wall temperature up to 33 and 40 K, respectively.

Here computational fluid dynamics (CFD) simulations are used to, first, predict the temperature and heat transfer rate to the combustor wall (called liner wall hereafter) by using a conjugate heat transfer method and, second, study quantities of convective and radiative heat transfer in this type of combustor. The analysis is carried out on a VT4400 LPP combustor developed by Volvo Aero Corporation. A slightly simplified geometry is used to simulate this combustor and some experimental data of inner and outer liner wall temperatures were provided to validate the simulation results.

#### 6.3.1 Combustor description and its modelling

The VT4400 LPP is a lean premixed combustor, which is fuelled by natural gas. In the case of measured data, the equivalence ratio has been set to 0.59. The supplied air from the compressor is divided into two parts. The primary air, after passing through a cooling duct, enters the swirl system and mixes with the natural gas and is then burnt. The height of the cooling duct is 8 mm. The primary air and swirl number are about 1.57 kg/s and 0.6, respectively. By using the geometrical data and definition of swirl number (see eqn (1)), the axial and tangential velocities at the inlet of the combustor can be set.

$$S_n = \int_0^{R_o} U_z U_\theta r^2 dr \,/\, R_o \int_0^{R_o} U_z^2 r dr \tag{1}$$

At the second inlet, the secondary air is mixed with the burnt gases before the entrance to the turbine. In the experiments, the combustor was equipped with two different cooling schemes; a simple duct and a ribbed duct with thermal barrier coating (TBC) on the inner side of the liner wall.

The thickness of the liner wall is 1.5 mm and its thermal conductivity is about 25 W/m K. In the second scheme, a TBC layer with thermal conductivity 1.3 W/m K has been used. The inlet temperature from the compressor is 662 K and according to the experiment this temperature is increased by 48 K at the outlet of the channel. The described combustor was modelled by a two-dimensional geometry (see Fig. 20). The model was meshed by two-dimensional (three-dimensional with one cell thickness) multi-block axisymmetric grids. A grid dependence study was carried out in the simple cooling duct case and 42,580 cells showed satisfactory accuracy. Then this meshed model was improved for the ribbed duct and TBC case and the number of cells reached 70,090. To capture the temperature distribution, the liner has been divided into 10 cell thickness. For boundary conditions, inlet and pressure boundaries were used for inlet and outlet, periodic and symmetry boundaries were used for the $r$–$z$ faces in the liner and cooling duct, respectively.

### 6.3.2 Governing equations and solution methods

To model the flow field, continuity and Navier–Stokes equations were solved. The turbulence was modelled by solving the transport equations for the turbulent kinetic energy and turbulent dissipation, which are implemented in the standard $k$–$\varepsilon$ model. The summarized governing equations are listed in Table 2.

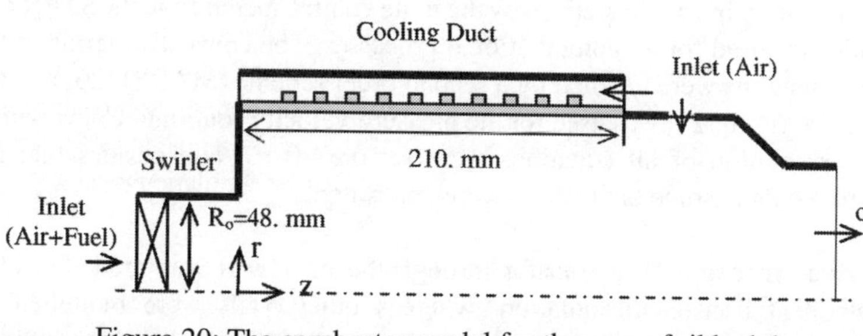


Figure 20: The combustor model for the case of ribbed duct.

Table 2: Governing equations in general form.

$$\frac{\partial(\rho\phi)}{\partial t}+\frac{\partial(\rho U_j\phi)}{\partial x_j}=\frac{\partial}{\partial x_j}\left(\Gamma_\phi \frac{\partial\phi}{\partial x_j}\right)+S_\phi$$

| $\varphi$ | $\Gamma\varphi$ | $S\varphi$ | Eqn |
|---|---|---|---|
| 1 | 0 | 0 | (2) |
| $U_i$ | $\mu+\mu_t$ | $-\frac{\partial P}{\partial x_i}+\frac{\partial}{\partial x_j}\left((\mu+\mu_t)\left(\frac{\partial U_j}{\partial x_i}-\frac{2}{3}\frac{\partial U_k}{\partial x_k}\delta_{ij}\right)-\frac{2}{3}\rho k\delta_{ij}\right)$ | (3) |
| $h$ | $\mu/Pr+\mu_t/Pr_t$ | $\frac{\partial P}{\partial t}-\frac{\partial q_{R,j}}{\partial x_j}$ | (4) |
| $Y_\alpha$ | $\mu/Sc+\mu_t/Sc_t$ | $\dot{\omega}_a$ | (5) |
| $k$ | $\mu+\mu_t/\sigma_k$ | $P_k-\rho\varepsilon$ | (6) |
| $\varepsilon$ | $\mu+\mu_t$ | $f_1C_{1\varepsilon}\frac{\varepsilon}{k}P_k-\rho\varepsilon f_2C_{2\varepsilon}\frac{\varepsilon}{k}$ | (7) |
| | | $\mu_t=\rho f_\mu C_\mu k^2/\varepsilon$ | |

*Note:* The indices in eqn (3) are not correct for the first velocity gradient.

The premixed turbulent combustion was modelled by a one-step reaction for burning methane. The reaction rate was approximated by using the Eddy Dissipation Concept [19] and implemented in the source term of the species transport equation. According to this model, the reaction rate is given by:

$$\dot{\omega}_a=-\frac{\rho\varepsilon A_{ebu}}{k}\min\left(Y_F,\frac{Y_O}{s_O},B_{ebu}\frac{Y_P}{s_P}\right) \tag{8}$$

where $A_{ebu}$ and $B_{ebu}$ are model constants, which were set to 4.0 and 0.5, respectively. The solution domain was discretized by the finite volume method and the STAR-CD CFD code was used for all computational processes. The convective terms of the transport equations were handled by a second order scheme (MARS) [20] and the SIMPLE algorithm [21] was used for the pressure-velocity coupling. Convergence criteria for solution of all equations were set on $1.0 \times^{-4}$ and besides that the temperature data at some boundaries were controlled.

**6.3.2.1 Heat transfer** Heat transfer through the liner wall was modelled by a conjugate heat transfer formulation, whereas other walls were modelled by their thermal resistance and environment temperature. Convective heat transfer on both hot and cold sides of the liner was modelled by the standard wall function relations [22], which are valid in the log-law region of the turbulent boundary layer. For this reason, the values of $y+$ in near the wall regions were kept in the range of 30–40. Also, to mitigate the effect of circulation zones in the case of the ribbed duct channel, a non-equilibrium wall function has been imposed, which can take the effect of pressure gradient into account. In the radiative heat transfer part, the radiative transfer equation (RTE) [22] in participating media has been solved by the S4 discrete ordinates method with 24 directions. This selection gives satisfactory accuracy [23] for absorption and emission and keeps the computational efforts as low as possible. Also, because of the low temperature in the cooling duct, the radiative heat transfer was only considered on the inside of the liner. Combustion of methane generates some $CO_2$ and $H_2O$ and these are the most important participating gases in the absorption and emission of thermal radiation. To take their effects into account, the spectral line weighted sum of grey gases method (SLW) [24] has been used, which is an accurate model. For calculation by the SLW method, five grey gases, optimised by a conjugate gradient method were selected. For each grey gas, the blackbody weight is calculated and then the grey gas absorption coefficient and its blackbody weight are used in solving the RTE. For the radiative boundary conditions, the liner wall emissivities have been set to 0.7 and updated wall temperatures are used during the solution process.

### 6.3.3 Results and discussion

**6.3.3.1 Flow and temperature fields** The axial velocity and temperature fields inside the liner and before mixing with the secondary air are shown in Fig. 21. As can be seen, the velocity changes sharply close to the swirler with negative values near the liner wall. The sharp variation is mitigated along the length of the liner. At $z/R_0 = 1.5$ (about 74 mm along the liner), the velocity profile starts to change direction near the wall and therefore a stagnation point is formed. By further increase in $z/R_0$, the velocity is stabilized in the new direction and its profile in the radial direction will be flattened.

Similar to the velocity field, the temperature field varies sharply close to the swirler. The main reason for the variation is the combustion of fuel in the region

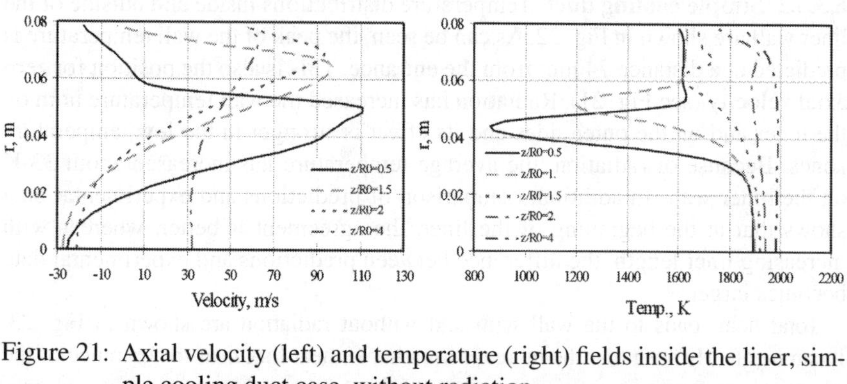


Figure 21: Axial velocity (left) and temperature (right) fields inside the liner, simple cooling duct case, without radiation.

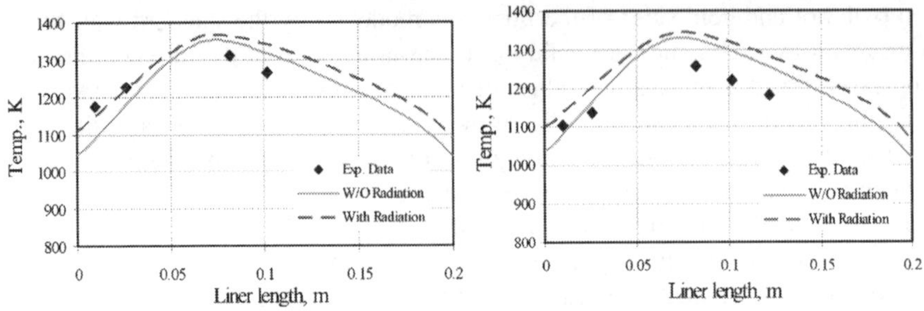


Figure 22: Temperature distribution at inner (left) and outer (right) liner wall in a simple cooling duct.

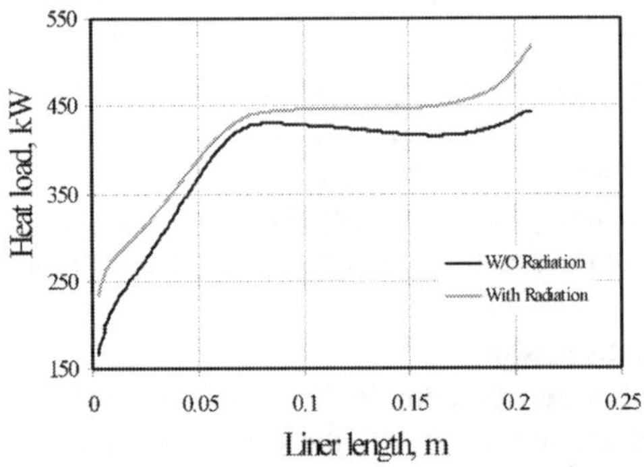


Figure 23: Prediction of heat load in a simple cooling duct.

and negative velocities at the centre and wall regions of the liner. By increasing $z/R_0$ up to 1, the temperature does not change considerably near the wall regions; however, by further increase, the temperature is decreased. This can be due to the tact that the combustion process has been completed and convective heat transfer.

**6.3.3.2 Simple cooling duct** Temperature distributions inside and outside of the liner wall are shown in Fig. 22. As can be seen, the peak of the wall temperature is predicted at a distance 74 mm from the entrance. This is also the position for zero axial velocity (see Fig. 21). Radiation has increased the wall temperature both on the inner and on the outer sides and its effect is stronger in the low temperature zones. Because of radiation, the average temperature has increased about 33 K on the inner wall. In addition, comparison of predictions and experimental data shows that at the beginning of the liner, the agreement is better, whereas with increasing liner length, the difference between predictions and experimental data becomes larger.

Total heat loads to the wall with and without radiation are shown in Fig. 23. It can be clearly seen that the increase in radiative heat flux near the entrance of the

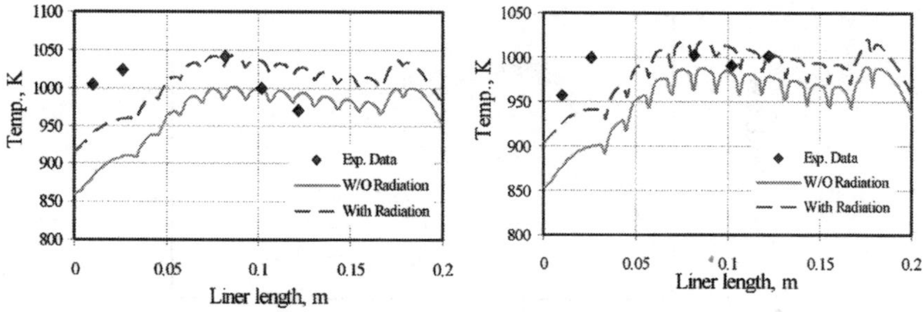


Figure 24: Temperature distribution at inner (left) and outer (right) liner wall for a ribbed cooling duct and TBC.

cooling duct is large and low radiative heat flux occurs at the hot region of the wall. In the mid part, the heat load with radiation is almost constant, 450 kW, which means that the temperature difference between the two sides of the wall for a short distance is almost constant. The average heat load without radiation is about 388 kW and radiation increases this value by 8%. Because of the small wall thickness, the heat load is very sensitive to the temperature difference between the two sides of the liner wall and it is noted that for 1 K difference, the heat load changes about 17 kW/m$^2$.

**6.3.3.3 Ribbed cooling duct and TBC** In Fig. 24, the temperature distributions on the inner and outer liner walls for the case of a ribbed duct with TBC are shown. In this case, the influence of radiation is higher than that for the simple duct case. The average temperature on the inner wall has increased by 40 K. The agreement between predictions and experimental data at the entrance of the liner is somewhat poor, but it is obvious that radiation is important. At the middle and end parts, the predictions for the outer face of the wall are very good, but the values of the inner face have been overpredicted.

This might be due to error in the experimental data, because the outer wall at the same position is well predicted. In that part of the liner, the flow and temperature near the wall are stabilized, so the sharp decrease in the inner wall temperature is doubtful. The predicted heat load is shown in Fig. 25. Similar to the simple cooling duct case, the radiative heat load is stronger near the entrance of the cooling duct. In this case, the average convective heat load is about 393 kW/m$^2$ and the radiative heat transfer increases the heat load by about 7%.

In summary, the wall temperatures and heat loads in a premixed combustor have been predicted. The results showed that in the mid part of the liner, the prediction of the wall temperature is good, but poorer agreement exists in other parts. In addition, radiative heat transfer has been included in the study. The results showed that radiative heat transfer for simple and ribbed duct cooling schemes can increase the average inner wall temperature by 33 and 40 K, respectively. As an extension of this study, the accuracy of the model in prediction of wall temperature and heat loads to the walls can be investigated by using different wall treatments such as a two-layer wall function approach or applying a low Reynolds model.

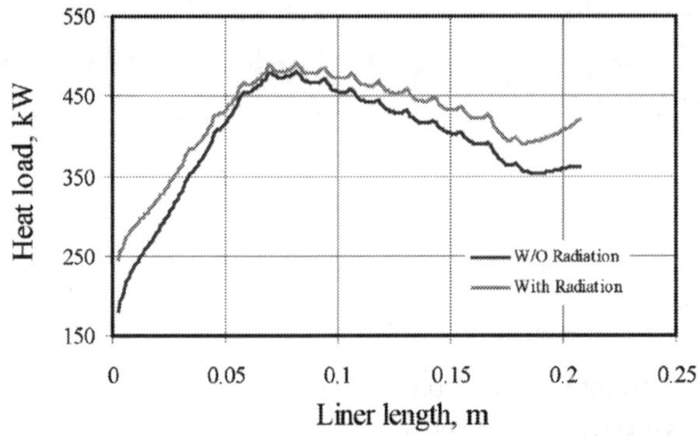


Figure 25: Prediction of heat load in a ribbed cooling duct and TBC.

### 6.4 CFD methods in analysis of thermal problems

CFD can be applied to heat exchangers in quite different ways. In the first way, the entire heat exchanger or the heat transferring surface is modelled. This can be done by using large scale or relatively coarse computational meshes or by applying a local averaging or porous medium approach. For the latter case, volume porosities, surface permeabilities and flow and thermal resistances have to be introduced. The porous medium approach was first introduced by Patankar and Spalding [25] for shell-and-tube heat exchangers and was later followed by many others.

Another way is to identify modules or group of modules, which repeat themselves in a periodic or cyclic manner in the main flow direction. This will enable accurate calculations for the modules, but the entire heat exchanger including manifolds and distribution areas are not included. The idea of streamwise periodic flow and heat transfer was introduced by Patankar *et al.* [26].

The finite volume method is a popular method particularly for convective flow and heat transfer. It is also applied in several commercial CFD codes. Further details can be found in [21, 27]. In heat transfer equipment like heat exchangers, both laminar and turbulent flows are of interest. While laminar convective flow and heat transfer can be simulated, turbulent flow and heat transfer normally require modelling approaches in addition. By turbulence modelling, the goal is to account for the relevant physics by using as simple a mathematical model as possible. This section gives a brief introduction to the modelling of turbulent flows.

The instantaneous mass conservation, momentum and energy equations form a closed set of five unknowns $u$, $v$, $w$, $p$ and $T$. However, the computing requirements, in terms of resolution in space and time for direct solution of the time dependent equations of fully turbulent flows at high Reynolds numbers (so-called direct numerical simulation (DNS) calculations), are enormous and major developments in computer hardware are needed. Thus, DNS is more viewed as a research tool for relatively simple flows at moderate Reynolds number. In the meanwhile, practicing thermal engineers need computational procedures supplying information about the turbulent processes, but avoiding the need to predict effects of every eddy in the flow. This calls for information about the time-averaged properties of the flow and temperature fields (e.g. mean velocities, mean stresses, mean temperature). Commonly, a time-averaging operation, called Reynolds decomposition is carried out. Every variable is then written as a sum of a time-averaged value and a superimposed fluctuating value. In the governing equations, additional unknowns appear, six for the momentum equations and three for the temperature field equation. The additional terms in the differential equations are called turbulent stresses and turbulent heat fluxes, respectively. The task of turbulence modelling is to provide the procedures to predict the additional unknowns, i.e. the turbulent stresses and turbulent heat fluxes with sufficient generality and accuracy. Methods based on the Reynolds-averaged equations are commonly referred to as Reynolds-averaged Navier–Stokes (RANS) methods.

### 6.4.1 Types of models

The most common turbulence models for industrial applications are classified as

- zero-equation models,
- one-equation models,
- two-equation models,
- Reynolds stress models,
- algebraic stress models and
- large eddy simulations (LES).

The first three models in this list account for the turbulent stresses and heat fluxes by introducing a turbulent viscosity (eddy viscosity) and a turbulent diffusivity (eddy diffusivity). Linear and non-linear models exist [28–30]. The eddy viscosity is usually obtained from certain parameters representing the fluctuating motion. In two-equation models, these parameters are determined by solving two additional differential equations. However, one should remember that these equations are not exact, but approximate and involves several adjustable constants. Models using the eddy viscosity and eddy diffusivity approach are isotropic in nature and cannot evaluate non-isotropic effects. Various modifications and alternative modelling concepts have been proposed. Examples of models of this category are the $k$–$\varepsilon$, and $k$–$\omega$ models in high or low Reynolds number versions as well as in linear and non-linear versions. A lately popular model is the so-called V2F model introduced by Durbin [31]. It extends the use of the $k$–$\varepsilon$ model by incorporating near-wall turbulence anisotropy and non-local pressure–strain effects, while retaining a linear eddy viscosity assumption. Two additional transport equations are solved, namely one for the velocity fluctuation normal to walls and another for a global relaxation factor.

In Reynolds stress equation models, differential equations for the turbulent stresses (Reynolds stresses) are solved and directional effects are naturally accounted for. Six modelled equations (i.e. not exact equations) for the turbulent stress transport are solved together with a model equation for the turbulent scalar dissipation rate $\varepsilon$. Reynolds stress equation models are quite complex and require large computing efforts and for this reason, are not widely used for industrial flow and heat transfer applications.

Algebraic stress models (ASM) and explicit ones such as EASM present an economical way to account for the anisotropy of the turbulent stresses without solving the Reynolds stress transport equations. One idea is that the convective and diffusive terms are modelled or even neglected and then the Reynolds stress equations reduce to a set of algebraic equations.

For calculation of the turbulent heat fluxes, most commonly a simple eddy diffusivity concept is applied. The turbulent diffusivity for heat transport is then obtained by dividing the turbulent viscosity by a turbulent Prandtl number. Such a model cannot account for non-isotropic effects in the thermal field but still this model is frequently used in engineering applications. There are some models presented in the literature to account for non-isotropic heat transport, e.g. the generalized gradient diffusion hypothesis and the WET (wealth = earnings × time) method. These higher order models require that the Reynolds stresses are calculated accurately by taking non-isotropic effects into account. If not, the performance may not be improved. In addition, partial differential equations can be formulated for the three turbulent heat fluxes, but numerical solutions of these modelled equations are rarely found. Further details can be found in [32].

The LES is a model where the time-dependent flow equations are solved for the mean flow and the largest eddies, while the effects of the smaller eddies are modelled. The LES model has been expected to emerge as the future model for industrial applications, but it still limited to relatively low Reynolds number and simple geometries. Handling wall-bounded flows with focus on the near wall phenomena like heat and mass transfer and shear at high Reynolds number present a problem due to the near-wall resolution requirements. Complex wall topologies also present problem for LES.

Nowadays, approaches to combine LES and RANS based methods have been suggested.

### 6.4.2 Wall effects

There are two standard procedures to account for wall effects in numerical calculations of turbulent flow and heat transfer. One is to employ low Reynolds number modelling procedures, and the other is to apply the wall function method. The wall functions approach includes empirical formulas and functions linking the dependent variables at the near-wall cells to the corresponding parameters on the wall. The functions are composed of laws of the wall for the mean velocity and temperature, and formulae for the near-wall turbulent quantities. The accuracy of the wall function approach is increasing with increasing Reynolds numbers. In general, the wall function approach is efficient and requires less CPU time and memory size, but it becomes inaccurate at low Reynolds numbers. When low Reynolds number effects are important in the flow domain, the wall function approach ceases to be valid. The so-called low Reynolds number versions of the turbulence models are introduced and the molecular viscosity appears in the diffusion terms. In addition, damping functions are introduced. Also, the so-called two-layer models have been suggested where the transport equation for the turbulent kinetic energy is solved, while an algebraic equation is used for, e.g. the turbulent dissipation rate.

### 6.4.3 CFD codes

Several industries and companies worldwide are nowadays using commercially available, general-purpose, so-called CFD codes for simulation of flow and heat transfer topics in heat exchangers, investigations on enhanced heat transfer, electronics cooling, gas turbine heat transfer and other application areas, e.g. fuel cells. Among these codes are: FLUENT, CFX, STAR-CD, FIDAP, ADINA, CFD2000, PHOENICS and others. Also many universities and research institutes worldwide apply commercial codes besides in-house developed codes. However, to apply such codes successfully and to interpret the computed results, it is necessary to understand the fundamental concepts of computational methods.

### 6.4.4 Ducts with bumps

As a duct with bumps is considered, this type of duct appears in some rotary regenerative heat exchangers. The basic idea with introduction of bumps is to design corrugated ducts as indicated in Fig. 26 for ducts with triangular cross section. The intention is that this corrugation should affect the flow field and introduce low Reynolds number turbulence and a swirling motion as sketched in Fig. 26. At a certain distance downstream the corrugation element, the turbulence and the swirling motion will be attenuated and gradually the intensity of the fluctuations will be reduced. Therefore, at a position upstream where the complex flow pattern (strong secondary cross-sectional flow and separated flow) has been significantly weakened or has disappeared, a new corrugation element is introduced to re-establish the violent and swirling like motion. CFD calculations have been performed and a non-orthogonal structured grid was employed. Periodic conditions were imposed in the main flow direction. About 40,000 control volumes (CVs) were used, 30 × 60 CVs in the cross-sectional plane. The existence of a secondary flow was revealed and a result is shown in Fig. 27. It is obvious that a swirling motion is created by the bumps and the triangular cross section. The Reynolds number corresponding to the flow in Fig. 27 is about 2000. In the simulations, a low Reynolds number $k$–$\varepsilon$ model was used. The secondary motion exists also for laminar cases, as it is partly geometry-driven. It is found that the heat transfer is enhanced compared to a smooth duct, but the pressure drop increase is high.

Further details of this investigation can be found in [33].

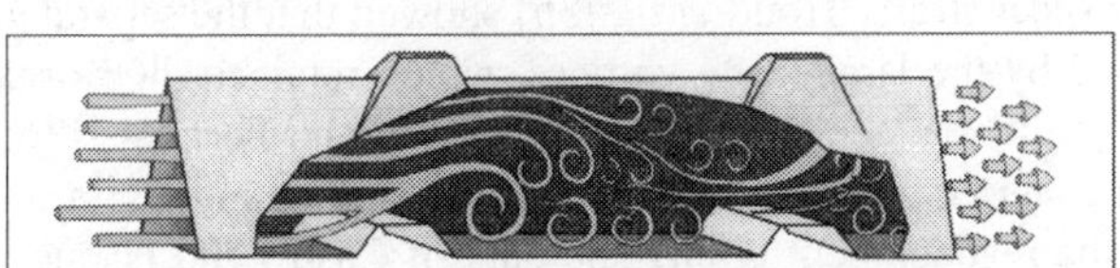

Figure 26: Conjectured flow pattern in a duct with bumps.

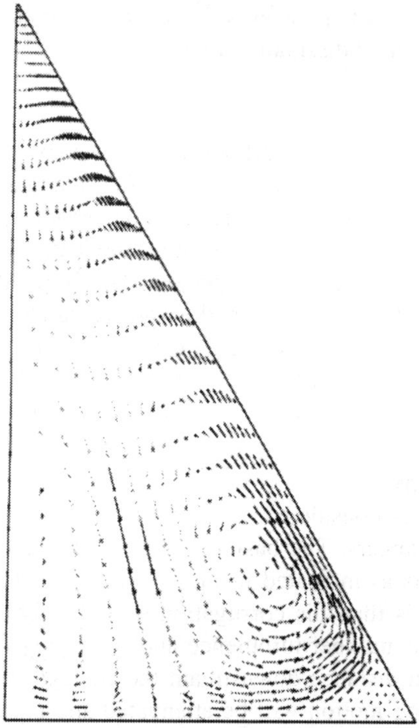

Figure 27: Secondary flow velocity vectors in a cross sectional plane midway over a corrugation element.

### 6.5 Flow structures in ribbed ducts

Ribbed duct flows are encountered in numerous engineering applications, e.g. turbine blades and combustor walls cooling. The flow behind a rib is typically characterized by flow separation and subsequent reattachment. Flow in a separated shear layer is complicated by the presence of reverse flow and a high level of the turbulence intensity. Despite the substantial progress in experimental and numerical studies on turbulent flows with s eparation, our understanding of this phenomenon is far from complete. The first review of the experimental data for separated flow was provided by Bradshaw and Wong [34] for flow over a backward-facing step. On the basis of single point measurement, they concluded that the shear layer split in two parts at the reattachment point and the bifurcation caused a rapid decrease in turbulence shear stress. Troutt *et al.* [35] showed that the separated shear layer was dominated by the large-scale vortices, which retained their organization far downstream of the reattachment region. Ruderich and Fernholz [36] indicated a self-similar behaviour for the mean and fluctuating quantities in a short region upstream of the reattachment point. The data of Castro and Haque [37] showed that the turbulent structure of the separated shear layer differed from that of a plane mixing layer between two streams. On the other hand, they argued that the flow close to the wall within the recirculation region had some features reminiscent of a laminar boundary layer in a favourable pressure gradient. Thereafter, Hasan [38] confirmed that the reattaching shear layer did split into two and a low-frequency flapping motion of the shear layer is observed. In the numerical simulation of turbulent flow over a backward-facing step, Le *et al.* [39] pointed out that the turbulent kinetic energy budget in the recirculation region is similar to that of a turbulent mixing layer.

Previous research mostly focused on the flow with separation induced by a backward-facing step, which is considered to be the benchmark to study this phenomenon. The flow past a rib is more complicated because it involves an additional separation in the upstream region of the obstacle. On the other hand, the geometry of rib has also an essential influence on the flow separation and reattachment. According to Fröhlich *et al.* [40], separation from continuous and curved surfaces displays a strong spatial and temporal fluctuation of the separation line; meanwhile, the mean location of reattachment is sensitively dependent on that of separation. These characteristics imply that the separation from contoured protrusions is more elusive than that from obstacles with sharp edge.

In our recent work, square-shaped, transversely placed ribs were employed to investigate the separated flow in a square channel. To highlight the physical mechanism of flow separation, only one wall of the channel is fitted with periodic ribs. The ribs obstruct the channel by 15% of its height and are arranged 12 rib heights apart. The inter-rib spacing is set such that the reattachment is allowed to take place on the portion between consecutive ribs and a distinct redevelopment region is introduced prior to a re-separation over the next rib.

Many numerical and analytical studies [41–44] were carried out to investigate the characteristics of flow separation in a ribbed channel based on the DNS or LES techniques. Corresponding to the numerous simulation works, very few experimental works, however, were executed to give high-resolution velocity measurements and turbulent properties. On the basis of literature review, Islam *et al.* [45] conducted an experimental study on the turbulent water flow in a rib-roughened rectangular channel by particle image velocimetry (PIV).

Given the limited body of experimental data, experiments are performed to study the unsteady turbulent flow inside a square, ribbed channel. In this study, two-dimensional PIV technique is implemented to measure the instantaneous velocity fields and turbulent statistical quantities. The research reported here was undertaken to fulfil two objectives, i.e. to gain sight into the physical process of separation and to provide experimental data of ribbed channel flows for validation of CFD models.

Wang *et al.* [46] examined experimentally the flow structures and turbulent properties associated with the flow separation in a square, ribbed channel by using PIV. The Reynolds number, based on the bulk-mean velocity and the channel hydraulic diameter, is fixed at 22,000. The ribs obstruct the channel by 15% of its height and are arranged 12 rib heights apart. Due to the flow periodicity, the investigated domain ranges from the sixth to seventh ribs. Two-dimensional velocity measurements are made in both the vertical symmetry plane and the horizontal planes.

The instantaneous velocity gives evidence that the separated shear layer is dominated by the coherent vortices, which are generated by the Kelvin–Helmholtz instability. Similar to the plane mixing layers, the growth rate of the separated shear layer is linear with respect to the streamwise direction. Moreover, it is noticed that the turbulence production, for both turbulent kinetic energy and shear stress, has a remarkable peak at $y/e = 1$, which approximately coincides with the inflexion point $d^2\langle u\rangle/dy^2$. Two distinct features near reattachment have been identified. First, the maximum shear stresses decrease rapidly just downstream of the reattachment. Second, the anisotropy parameters deviate to a small extent from unity at reattachment. Further downstream of the reattachment, the acceleration in the inner part coupled with deceleration in the outer part makes the redevelopment of the boundary layer different from the behaviour of an equilibrium boundary layer.

## 7 Conclusions

To develop sustainable energy systems, one must minimize the final use of energy, improve the efficiency of energy conversion and use renewable energy sources. In all these aspects, heat transfer and heat exchangers play a significant role, and this fact has been reviewed and illustrated throughout this article. In addition, heat transfer and heat exchangers also have great influence on the reduction of emissions and pollutants directly and indirectly. Therefore, the attempt to provide efficient, compact and cheap heat transfer methods and heat exchangers is a real challenge for research. Both theoretical and experimental investigations must be conducted, and modern scientific techniques must be adopted, such as CFD, laser techniques, liquid crystal thermography. By doing so, sustainable energy systems can be established, and this will contribute to global sustainable development.

## Acknowledgement

Financial support has been received from Swedish Energy Agency, Swedish Scientific Council as well as EU. Several graduate students have been involved in performing the various projects.

## References

[1] http://www.un.org/esa/sustdev/agreed.htm

[2] Woudstra, N., Towards sustainable energy systems. *International Conference on Efficiency, Cost, Optimisation, Simulation and Environmental*, Enschede: the Netherlands, pp. 1381–1392, 2000.

[3] Grossmann, I.E., Mixed-integer programming approach for the synthesis of integrated process flowsheets. *Comp. Chem. Eng.*, **9(5)**, pp. 463–482, 1985.

[4] Wang, L. & Sundén, B., On heat exchanger technologies in process integration. *Process Innovation and Process Intensification* $(PI)^2$ *Conference*, Edinburgh, Scotland, UK, 9–13 September 2002 (accepted).

[5] Shah, R.K., Compact heat exchanger surface selection methods. *Proc. Sixth Int. Heat Transfer Conf.*, Vol. 4, pp. 193–199, 1978.

[6] Linnhoff, B., Townsend, D.W., Boland, D., Hewitt, G.F., Thomas, B.E.A., Guy, A.R. & Marsland, R., *User Guide on Process Integration for the Efficient Use of Energy*, Institution of Chemical Engineers, Pergamon Press: Oxford, 1982.

[7] Wang, L., Performance Analysis and Optimal Design of Heat Exchangers and Heat Exchanger Networks, PhD Thesis, Lund Institute of Technology, Sweden, 2001.

[8] Wang, L. & Sundén, B., Performance analysis of some tube inserts. *Int. Comm. Heat Mass Transfer*, **29(1)**, pp. 45–56, 2002.

[9] Cengel, Y.A. & Boles, M.A., *Thermodynamics: An Engineering Approach*, 3rd edn, McGraw-Hill: Boston, 1998.

[10] Larminie, J. & Dicks, A., *Fuel Cell Systems Explained*, Wiley: New York, 2000.

[11] Selimovic, A., Modelling of Solid Oxide Fuel Cells Applied to the Analysis of Integrated Systems with Gas Turbines, PhD Thesis, Lund Institute of Technology, Sweden, 2002.

[12] Arcoumanis, C., Nagwaney, A., Hentschel, W. & Ropke, S., Effect of EGR on spray development, combustion and emissions in a 1.9L direct-injection diesel engine, SAE paper 952356, 1995.

[13] Cohen, H., Rogers, G.F.C. & Saravanamuttoo, H.I.H., *Gas Turbine Theory*, 4th edn, Longman Group Limited: Essex, England, 1996.

[14] Selimovic, F., Sunden, B., Assadi, M. & Selimovic, A., Computational analysis of a $CO_2$ separating membrane for a $CO_2$-emission-free power process, ASME, IMECE 2004-59382, 2004.

[15] Heck, R.M. & Gulati, S.R.F., The application of monoliths for gas phase catalytic reaction. *Chemical Engineering Journal*, **82**, pp. 149–156, 2001.

[16] Williams, J., Monolith structures, materials, properties and uses. *Catalysis Today*, **69**, pp. 3–9, 2001.

[17] Selimovic, F. & Sunden, B., Computational analysis of gas flow and heat transport phenomena in monolithic structures for high temperature processes, ASME HT 2005-72183, 2005.

[18] Rafidi, N. & Blasiak, W., Thermal performance analysis on a two composite material honeycomb heat regenerator used for HiTAC burners. *Applied Thermal Engineering*, **25(17–18)**, pp. 2966–2982, 2005

[19] Magnussen, B.F. & Hjertager, B.H., On mathematical modelling of turbulent combustion with special emphasis on soot formation and combustion. *16th Symposium (International) on Combustion*, The Combustion Institute, Pittsburgh, pp. 719–729, 1976.

[20] STAR-CD V3.20 Documentation (Methodology Part), CD-adapco, 2005.

[21] Versteeg, H.K. & Malalasekera, W., *An Introduction to Computational Fluid Dynamics*, Prentice Hall, 1995.

[22] Modest, M.F., *Radiative Heat Transfer*, McGraw-Hill, 1993.

[23] Bahador, M., *Radiative Heat Transfer Analysis of Relevance for Combustors*, Licentiate Thesis, ISSN 0282-1990, Lund University, 2005.

[24] Denison, M.K. & Webb, B.W., The spectral-line weighted-sum-of-gray-gases model for $H_2O/CO_2$ mixtures. *Journal of Heat Transfer*, **117**, pp. 788–792, 1995.

[25] Patankar, S.V. & Spalding, D.B., A calculation procedure for transient and steady state behavior of shell-and-tube heat exchanger. *Heat Exchanger Design and Theory Source Book*, eds N.F. Afgan & E.U. Sclunder, McGraw-Hill: New York, 1974.

[26] Patankar, S.V., Liu, C.H. & Sparrow, E.M., Fully developed flow and heat transfer in ducts having streamwise-periodic variations of cross-sectional area. *ASME J. Heat Transfer*, **99**, pp. 180–186, 1977.

[27] Patankar, S.V., *Numerical Heat Transfer and Fluid Flow*, McGraw-Hill Book Company: USA, 1980.

[28] Pope, S., *Turbulent Flows*, Cambridge University Press, 2000.

[29] Wilcox, D.C., *Turbulence Modeling for CFD*, 2nd edn, DCW Industries, Inc.: La Canada, California, 2002.

[30] Durbin, P.A. & Shih, T.I.-P., An overview of turbulence modeling. *Modeling and Simulation of Turbulent Heat Transfer*, eds B. Sunden & M. Faghri, WIT Press: Southampton, UK, pp. 1–26, 2005.

[31] Durbin, P.A., Separated flow components with $k$-$\varepsilon$-$v^2$ model. *AIAA Journal*, **33(4)**, pp. 659–664, 1995.

[32] Launder, B.E., On the computation of convective heat transfer in complex turbulent flows. *Journal of Heat Transfer*, **110**, 1112–1128, 1988.

[33] Sunden, B., On partially corrugated ducts in heat exchangers. *ASME/IMECE 2002 CD-ROM proceedings*, IMECE2002-39651, 2002.

[34] Bradshaw, P. & Wong, F.Y.F., The reattachment and relaxation of a turbulent shear layer. *Journal of Fluid Mechanics*, **52**, pp. 113–135, 1972.

[35] Troutt, T.R., Scheelke, B. & Norman, T.R., Organized structures in a reattaching separated flow field. *Journal of Fluid Mechanics*, **143**, pp. 413–427, 1984.

[36] Ruderich, R. & Fernholz, H.H., An experimental investigation of a turbulent shear flow with separation, reverse flow, and reattachment. *Journal of Fluid Mechanics*, **163**, pp. 283–322, 1986.

[37] Castro, I.P. & Haque, A., The structure of a turbulent shear layer bounding a separation region. *Journal of Fluid Mechanics*, **179**, pp. 439–468, 1987.

[38] Hasan, M.A.Z., The flow over a backward facing step under controlled perturbation: laminar separation. *Journal of Fluid Mechanics*, **238**, pp. 73–96, 1992.

[39] Le, H., Moin, P. & Kim, J., Direct numerical simulation of turbulent flow over a backward-facing step. *Journal of Fluid Mechanics*, **330**, pp. 349–374, 1997.

[40] Fröhlich, J., Mellen, C.P., Rodi, W., Temmerman, L. & Leschziner, M.A., Highly resolved large-eddy simulation of separated flow in a channel with streamwise periodic constrictions. *Journal of Fluid Mechanics*, **526**, pp. 19–66, 2005.

[41] Miyake, Y., Tsujimoto, K. & Nagai, N., Numerical simulation of channel flow with a rib-roughened wall. *Journal of Turbulence*, **3**, 035, 2002.

[42] Cui, J., Patel, V.C. & Lin, C.L., Large-eddy simulation of turbulent flow in a channel with rib roughness. *Int. J. Heat and Fluid Flow*, **24**, pp. 372–388, 2003.

[43] Nagano, Y., Hattori, H. & Houra, T., DNS of velocity and thermal fields in turbulent channel flow with transverse-rib roughness. *International Journal of Heat and Fluid Flow*, **25**, pp. 393–403, 2004.

[44] Leonardi, S., Orlandi, P., Djenidi, L. & Antonia, R.A., Structure of turbulent channel with square bars on one wall. *International Journal of Heat and Fluid Flow*, **25**, pp. 384–392, 2004.

[45] Islam, M.S., Haga, K., Kaminaga, M., Hino, R. & Monde, M., Experimental analysis of turbulent flow structure in a fully developed rib-roughened rectangular channel with PIV. *Experiments in Fluids*, **33**, pp. 296–306, 2002.

[46] Wang, L., Hejcik, J. & Sunden, B., PIV measurement of separated flow in a square channel with streamwise periodic ribs on one wall. *Journal of Fluids Engineering*, **129**, pp. 834–841, 2007.

# CHAPTER 2

# Advanced technologies for clean and efficient energy conversion in power systems

A.K. Gupta
*Department of Mechanical Engineering,*
*University of Maryland, College Park, MD, USA.*

## Abstract

Efficient fossil energy use with low pollution in all kinds of power plants is of growing interest in all industrial sectors. The use of energy is growing steadily by many countries worldwide due to the greater desire to enhance standards of living and increase productivity. Efficient energy use is favorable for better productivity, product quality, costs, and quality of human life but the energy use adversely impacts our environment. During the past decade, energy demands from China and India have grown significantly and are projected to grow even more in the coming decades as they enhance their living standards and productivity. This clearly requires close examination of the available methods of fossil fuel energy conversion as well as advanced methods that will provide increased efficiency and pollution reduction. After a brief review of the historical perspectives on energy conversion, this chapter reviews the various methods used for energy conversion in industrial power plants as well as new innovative methods that are now becoming available with significant fuel saving, compact size of the equipment and low pollution. As an example, high-temperature air combustion technology has seen wider acceptance in many kinds of industrial furnaces, in particular for use in the steel industry and fuel reforming furnaces with demonstrated energy savings of about 30% (on average), about 25% reduction in the size of the equipment (compact size), about 50% reduction in pollution and better quality of the product produced. Such significant energy savings were only dreams of engineers in the past. This technology has been adopted by many countries worldwide. In addition, advanced technologies are developed or being developed for use with coals (and solid fuels) since its cost based on energy content is low compared to gaseous fuels such as natural gas. Several commonly used methods in power generation are presented along with integrated and hybrid systems for increased efficiency.

A clear representation is provided of the output energy available from the known input energy. Some energy conversion techniques provide *in situ* capture of pollutants, while others must use air pollution control devices for pollution capture and/or pollution reduction prior to discharge of waste gases into the atmosphere. The information provides a quick guide on the commonly used methods of energy conversion in USA and Europe, in particular when using solid fuels such as coal. It is expected that future systems will be more energy-efficient and emit even lower pollution via the use of combined cycle systems and integrated energy conversion systems.

## 1 Introduction

Energy conversion is one of the important issues in the utilization of the chemical energy contained in the fuel in the desired form of energy for a specified need [1–30]. This has been the prime theme (formerly also known as the heat/ power engineering) for the numerous technological developments worldwide. Energy conversion deals with the conversion of energy (from such sources as fossil fuel, nuclear, biomass, various renewable and alternative energy sources) into a desired and convenient usable form of thermal, mechanical, or electrical energy. The resulting energy is used for various practical applications, including electricity, power, propulsion, transportation, heating and cooling. In some cases, the chemical energy desired is of high energy density and desired octane rating, while in other cases it is not of critical importance. But in all cases, it must be environmentally benign and of high efficiency. In the past, the main focus was on increased power and not as much on efficiency and pollution because of the relatively inexpensive costs of energy. However, the oil embargos of the 1970s, 1980s and 2000s together with the limited earth's resources and observed global warming caused by the increased carbon dioxide in the atmosphere have caused greater awareness among engineers to develop environmentally benign energy and even more energy-efficient devices. Innovative research and development efforts in an academic and research laboratory environments together with the entrepreneurial needs provide much of the new products. In most cases, the basic support has been provided by federal funds, which are then used by the industry for product development for specific needs. Examples to support this include development of commercial aircraft engines from military engines, automobile engines for increased fuel efficiency and flexibility with lower emissions, electric power from fossil, alternative and fossil fuels, micro-power generation for local use, portable electrical energy. Many times the budgetary issues on research and development efforts may impose an adverse effect on specific product development due to priorities for short-term issues over long-term issues. With the recent increased interest in energy, greater emphasis is placed on clean energy conversion from the limited fossil, alternative or nuclear energy sources. New as well as the more traditional energy conversion systems are being considered or reconsidered for increased efficiency and reduced pollution. A few examples include advanced designs for internal combustion engines, turbocharged engines, hybrid vehicles, combined cycle steam and gas turbine engines, advanced gas turbine cycle engines, fluidized bed combustors, advanced incinerators, coal gasification advanced power plants, zero emission power systems and miniaturized power systems. In fact, we are living in a rapidly changing world that requires rapid and continuous improvement of old technologies with new ones for efficiency improvement and reduced or sometimes even negative environmental burden. Energy conversion is a more complex, stimulating and viable field today than ever before because of the changed trend for significantly increased efficiency and negligible environmental impact.

Although we have made much progress in the energy conversion system, there is no doubt that progress in the electronic industry has been significant during the immediate past five decades. Table 1 gives a direct comparison in the energy conversion sector and its comparison with the electronic sector during the latter half of the 20th century. This table shows that although major advances have been made in both areas, the advances in electronic industry are more significant with laboratory on the computer chip using very large to super large-scale integration so that the size of the system is much smaller and simultaneously provides a much larger processing capability. In contrast, the progress in the energy/power industry has been primarily focused on larger-scale units with high energy efficiency and lower pollution until the early 1990s when a new concept of combustion was introduced, called high temperature air combustion (HiTAC), to provide simultaneous and significant benefits of energy conservation, lower emission (including emission of $NO_x$, $CO_2$, unburned hydrocarbons, carbon monoxide and soot), compact size of equipment, near uniform thermal field in the entire burning zone to be called as an isothermal reactor and low noise emission levels.

Table 1: Major advances in the electronics and the energy industry during the latter half of 20th century [7].

| Decade | Electronics industry | Energy/power industry |
|---|---|---|
| 1950s | Vacuum tubes | High energy conversion |
| 1960s | Transistors | Large-scale units |
| 1970s | Integrated circuits | Environmental pollution |
| 1980s | Printed circuits and large-scale integration | High efficiency |
| 1990s | Very large-scale integration | Higher efficiency, compactness, low pollution including, $CO_2$ and noise |
| 2000+ | Super large-scale integration, compact and high-density systems | Energy and environment conservation (zero emission) from variety of fuel sources |

## 2 Brief history of energy conversion

Before proceeding to the major revolutionary methods of energy conversion, it is appropriate to consider a few of the significant milestones on inventions and ideas on energy conversion. These are summarized in Table 2. It should be noted that Table 2 does not include the entire history, but only some major inventions so that a reader is referred to seek other references for gaining a historical perspective. Some of the omissions include V8 engine, ramjet, scramjets, rockets, space launch vehicles, micro-satellites. The struggling efforts of many heroes on their ideas to produce a practical hardware with the limited tools and resources available were quite challenging to produce machines for people and industry. In addition, the fuels available for these machines were not as well defined as

Table 2: Major significant invention milestones in various energy conversion areas.

| | | |
|---|---|---|
| Atmospheric engine using steam (first widely used heat engine) | Thomas Newcomen | 1700 |
| Steam condenser idea | James Watt | 1765 |
| | Boulton and Watt steam engine | (1775) |
| Gas turbine idea (patent) | James Barber | 1791 |
| Invention of internal combustion engine using hydrogen/oxygen fuel mixture | Francois Isaac de Rivaz (Switzerland) | 1807 |
| Two-stroke engine (patented) | Nikolaus August Otto | 1861 |
| Four-stroke engine operation | Nikolaus August Otto | 1876 |
| Diesel fueled engine (patent issued) | Rudolf Diesel | 1892 (1898) |
| Gas turbine turbo jet idea (patent issued) | Sir Frank Whittle | 1929 (1932) |
| Turbo jet concept (patent issued) | Hans Von Ohain | 1933 (1935) |
| Turboprop idea for long range transports | Herbert Wagner | 1934 |
| Regenerative combustion in engines | L. Jiang and A.K. Gupta | 1990 |
| High temperature air combustion (HiTAC). Also called Green flame and Colorless distributed combustion | NFK Ltd and A.K. Gupta | 1990 |

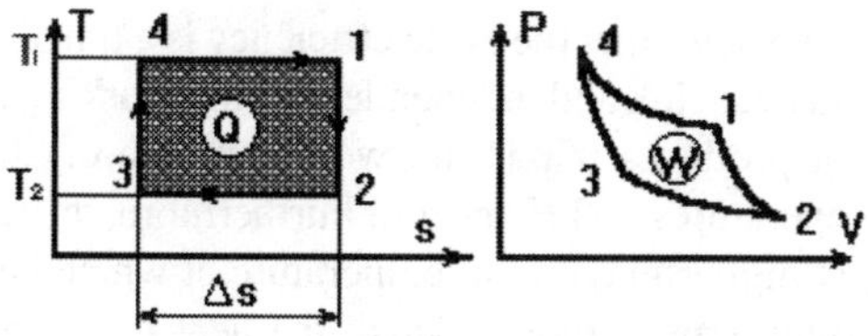


Figure 1: *T–S* and *P–V* diagrams for the Carnot cycle.

they are today. The progress on technology and industry was slow until a sound knowledge on thermodynamics became established in the 19th century. The development of nuclear power was in the second half of the 20th century based on the theoretical and experimental knowledge gained during the first half of the last century. The Table reveals that fundamental R&D has made a significant impact on the energy conversion needs in a range of areas for power, propulsion and transportation.

## 2.1 Basic energy conversion concepts

We now consider the basic concept of energy conversion using steam and air as the working media. The Rankine cycle is accepted as the basic standard for steam power plants. Before considering Rankine cycle, we consider Carnot cycle, as this cycle provides maximum possible efficiency. The ideal diesel cycle is a gas cycle and the Carnot cycle is a cycle for all fluids. The Carnot cycle provides the foundation for the second law of thermodynamics and the concept of irreversibility. The temperatures of the heat source and heat sink in this cycle provide thermal efficiency of a reversible cycle. Carnot cycle is hypothetical, as one cannot build a reversible engine. Figure 1 shows Carnot cycle on a *T–S* diagram consisting of four processes. Process 1 is reversible adiabatic compression (3-4). Process 2 consists of reversible constant temperature heat addition (4-1). Process 3 consists of reversible adiabatic expansion (1-2).

Finally, the process 4 consists of reversible constant temperature heat rejection (2-3). The thermal efficiency of the Carnot cycle can be determined. One must note that for the reversible case the magnitude of entropy change during heat addition and heat rejection are equal so that $T_H = T_1$ (or $T_4$) and $T_L = T_2$ (or $T_3$) since the heat transfer between the heat source and the working fluid occurs at no temperature difference. The Carnot cycle thermal efficiency is given by

$$\begin{aligned}\eta_{th} &= (Q_A - |Q_R|)/Q_A \\ &= \Delta W_{net}/Q_A \\ &= T_H(S_2 - S_3) - T_L(S_2 - S_3)/T_L(S_2 - S_3) \\ &= (T_1 - T_2)/T_1 = 1 - (T_2/T_1)\end{aligned}$$

The above relationship shows that the cycle efficiency is a function of temperatures of the heat source and heat sink and independent of the working fluid. Because this cycle is reversible, it produces maximum work possible of all cycles operating between the two temperatures of TH and TL. Furthermore, the real cycle will have a higher efficiency at higher maximum temperature at which heat is received and also at lower minimum temperature at which the heat is rejected. The reversible cycle operating between two given temperatures has the highest possible thermal efficiency of all cycles operating between these two temperatures. Carnot cycle is reversible and therefore has the highest possible thermal efficiency between any two given temperatures.

## 3 Energy and power generation

The amount of energy used as well as the distribution of different kinds of energy used to produce power in the world is given in Fig. 2. Coal oil and gas remain the major sources worldwide. Renewable energy accounts for about 11.5% of the world primary energy and presently provides only about 2.1% of the power generation. Since the amounts of fossil energy available are limited, we must make all the efforts to utilize the available energy at the highest efficiency possible. Most simple cycle power plants have efficiency of the order of about 30–35%. This is because all practical cycles are not reversible and therefore have some efficiency (or entropy production) associated with them. As an example, all turbines and pumps are never isentropic and so we must seek novel means to minimize the isentropic losses. Therefore, a design engineer always has a goal to maximize the efficiencies (or minimize the isentropic losses) by making some modifications to

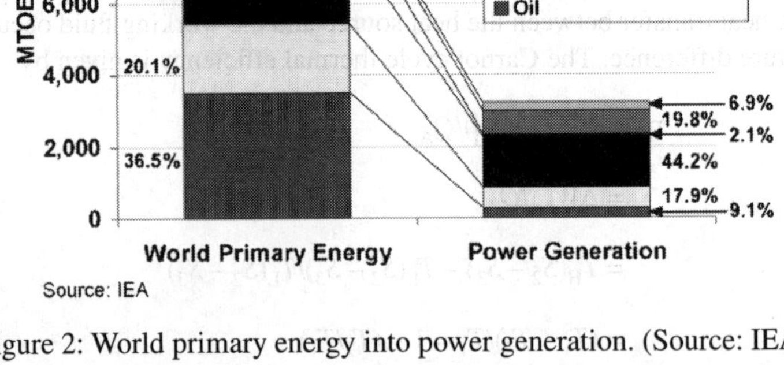


Figure 2: World primary energy into power generation. (Source: IEA.)

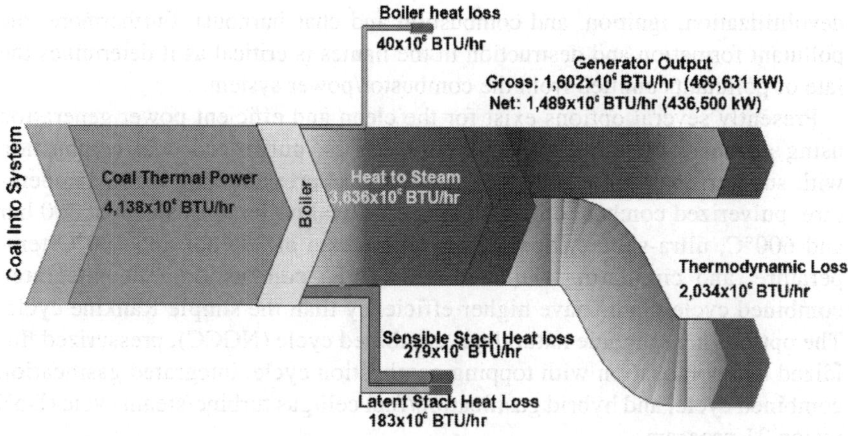


Figure 3: Major thermal energy power usage per unit time from a pulverized coal fired unit with no carbon capture.

the cycles. A schematic diagram of the various losses associated with a power plant is given in Fig. 3. A number of methods have been used to minimize such losses. In no case has the irreversibility and other losses present in power plant been completely eliminated.

### 3.1 The steam cycle

In the steam cycle, called Rankine cycle, steam at high temperature and high pressure produced from a boiler is expanded in a steam turbine. The work output is used to drive an electrical generator. The low pressure steam rejects heat in a condenser (heat sink) and the condensate (water) is pumped back to the boiler since the water treatment for use in a boiler must be free of unwanted dissolved and other materials that can adversely build up scaling on the inside surfaces of the boiler and boiler tubes. The thermal energy gained by water, to cool the condensate, is dissipated by the use of cooling towers. Thus, the coolant and condensate are kept separate to maintain the desired high quality water in the steam cycle. Steam cycle can use gas, oil, coal, biomass or their mixtures. The addition of small amounts of biomass (about 5–10%) to coal as fuel is particularly attractive because it does not require much hardware changes with significant benefits on reduction in fuel costs.

The most traditional approach to burn coal for steam power generation has been via pulverized coal combustion. In this approach, the burners utilize very small size coal particles that are mixed with some air and injected into the combustion chamber for ignition and complete combustion within the combustion chamber. However, the fly ash leaving the combustion chamber at high temperatures and entering the convective section of the boiler must be cooled down to temperatures below its softening temperature to avoid fouling of the heat exchange surfaces. It is therefore important to consider important design considerations associated with coal combustion (that includes the process of coal devolitilization, ignition, and combustion and char burnout). Furthermore, the pollutant formation and destruction in the flames is critical as it determines the fate of pollutants emitted from the combustor/power system.

Presently several options exist for the clean and efficient power generation using advanced Rankine cycle plants, such as, pulverized coal combustion with supercritical boiler with steam at 245 bar pressure and 565°C temperature, pulverized combustion in ultra supercritical boiler with steam at 300 bar and 600°C, ultra-supercritical boiler with steam at 375 bar and 700°C temperature, and circulating fluidized bed (CFB) combustion. The integrated combined cycle plants have higher efficiency than the simple Rankine cycle. The options here include natural gas combined cycle (NGCC), pressurized fluidized bed combustion with topping combustion cycle, integrated gasification combined cycle, and hybrid gasification/fuel cell/gas turbine/steam cycle (DoE vision 21 program).

## 3.2 Pulverized-coal firing system

Pulverized coal combustion in the Rankine steam cycle mode has been the frequently used mode of burning coal in power plants since the 1920s worldwide, with characteristic subcritical steam parameters of 163 bar and 538°C. The lower heating value (LHV) efficiency as high as 40% can be obtained. Pulverized coal combustion in supercritical boilers has been in operation since the 1930s with plant efficiency as high as 45% as illustrated by Schilling [2]. The first two are waste gas heat loss from the boiler that can be as much as 6–8% and represents the largest heat losses from a boiler. The exit gas temperature from the boiler can be reduced from the dew point design considerations of the flue gases. The role of subcritical, ultra-critical and supercritical steam (referring to the steam conditions given above) has a significant effect on the efficiency and percentage of $CO_2$ emission reduction. The plant efficiency increases by about one percentage point for every 20°C increase in superheat steam temperature. Figure 4 shows the net plant efficiency gains and carbon dioxide emission reduction obtained from going from subcritical steam plant to ultra-supercritical pulverized coal plant [3]. It is anticipated that with the new construction of ultra-supercritical plant in the next decade it will provide efficiency of about 50% for coal-fired plant (LHV), which can also provide reduction of about 25% in carbon dioxide emission as well as significant reduction of other pollutants.

Relationship exists between the excess air of combustion and low limit of exit gas temperature using sulfur-containing fuel. Excess air favors the oxidation of $SO_2$ to $SO_3$, which is then converted to sulfuric acid ($H_2SO_4$) in the combustion products. Sulfuric acid vapors increase the dew point of the flue gases and so the permissible minimum exit temperature of the flue gases decreases (i.e. high exhaust gas temperatures). Only about 1 ppm of $SO_2$ in the gas stream raises the due point of the flue gases by about 62°C. This effect is not liner, if the concentration of $SO_2$ increases to higher values. At stack gas exit temperature of 130°C, the plant efficiency increases by about 0.3% for every 10°C decrease in boiler exit temperature.

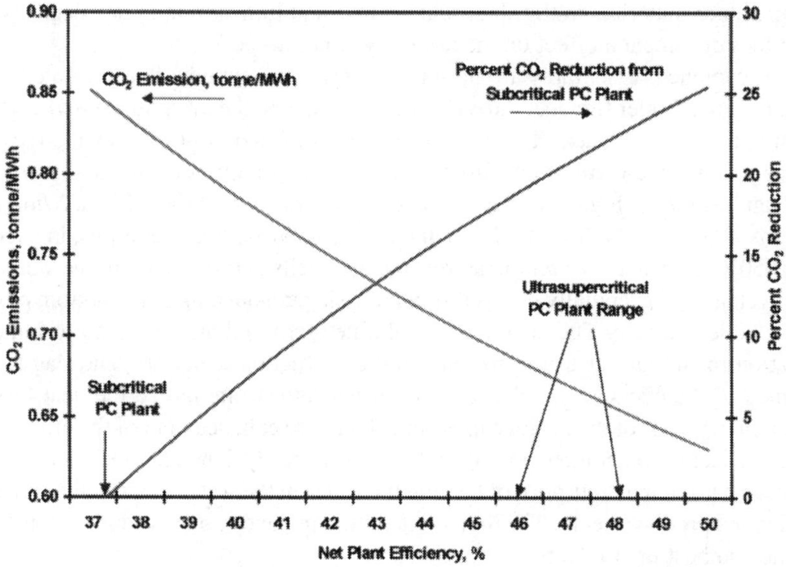


Figure 4: Net plant efficiency vs. carbon dioxide emission (based on firing Pittsburgh #8 Coal) [3]. PC: pulverized coal.

The pulverized coal system consists of coal pulverizer, fuel delivery and burning equipment to produce steam. A pulverized coal burner receives dried pulverized coal in suspension in the primary air which mixes with the main combustion air from the steam generator's air preheater. The surface to volume ratio pulverized coal requirements vary from the coal type (the higher the fixed carbon the finer the coal). For example, pulverized coal with 80% passing through a 200-mesh screen and 99.5% passing through 50-mesh screen have a surface area of approximately 1500 $cm^2/g$ with more than 97% of the surface area passing the 200-mesh screen.

The Rankine cycle efficiency is directly proportional to the pressure and temperature of the steam with which heat is added to the cycle and inversely proportional to the condenser pressure and thus the temperature of the cooling medium. The common design pressure is 2 inches of Hg absolute (about 67 mbar) in USA, while it is 1 inch of Hg absolute in Europe. This reduced pressure operation can provide net efficiency gains of about 2% in the power plant.

### 3.3 Cyclone furnaces

Cyclone furnace firing, developed in the 1940s, represents the most significant step in coal firing since the introduction of pulverized coal firing in the 1920s. It is widely used to burn low-grade coals with high as content (6–25%), high volatile matter (above 15%) and wide range of moisture content with pre-drying. One limitation is that the ash should not contain high sulfur content or high ratio of ($Fe_2O_3/CaO + MgO$). Such a coal has the tendency to form high ash fusion temperature materials in the slag, such as iron and iron sulfide in the slag, which provides detrimental effect on the use of cyclone furnaces.

The cyclone is essentially a water-cooled horizontal cylinder that is located outside the main boiler furnace. Pulverized coal is fed into the cyclone to provide very high rates of heat release. The combustion is completed prior to allowing the hot gases to enter the main boiler furnace. The swirling motion of the coal particles and air results in high volumetric heat release rates of 4700–8300 kW/m$^2$ (or approx. 450,000–800,000 Btu/h.ft$^3$) and high combustion temperatures, in excess of 1650°C. At such high temperatures, the ash melts and flows along the walls of the cyclone and eventually drains through the slag tap opening at the bottom of the main boiler furnace. The slag is then solidified, crushed and removed to remote location for its further use in roadbed and construction materials. The slag layer formed in the cyclone chamber acts as an insulator to minimize the heat losses through the walls of the cyclone thus contributing to enhancement of the efficiency of the system. The combustion of fuel occurs in the cyclone chamber. The boiler furnace has the main purpose of heating the water in the water tubes in the furnace boiler and produce steam. The high temperature in the cyclone produces high $NO_x$ in the combustion products.

## 3.4 Fluidized bed combustion

In a fluidized bed, combustion occurs at lower temperatures compared with cyclone and pulverized coal boilers. The fluidized bed contains solid particles (sand) and fuel particles (coal) that are in intimate contact with a fluid passing through at a velocity sufficiently high to cause the particles to become freely supported by the fluid. In contrast, the fixed bed has too low of a velocity to cause fluidization. The minimum fluid velocity necessary for fluidization can be calculated by equating the drag force on a particle due to the fluid motion to the weight of the particle. The fluidized bed provides intimate contact between the fuel and air and the combustion of coal particles occur in a hot bed of sorbent particles that are in the fluidized state. Enhanced heat transfer from the bed to the heat exchanger tubes that are embedded in the fluidized bed occurs in the fluidized combustor. Another advantage of the fluidized bed combustion is the removal of sulfur during the combustion process. CFB is the most common fluidized bed design today. In this design, high gas velocities (about 4–10 m/s) entrain large portion of the solids, which are then separated from the flue gases and recirculated to the lower part of the furnace to achieve good carbon conversion efficiency and $SO_2$ sorbent utilization. A cyclone is used for the separation of particles and hot gases at the furnace exit.

Removal of sulfur dioxide during the combustion process is achieved by adding limestone to fluidized bed along with the crushed coal. The limestone is converted to free lime, which reacts with $SO_2$ to form calcium sulfate. Under steady state conditions, the fluidized bed consists of unburned fuel, limestone, free lime, calcium sulfate, ash and sand bed material. The temperatures in the bed are fairly uniform (850–900°C) due to very good mixing and the residence times are very long so that carbon conversion efficiency is very high. The above combustion temperature range in the fluidized bed is optimum for *in situ* capture of $SO_2$ by the available free lime in the bed. In addition, the bed temperature is low for thermal $NO_x$ formation. Therefore, fluidized bed is also good from the point of view of $NO_x$ emission. However, low temperature combustion in CFB results in higher levels of $N_2O$ emission (in the range of 40–70 ppm) since it survives at temperatures below 1095°C. $N_2O$ is a green house gas with global warming potential of 296 times higher than $CO_2$ so that the characteristic $N_2O$ emission from CFB corresponds to about 15% increase in $CO_2$ emissions.

# 4 Efficiency improvements in power plants

## 4.1 Combined cycle power plants

Fossil fuels constitute 85% of the total energy consumption. It is anticipated that this consumption will increase in the next 25 years. It is well known that the supplies and reserves of fossil fuels are limited, lasting about 50–300 years depending on the fuel type, use, recovery rate and exploration. The use of fossil fuels suggests good evidence of increase in $CO_2$ emissions, global temperature and long-term climate change. High efficiency systems, as well as increased utilization of carbon neutral renewable energy (also nuclear fuels), offer the benefits of controlling the rate of $CO_2$ increase and global temperature rise. The current market share of renewable energy consumption is about 11.5%, while that of electricity production is only about 2.1%. Therefore, efforts must be made to enhance power plant efficiency using fossil and renewable fuels.

Combined cycle plants that incorporate gas turbine and steam cycle plants offer good advantages of enhancing the plant efficiency. This is because the temperature range of the Brayton (gas turbine cycle) and Rankine (steam cycle) are in successive ranges for the two cycles and so the waste energy from the gas turbine Brayton cycle (operational range 900–1600 K) can be used for operating the steam Rankine cycle (operational range 288–850 K). Therefore, combined GT and stream cycle can provide a greater thermodynamic cycle efficiency compared with single cycle alone, see Fig. 5 [1]. In a characteristic gas turbine cycle, the thermodynamic efficiency from gas turbine can be about 38% and the rest of the energy (about 61.5%) is lost in the exhaust due to high temperature of the exhaust gases. The radiation losses account for only about 0.5%. However, if exhaust gases are used for operating steam turbine cycle, the efficiency of this cycle can be about 21% of the input fuel energy. The condenser losses in the steam turbine cycle account for about 30%, while stack and radiation losses account for 10% and 0.5%, respectively. Therefore, the net efficiency of the combined cycle plant is enhanced to near 59%. This net gain in efficiency has a significant effect on energy use, $CO_2$ and other pollutant's emission, and global temperature rise.

High carbon to hydrogen ratio of natural gas makes NGCC power plants environmentally benign, thus allowing them to be located in areas of high population density. These plants are primarily used as distributed power generation or heat and cogeneration plants. However, due to high price of natural gas (compared to coal on

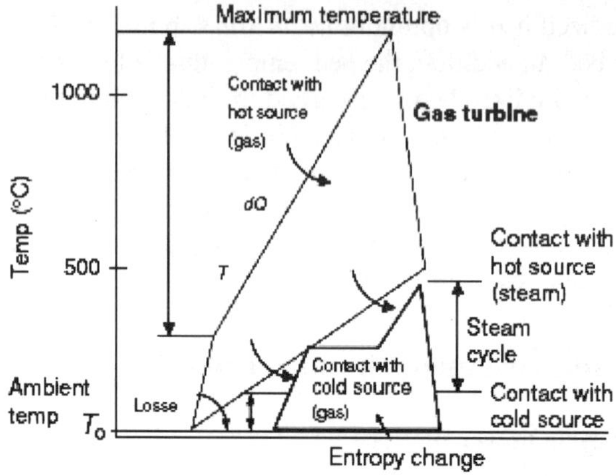


Figure 5: Temperature–entropy (*T*–*S*) diagram of combined cycle having Brayton cycle gas turbine and Rankine steam turbine cycle [1].

the basis of unit energy content), use for central power production is not economical. The use of coal is favorable due to its abundant supply in coal rich countries (e.g. USA) and availability of clean coal technologies for efficient and clean power production. It is, therefore, favorable to use coal-based technologies to produce power, as coal is much cheaper than natural gas or oil. It is expected that this cost issue of high natural gas price compared to coal will not change in the future too.

## 4.2 Hybrid integrated power plants

Combined cycle power plants use more than one unit to maximize overall efficiency by operating each unit within its own best designed parameter. In the case of combined cycle, the wasted energy from the Brayton cycle can be effectively used in the bottoming Rankine cycle. Many other examples exist on the use of integrated system, such as waste heat from the power plant for district heating, superheat and reheat cycles, hybrid power, energy conversion and propulsion units integrated through a storage unit. In the transportation sector, some specific examples of such hybrid systems include internal combustion engine/electric motor automobiles, fuel cell/batteries/internal combustion engines. Integration of conversion and storage units has been shown to be very promising in bridging the gap between demand load and available energy source when using intermittent sources. Advanced concept plants that use multiple energy sources, such as, fossil fuel and solar thermal energy, overcome the intermittency and maximize overall thermal efficiency. Combined power and heat units have been used in minimizing waste heat in decentralized power thus increasing overall efficiency of the plant. Integration can be in several forms that satisfy the specific criteria. We expect that integrated systems will continue to evolve and grow further. This is true of larger electricity plants, with a characteristic lifetime of about 50 years, so that they can reach efficiency of over 60% and capture all the $CO_2$ released from the plant for its sequestration or further use. Power plants and refineries can be integrated optimally to exploit the synergy between the processes of simultaneous electricity production and fuels transportation. Hybrid power trains that integrate high temperature fuel cells and thermoelectric devices with the internal combustion engine and battery will result in higher fuel mileage. Efficient and economical expansion in the use of renewable resources (since fossil fuels are limited) including biomass, wind, solar will greatly benefit from integration with fossil fuel systems. The effort to achieve maximum conversion efficiency through integration will not be successful unless it is followed by a gradual transition strategy that fosters economic and social issues. The examples of combined cycle power plants and hybrid power trains are good records of integrated energy systems as they are now commonly used and widely accepted.

## 4.3 Other methods to increase efficiency

The area under the *P–V* curve (see Figure 1) provides the net power output from the plant. Therefore, operation at higher pressures and heat rejection at lower temperature provides an increase in net power output. In Rankine cycle, this has been used much in the industry. Many of the European power plants operate at supercritical steam temperatures, while the North American plants are typically operated at lower steam temperatures. The increase in power plant efficiency associated with higher pressures and temperatures, as well as reducing the discharge at lower pressures and temperatures, results in higher efficiency. This is illustrated in Fig. 6. Plant operation at higher temperatures and low back pressure can increase the plant efficiency significantly.

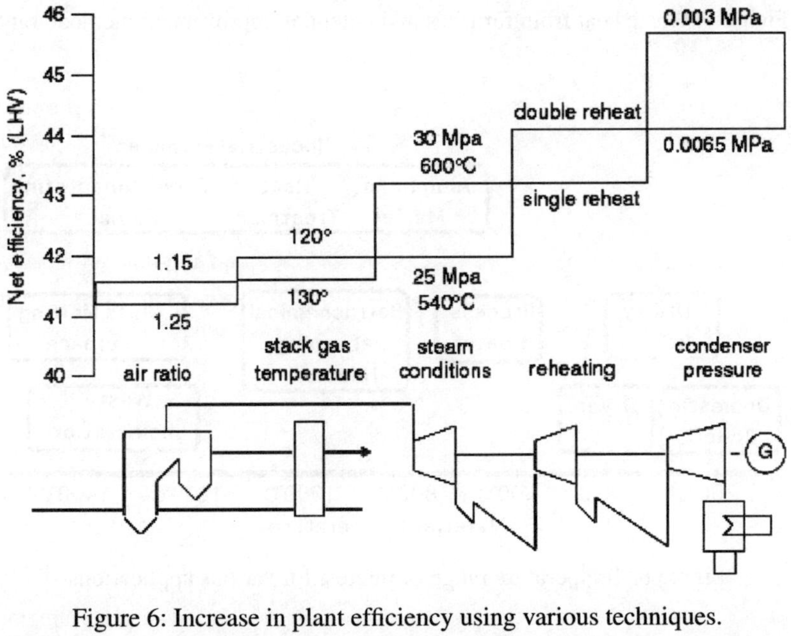


Figure 6: Increase in plant efficiency using various techniques.

## 4.4 Waste heat recovery for improving efficiency

Plant efficiency increases with decrease in heat losses from the stack, operation at higher temperature and pressures, combined heat and power systems, use of integrated energy conversion systems.

In the combustion zone, not all the heat is used to produce power. In practical power plants, much of the heat is lost via the stack or tail end of the boiler or power plant, see Fig. 7. By utilizing much of this waste heat, significant gains in the thermal efficiency can be obtained. The characteristic efficiency of a power plant is only about 30%.

However, by utilizing the waste heat, significant increase in plant efficiency can be achieved. The extent of waste heat from a plant varies with the type of plant. This is because, for each application, there is certain temperature range of interest, see Fig. 8.

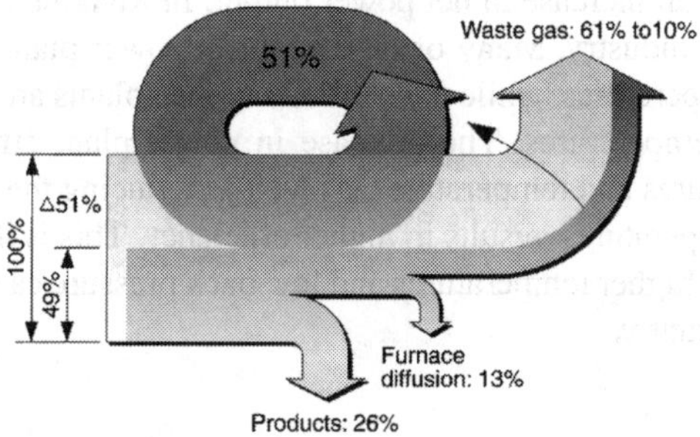


Figure 7: Waste heat from furnaces and potential improvement methodology.

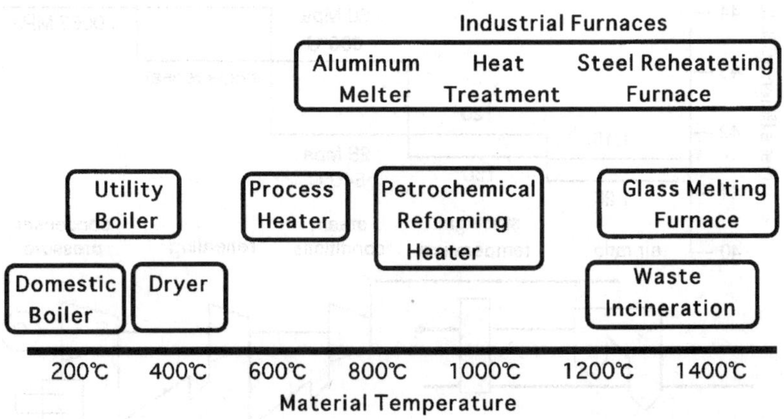


Figure 8: Temperature range of interest for various applications.

# 5 High temperature air combustion technology

Recent advances on high temperature air combustion (HiTAC), also called colorless distributed combustion or flameless oxidation, have demonstrated significant energy savings, higher and uniform thermal field, lower pollution including noise, and smaller size of the equipment for a range of furnace applications. In HiTAC, combined heat and gas recirculation are used to provide high temperature and low oxygen concentration so as to enlarge and control the flame thermal behavior. This technology has shown promise for much wider applications in various process and power industries, energy conversion and waste to clean fuel conversion. For each application, the flow, thermal and chemical behavior of HiTAC flames must be carefully tailored to satisfy the specific needs. Sample results are given from a few diffusion flames using high temperature combustion air. A specially designed regenerative combustion test furnace facility, built by Nippon Furnace Kogyo, Japan, was used to preheat the combustion air to very high temperatures. The flames with highly preheated combustion air were significantly more stable and homogeneous (both temporally and spatially) compared with the flames with room temperature or moderate temperature combustion air. The global flame features showed the flame color to change from yellow to blue to bluish-green to green to colorless or flameless distributed in the entire combustion zone over the range of conditions examined. In some cases, hybrid color flame was also observed. The flameless or colorless oxidation of the fuel was observed only under certain conditions. Very low levels of $NO_x$ along with negligible levels of CO and hydrocarbons (HC) have been obtained using HiTAC. The thermal and chemical behavior of HiTAC flame depends on air preheats temperature, oxygen concentration of air, fuel chemical property and fuel–air mixture preparation. Waste heat from a furnace in HiTAC technology is retrieved and introduced back into the furnace using a regenerator. These features help save energy, which subsequently also reduce the emission of $CO_2$ (greenhouse gas) into the environment. Flames with high temperature air provide significantly higher and uniform heat flux than normal air, reduce noise, reduce the equipment size or increase the material throughput for the same size of the equipment. The HiTACHiTAC technology has shown practical demonstration of significant energy savings (up to about 60%), downsizing the equipment (about 25%), pollution reduction (about 50%), and low noise (about 7 dB) in practical systems. Fuel energy savings directly translate into reduction of $CO_2$ and other greenhouse gases to environment. HiTAC has been used to demonstrate fuel reforming at high efficiency and also clean conversion of wastes to hydrogen-rich syngas and also liquid fuels.

## 5.1 Background on HiTAC

High temperature air combustion (also called colorless distributed combustion, flameless oxidation) of fuels is a new and innovative method for the conversion of chemical energy to thermal energy from hydrocarbon fuels. The concept of excess enthalpy combustion originated almost three decades ago for the combustion of low heating value fuels [4]. In excess enthalpy combustion, the thermal energy released is fed back to the fresh reactants so that the temperatures obtained with excess enthalpy combustion are much higher than its counterpart with normal temperature air. However, colorless oxidation of fuels is obtained by using the design and fundamental principles of HiTAC technology in which high temperature combustion air at low oxygen concentration (obtained by heat and gas recirculation) is mixed with the fuel prior to the combustion of fuel. High temperature air is obtained by preheating the air with the exhaust gases from the furnace or reactor. The peak temperature in the flame zone is much reduced with the use of diluted low oxygen concentration combustion air even though the air is preheated to high temperatures. This low oxygen concentration (or diluted) air is obtained by mixing portion of the exhaust gases by recirculating part of the combustion products into the incoming hot combustion air. The combustion air is preheated to temperatures in excess of 1000°C, depending on the application, using honeycomb type or ceramic ball type heat exchangers. Thus, HiTAC uses fresh air for combustion, but this air is prepared such that the temperature and chemical composition is appropriate for the specific application to provide colorless distributed combustion. Honeycomb type heat exchangers are more effective than the ceramic ball type heat exchangers [5]. Most of the previous research activities have focused on gaseous fuels such as methane, propane, liquefied petroleum gas and process gases [5–37]. The HiTAC technology has also been demonstrated to provide significant benefits with other gas, liquid, solid and waste fuels [7, 38] for applications to various kinds of furnaces, processes, and industrial and power systems.

Combustion research works are always concerned with meeting various industrial, societal, national and global needs. An example of this is excess enthalpy concept that was used for the combustion of low heating value fuels without using any auxiliary fuel. Global environmental problems, including acid rain, particulate, ozone layer destruction, and global warming from greenhouse gases, such as $CO_2$ and methane, are now recognized to cycle between atmosphere and the earth. Simultaneous requirements of both environment protection and energy conservation require rapid development of combustion science and technology for various applications. The science and technology of combustion has made a significant progress during the past century, in particular, after the World War II. For the furnaces area, HiTAC technology has shown to be revolutionary for energy conservation, reduction of pollution generation including $NO_x$ and $CO_2$, far uniform thermal field and higher heat flux than previously possible, to give superior quality product [7, 13, 33]. Higher heat flux is from increased radiative heat transfer. This increase is from higher and superior uniform temperatures in the entire combustion chamber.

Recent progress in HiTAC is first reviewed here followed by hydrocarbon fuel reforming to hydrogen and more recent application of HiTAC for the gasification of coal, wastes and low-grade fuels to produce hydrogen-rich syngas. Data are obtained from laboratory scale investigations as well as pilot plant studies in Japan [33]. Uniform thermal field in HiTAC furnace has also been used for fuel reforming of hydrocarbon fuels to hydrogen [37]. The gasification studies provide innovative technology to convert wastes to hydrogen-rich syngas [38–40].

Fossil fuels, such as, coal, oil and gas, have been used by the society for millions of years. These fuel sources have never been considered exhaustible until recently. Rapid industrialization and growth of human society have resulted in rapid utilization of natural resources on the earth. This then leads to the reality that in the very near future, the fossil fuels, which encompassed our main energy consumption cycle, will run out. Although the nuclear and solar energy may provide more promise for the future, fossil fuel cannot be quickly replaced for all applications, at least in the near future because of their unique advantages including those of mature utilization technologies, non-radioactivity, safety, high conversion efficiency and cost effectiveness. However, when one considers the negative impact of fossil fuels, besides the limited reserves, concerns over environmental issues are quite serious.

The HiTAC technology is one of the most promising combustion techniques that provide solution for energy savings and pollution formation and emission. HiTAC technology is particularly attractive for processes that require uniform temperature, and higher and uniform heat flux distribution in the combustion zone, and low noise. Controlled flame behavior results in uniform thermal field, significantly lower emissions including combustion noise, and smaller chamber size for processing the same material or increased material throughput for the same furnace size. However, to explore further the full potential of HiTAC technology, one must seek a greater understanding on the detailed structure of highly preheated air flames so that the results can be translated into wide applications.

## 5.2 Benefits of HiTAC technology

### 5.2.1 Energy saving

HiTAC technology uses regenerative heat exchangers to extract thermal energy from the waste gases in the combustion products. Ceramic honeycomb or balls have been shown to provide good energy storage heat media heat exchangers. Honeycomb type heat exchanger is more effective than ceramic ball type, as this provides larger surface area, low-pressure drop and high efficiency. Exhaust gases from industrial furnaces and processes represent one of the major heat losses from the system. The regenerative media used in the HiTAC devices recovers large amounts of thermal energy from the exhaust gases, which can then be transported back to the combustion zone. The amount of energy recovered translates directly into fuel saving. In the ‘High Performance Industrial Furnace Development’ project [2, 30], the objective was to demonstrate significant energy savings (about 30%) using regenerative media, reduce physical size of the equipment by about 25%, and pollutants emission reduction (about 50%). Indeed the HiTAC technology has demonstrated average energy savings of about 30% from industrial furnaces. This goal was successfully demonstrated using 167 field demonstration tests [7, 33] with energy savings of 150,000 kiloliters of fuel. Now this technology is used in many countries worldwide.

### 5.2.2 $CO_2$ reduction

The role of $CO_2$ in global warming phenomena is now widely recognized. The demands for reducing $CO_2$ emission are higher than ever before. All fossil fuels contain carbon, which generates $CO_2$ as a byproduct during the combustion of carbon so that any efforts to reduce energy consumption will directly translate into reduction of $CO_2$ emission. Good correlation between fuel consumption and $CO_2$ production suggests that $CO_2$ reduction should be nearly the same as energy saving [33].

### 5.2.3 $NO_x$ reduction

Emission of $NO_x$ is now known to be responsible for the destruction of ozone layer in the upper atmosphere. In addition, $N_2O$ emission is of concern from combustion systems (for example in low temperature combustion devices, such as fluidized bed, that produce several tens of ppm of $N_2O$. This gas has 296 times more greenhouse gas potential than $CO_2$. $NO_x$ (primarily NO and $NO_2$) involves the complicated reaction mechanisms, which result in accelerating the ozone depletion in the oxygen cycle on earth. Therefore, combustion engineers have focused their attention on developing various strategies to reduce $NO_x$ emission and improve the combustion process. HiTAC is one of the most advanced techniques because of low levels of $NO_x$ formation and emission, significant energy savings and uniform thermal properties in the entire combustion space [7, 22, 24, 33].

### 5.2.4 Equipment size reduction

Higher and far uniform thermal field and heat flux distribution is obtained with HiTAC technology compared to any other known method for combustion using normal temperature air [16, 17, 33], air preheated to some moderate temperatures using recuperators or with the use of air preheaters. The ignition delay time is different from that obtained with room temperature air for combustion [8, 11]. This means that many industrial combustion equipment has a potential for reduced size, better product quality from the process and material's conservation with high temperature combustion air technology [18, 24, 33].

### 5.2.5 Noise reduction

Recent studies on HiTAC have shown a significant reduction (over 7 dB) in combustion noise from HiTAC flames. This is attributed to smaller volumetric expansion with the HiTAC flames caused by distributed combustion features. Results on noise reduction under colorless distributed combustion conditions are given in refs. [33–36].

### 5.2.6 Uniform thermal field

The HiTAC combustion technology has shown temperature distribution uniformity in the entire combustion to be better than 30 K. This has never been possible before so that one can classify HiTAC to be an isothermal reactor. Simultaneous realization of the above benefits was combustion engineers' dream and considered impossible before. The unique flame features associated with HiTAC assist in utilizing this technology for wider range of applications than initially envisioned [7, 13, 33].

## 5.3 Basic principle of HiTAC technology

### 5.3.1 Thermodynamic consideration of combustion process

Flame temperature is one of the important factors for considering combustion efficiency and energy conversion efficiency. Weinberg [4] provided the initial concept of excess enthalpy combustion. In his study, limitations on combustion temperature were discussed, including both positive and negative factors associated with combustion temperatures in a certain range. Heat circulation from the exhaust gases, using high-efficiency heat exchanger, was adopted to increase the combustion temperature thus allowing the combustion of low heating value fuels that could otherwise not be combusted without using auxiliary fuel. However, combustion engineers have to pay attention to the upper limit of the combustion temperature because of material constraints used in the equipment and/or pollution formation at higher temperatures, in particular $NO_x$. Various possibilities on enthalpy intensification have been described [5]. From the economic point of view, it is of course better to use thermal energy generated by combustion process itself to heat-up the oxidant or fuel, which is often of low thermal energy, than via the use of electrical or mechanical energy.

The amount of combustion-generated energy circulated into the combustion process is given as [4]:

$$\int_{T_0}^{T_t} C_p dt = Q_c + Q_a = H_t - H_0$$

where $T_f$ is the final temperature, $T_0$ is the initial temperature, $Q_c$ is the heat release by chemical energy conversion, $Q_a$ is the energy added, $H_t$ and $H_0$ are the enthalpies at two states. The circulation part of thermal energy from combustion-generated products will increase the combustion temperature and so the enthalpy of the reaction zone will be above the conventional combustion level. This has resulted in the use of the term called "excess enthalpy combustion."

Increase in thermodynamic efficiency must be coupled with other desirable characteristics, such as low $NO_x$ formation, reliability of the equipment and refractory material, and spatial uniformity of temperature. Nonetheless, the heat circulation and excess enthalpy methods provide new light on next generation of advanced energy conversion technology and combustion chamber design. This method provides new ideas to control temperature in the combustion zone. The method is independent of the fuel composition, and can simultaneously satisfy the demands of high combustion intensity and reduced pollutant formation from fuels, including low-grade fuels.

### 5.3.2 Basics of excess enthalpy combustion design

Excess enthalpy combustion can be realized by internal or external circulation or their suitable combination. Most designs for internal heat circulation use bluff body, porous media or swirl to the flow. Most of these are aimed at flame stabilization [19] and are very effective. Internal heat circulation relies on heat convection and species circulation so that pool of hot and active radicals in reaction zone can be maintained, which also assists in flame stabilization.

The external circulation methods use heat exchanger to transfer the thermal energy via conduction between combustion products and solid media so that heat can be exchanged to the cool reactants. Most of the enthalpy contained in hot combustion products can be recirculated back to the combustion process. This allows one to utilize much of the waste energy for flame stabilization.

### 5.3.3 Excess enthalpy combustion and high temperature air combustion

Thermodynamic considerations suggest that preheating the oxidant to very high temperatures (not fuel due to possible fuel decomposition and safety) increases the cycle thermal efficiency. Air preheats add enthalpy to the combustion zone. From the point of pollutant formation, especially $NO_x$, high temperature of combustion has been recognized as one of the most important parameters for high levels of $NO_x$ emission. This then requires method to control the emission of $NO_x$. One method of maintaining the same temperature in the combustion zone is to dilute the incoming combustion air with hot combustion products to cause mild combustion condition. Partial recirculation of hot products into the incoming combustion air dilutes the inlet concentration of oxygen in the air. This diluted air lowers the peak flame temperatures with negligible oxidation of $N_2$. Further examples of this are given in refs. [5, 7, 20, 33].

The goal of HiTAC was to demonstrate a significant pollution reduction from industrial furnaces (by about 25%), to enhance efficiency and achieve fuel energy savings (by about 30%), to reduce equipment size (by about 25%), and to reduce noise (by about 7 dB) using honeycomb type regenerative burners that can be classified as operating under excess enthalpy principles. In the North American design, ceramic balls have been used in the regenerators to preheat the combustion air with hot gases exiting the furnace. The use of ceramic balls in a packed bed provided higher combustion air temperatures; however, the regenerator efficiency is not as high as that of the honeycomb type regenerator. Nonetheless, the air preheat temperatures were much higher than those achieved previously by using recuperators. In the HiTAC technology of NFK, a honeycomb type regenerator is used. The honeycomb regenerator is much more compact than a bed packed with ceramic balls because of high specific surface area, low thermal inertia and gives very low pressure drop [5, 18]. In the HiTAC technology, low oxygen concentration air at high temperatures is used for combustion air. In conventional burners, increasing the air preheat temperature increases $NO_x$ emission levels. However, with high temperature combustion, the temperature increase in the furnace or reactor is small (only about 30–100°C above the incoming high temperature combustion air). The oxygen concentration in the combustion air is very low (only about 2–5% by volume, depending on the application). Under these conditions, the thermal field in the combustion zone is very uniform [17, 32, 33]. The peak temperatures in the combustion zone are suppressed to result in very low $NO_x$ emission levels. The heat flux from the flame with high temperature combustion air is also very high [16, 17, 33]. A schematic diagram of the flame and heat flux distribution with

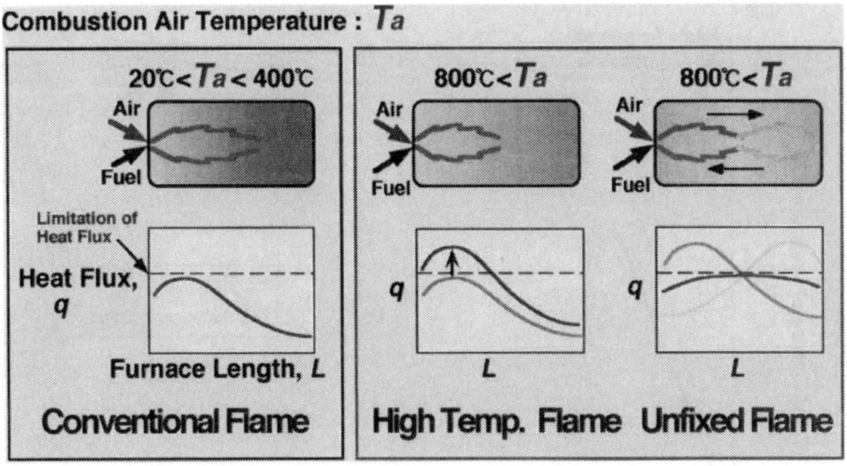


Figure 9: A schematic diagram of flame and heat flux distribution in a furnace with low temperature combustion air, high temperature air, and high temperature and low oxygen concentration air HiTAC condition.

normal temperature air and high temperature combustion air (called HiTAC flame) is shown in Fig. 9. The HiTAC flame results in a very uniform thermal field and high and uniform heat flux in the combustion zone.

## 5.4 Flame characteristics and energy savings with HiTAC

### 5.4.1 Flame stability and characteristics

The flame characteristics of propane fuel and high temperature combustion air have been examined in the test furnace facility using several advanced diagnostics. The flame stability limits of propane fuel as a function of air-preheat temperature and oxygen concentration in air is shown in Fig. 10. The flame is stable above the curve as shown in the figure. The flame stability limits increase significantly at high air preheats. Note that very wide flame stability limit occurs even with low oxygen concentration air. Under HiTAC conditions (high temperature and low oxygen concentration air), the flame stability limits are essentially infinite. The results also suggest that it is possible to use exhaust gases from a furnace as oxidant since these gases often contain several percent of oxygen.

The test regenerator facility utilized could provide high temperatures with any gas mixtures to about 1100°C. The size and color of these flames depend on air preheat temperature and oxygen concentration (or the amount of gas recirculation) in the combustion air and fuel property [7]. All flames showed very different flame structure as the air-preheat temperature was increased and the oxygen concentration in air was reduced from 21% to <5%. The flame volume was found to increase with increase in air preheat temperature and decrease in $O_2$ concentration in the combustion air. At any fixed temperature, the total flame volume

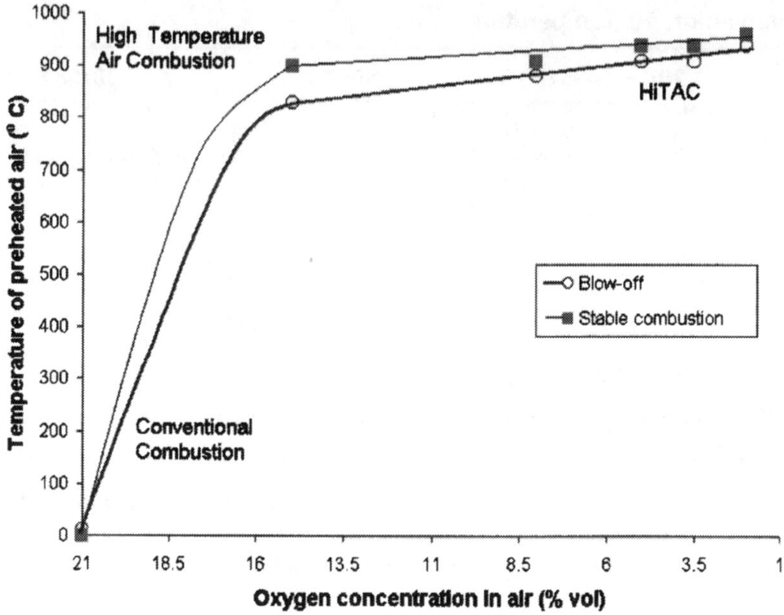


Figure 10: Stability limits of the propane flame as a function of air preheat temperature (heat recirculation) and oxygen concentration in air (flue gas recirculation).

decreased with increase in oxygen concentration from 2% to 21%. No yellow color flame was found at temperatures below 950°C and oxygen concentrations below 15%. The size of blue color region in the flame decreased with increase in oxygen concentration (up to about 15%) and temperature. The flames were of blue color for air preheats temperature between 900°C and 950°C and $O_2$ concentrations between 5% and 15%. For very fuel-lean mixtures at high air preheat temperature (1100°C), the luminosity of the flame (and hence the heat flux) was found to be very high. Further discussions on flame features are given in refs. [4, 7, 33]. At high air preheat temperatures and low oxygen concentration of about 2–5% in air, the flame was found to be of green color. The green flame color, which is considered environmentally benign, pronounced at higher air preheats and low oxygen concentration in the combustion air. This is due to high levels of $C_2$ species (swan band) produced from within the flames under these conditions. Low reaction temperatures cause increased production of swan band $C_2$ species in flames while normal flames cause blue color at high reaction temperatures. A dramatic increase in the flame volume was obtained with low oxygen concentration and high temperature combustion air. No flame color could be observed at very low $O_2$ concentration in air (about 2% or less). We describe this condition as colorless oxidation of fuel (also called Flameless oxidation of fuels). Sample photograph of the colorless combustion is shown in Fig. 11.

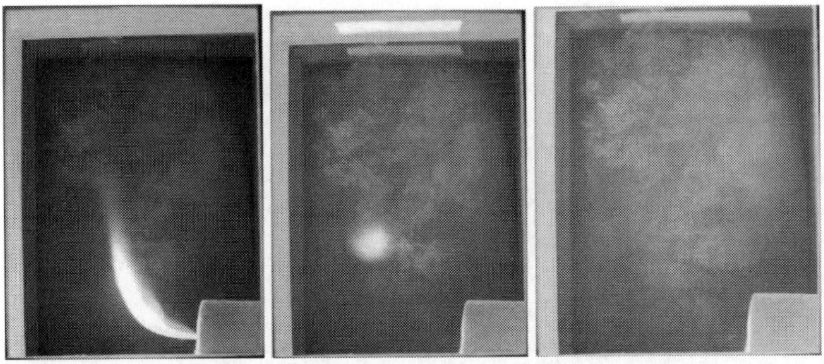

Figure 11: Methane flame photographs with combustion air temperature of 1000°C and oxygen concentrations (from left to right) of 21%, 8% and 2%, respectively. Note the colorless combustion in the extreme right side flame photograph.

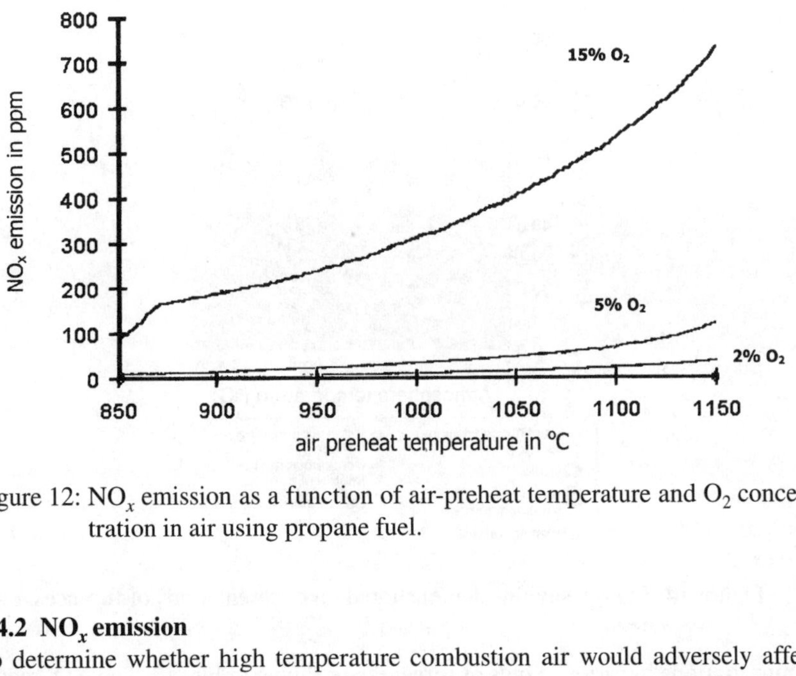


Figure 12: $NO_x$ emission as a function of air-preheat temperature and $O_2$ concentration in air using propane fuel.

### 5.4.2 $NO_x$ emission

To determine whether high temperature combustion air would adversely affect pollution emission, measurements were made of various gaseous species. Figure 12 shows the $NO_x$ emission levels for propane flame as a function of air preheats at 15%, 8% and 2% $O_2$ concentration in air. The emission of $NO_x$ increases with air preheats temperature under normal combustion conditions. This was also true for HiTAC conditions, but the rate of increase was very small. Very low $NO_x$ emission is observed under high temperature and low oxygen concentration combustion conditions. $NO_x$ emission at air preheats temperature of 1150°C decreased from 2800 ppm at 21% $O_2$ to 40 ppm at 2% $O_2$. Results obtained from field test

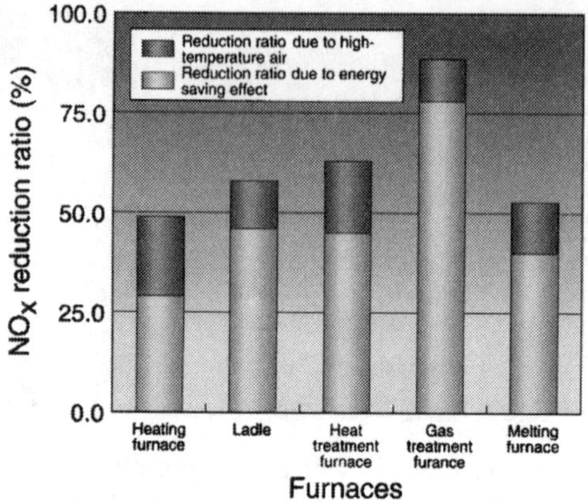


Figure 13: $NO_x$ reduction obtained from the field test demonstration using HiTAC technology.

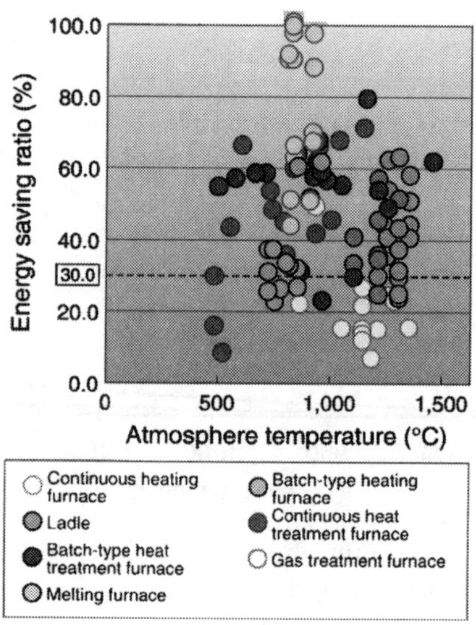


Figure 14: Energy savings demonstrated on different kinds of furnaces.

demonstrations in various kinds of furnaces are shown in Fig. 13. The $NO_x$ reductions are from two sources of energy savings and thermal field uniformity in the combustion zone. The results clearly show $NO_x$ reductions of nearly 50%.

### 5.4.3 Energy savings

Field tests demonstrations in actual plants were carried out to determine the extent of energy savings from the use of HiTAC in furnaces. Several different kinds of industrial furnaces were used in the field test demonstration. These field test demonstrations were carried out on 167 industrial furnaces over a duration of 2 years. The results are shown in Fig. 14. The energy savings are significant and depend on the type of application and furnace. The results clearly show average energy savings of 30% during the time duration. In some cases, nearly 100% energy saving is achieved as in the old furnace the low calorific value fuel gases could not be burnt with the result of nearly 100 energy recovery under HiTAC conditions. As a result of this demonstration, several other countries worldwide have accepted this technology as the most advanced technology demonstrated in the 21st century for significant energy savings and low pollution.

### 5.5 Diagnostics for colorless distributed combustion (flameless oxidation) in HiTAC

Flameless oxidation of fuels means no visible flame and so the diagnostics used must be carefully selected to observe the flame features. The radiation from the HiTAC flame is not luminous as it is primarily from water vapor (steam) and carbon dioxide. In addition, the temperature rise in the HiTAC flame is small. Most of the required heat is provided from furnace wall as solid radiation. Several diagnostic techniques may be considered to monitor the fate of flames in furnaces. They include PLIF, optical measurement of chemiluminescence spectra, FTIR, GC/MS via gas sampling and ion concentration measurement. Some of these techniques are expensive and not easily adaptable to hostile industrial size furnace environments. Ion concentration measurement with electrostatic probe has the advantages of high sensitivity response; it can withstand hostile environment, and it is of low cost. Electrostatic probe utilizes conductive characteristics of the combustion reaction zone. During the combustion of hydrocarbon fuels, a high level of ion concentration is formed by the chemical ionization reaction. A new practical diagnostic technique, based on the modified electrostatic probe, has been successfully developed and used to monitor ion concentration in the flame zone. The technique has been applied to monitor invisible-HiTAC-flame for improving heating performance in a test steam reformer furnace [35]. The diagnostics was first developed using a 1.3 kW HiTAC furnace facility and then applied on a pilot plant fuel reformer facility.

Figure 15 shows a comparison of the radial distribution of flame ion concentration at selected axial positions in the flame measured with the ionization probe (units in mV) from three different flames (ordinary flame, (left), high temp. air flame (middle), and HiTAC flame (right)) using the 1.3 kW furnace facility. The results are shown on the global flame behavior superimposed on the ionization probe data with the probe traversed horizontally at three different heights of $h = 45$, 130, and 235 mm. The results are: normal air (25°C temperature, Fig. 15a), high temperature air (21% $O_2$, in excess of 1000°C air preheats, Fig. 15b), and diluted and high temperature air with 3% $O_2$ (HiTAC case, Fig. 15c). For the case of ordinary flame with ambient temperature combustion air, shown in Fig. 15a, although some level of output signal is obtained from the ionization probe at $h = 45$ mm, no signal is obtained at $h = 130$ mm.

The flame ion concentration is detected only in the areas where the flame is present, see Fig 15. No signal is detected with the probe outside the flame areas where no visible flame is present. The results clearly show high level of ion concentration signal obtained with the high temperature air combustion case. Relatively low level of ion concentration obtained for the HiTAC case, shown in Fig. 15c, suggests that the technique is capable of providing information under flameless oxidation conditions so long as the chemical reactions continue to occur in the flame. The results showed that the measured ion concentration is proportional to the probe size. A larger size probe was developed at NFK for use in full-scale reformer furnace (hydrocarbon to hydrogen transformation reformer facility) with good success. The results obtained showed that the flame ionization probe is a good diagnostic tool for application under distributed colorless combustion conditions.

### 5.6 Waste fuel gasification and fuel reforming using HiTAC

Most of the HiTAC technology efforts have been for furnaces, boilers and waste fuel reforming [33]. Some of the near-term applications include the following:

1. Industrial furnaces used in the steel industry for melting, reheating, heat treatment and soaking, and for process heater, aluminum melters, petrochemical reforming, glass melting, drying, boilers, ceramic heater, and domestic boilers and heater.
2. Waste incineration/waste thermal destruction of solid and liquid wastes including wastes produced from municipal, industrial, yard, chickens, animals and farms. The technology is attractive for gasifying wastes, biomass fuels, and mixed wastes so that the gaseous fuel produced is clean and environmentally benign form of energy. High temperature steam gasification provides further benefits of higher heating value, higher hydrogen content, and high efficiency with minimum residue.
3. Fuel reforming and energy transformation to cleaner fuels for use in fuel cells and other power, propulsion and energy conversion applications.
4. Power production using micro-gas turbines with the combustion of low to medium gas produced from biomass and wastes.
5. Destruction of odors and certain pollutants.

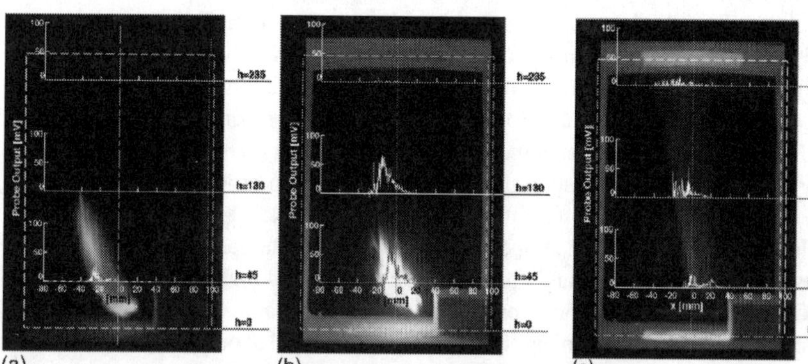


(a) (b) (c)

Figure 15: Direct flame photographs and flame ion concentration from (a) ordinary, (b) high temperature and (c) HiTAC flames.

6. Production of nano-powders and materials, nano-catalysts, nano-fluids, nano-sensors, and nano-related products, including energetic materials.

### 5.6.1 High temperature steam gasification of wastes

The HiTAC technology can also be used to produce very high temperature steam that can be used for coal, biomass and wastes and low grade fuels to produce clean syngas. Under favorable conditions, very high molar concentrations of hydrogen can be produced from biomass and wastes. The tar produced with the use of high temperature steam is minimal. Sample results of the gas composition with paper as the solid fuel are given in Fig. 16. The experimental results provided the influence of elevated gasifying agent temperature on hydrogen yield in the gas composition. In general, the higher the steam temperature, the higher was the amount of hydrogen produced. This trend was true for all the solid fuels examined, including solid wastes, biomass, coal and agricultural wastes (rice husk, rice straw and corn cob). Methane gas production is an important component of syngas in the steam-assisted gasification. Steam gasification at low temperatures favors methane production. In contrast, at high gasification temperatures, hydrogen yield increases as shown in the sample results in Fig. 16. The HiTAC has been successfully demonstrated in Japan on a pilot plant fuel reformer to convert hydrocarbon fuels to hydrogen using high temperature steam.

## 6 Practical aspects of power generation

A by-product from all fossil fuel power generation is the emission of pollutants including the emission of $CO_2$ greenhouse gas. Historic and geographic data indicate that temperature of the earth has been increasing since the onset of the industrial revolution in the 1800s. The concentration of $CO_2$ is high compared to long-long- term historic levels and has been increasing very rapidly with the onset of industrial revolution, see Fig. 17. This correlates very well with the increased combustion of fossil fuels. The global temperature and $CO_2$ concentration correlate quite well as seen in Fig. 18. Change in global temperature as small as 1 K has a profound effect on weather patterns, distribution of flora and fauna, and such geographical phenomena as ocean currents and the sea level. Some of the $CO_2$ capture approaches for power include: (a) post capture (such as adsorption, absorption,

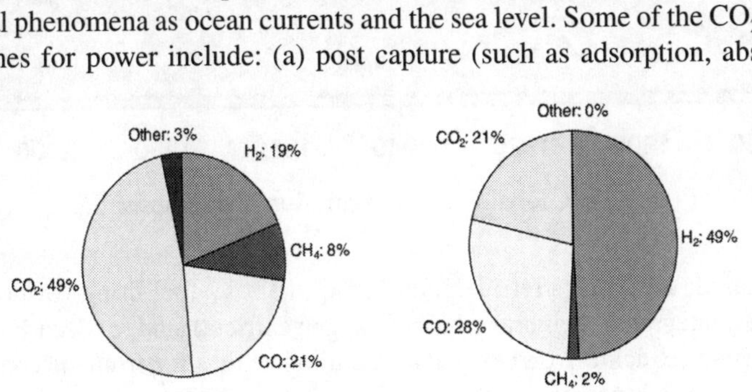


Figure 16: Chemical composition of the syngas produced by high temperature gasification of paper using steam at 700°C (left) and 1000°C (right).

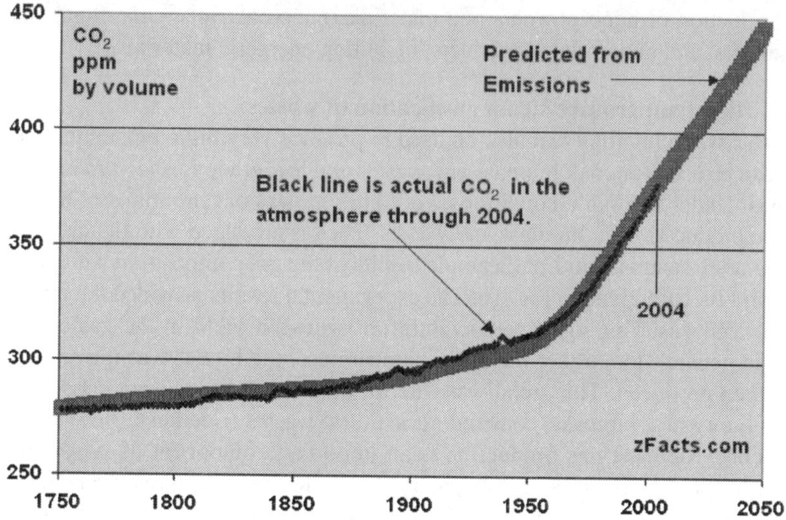


Figure 17: Increase in $CO_2$ emission levels since 1970 and projected to 2050.

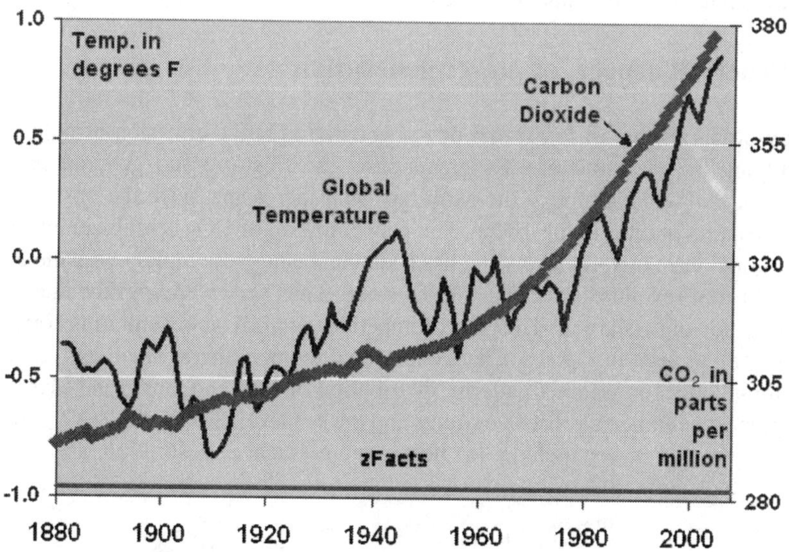


Figure 18: Global $CO_2$ concentration and temperature.

hydrate based, cryogenics/refrigeration based), (b) oxy-fuel firing (external oxygen supply, integrated membrane-based, oxygen carriers), and (c) decarbonization (fuel reforming to achieve carbonized fuel, carbonate reaction from via combustion decarbonization). Innovative options will evolve with the focus on the carbon-constrained world and from our focus on the means to reduce the carbon emissions. The $CO_2$ mitigation options include fossil fuel energy conservation and increase in efficiency of the plant. Fuel switching from fossil to nuclear, renewables, and

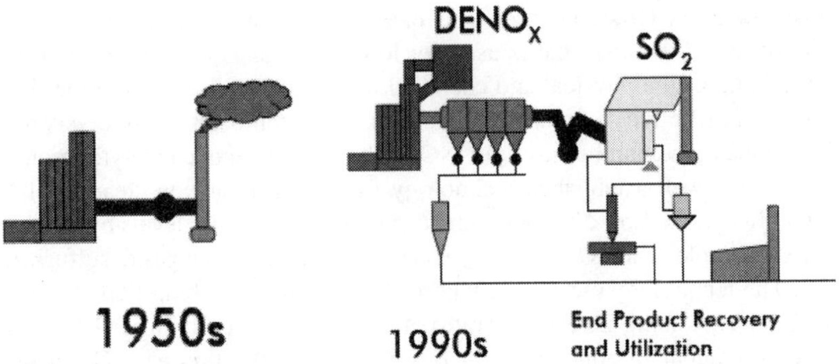


Figure 19: A schematic diagram of the power plant of the 1950s and 1990s.

natural gas is favorable for $CO_2$ mitigation. In case the fuel switching and other $CO_2$ mitigation options are not favorable, the $CO_2$ generated from a power plant can be sequestrated by capturing it from the tail end of the power plant and then sequestering it using either or some combination of sequestration approaches of: terrestrial ecosystems, geological storage in active oil beds, coal beds, depleted oil and gas wells, deep aquifers, mined caverns and salt domes.

In 2006, global $CO_2$ emission from fossil fuel use has shown an increase of about 2.6%. China topped the international $CO_2$ emission table in 2006 for the first time in history. Total $CO_2$ emission in China increased by 8.7% in 2006 compared to 1.4% in USA and almost unchanged in the EU-15 countries (the EU-15 had a decrease of 0.8% $CO_2$ emission in 2005). Even with this increase, the average Chinese emits 3.5 tons of $CO_2$ per year, whereas Britons emit nearly 10 tons and Americans 20 tons.

### 6.1 Pollutants emission

The emissions from a power plant depend upon the type of fuel used and the design and operational conditions of the power system. The older units in the 1950s had no pollution control device, while the 1990s unit had pollution control for particulate mater, $NO_x$, and $SO_2$. A schematic diagram of the 1950s and 1990s power plant is given in Fig. 19.

Fluidized bed uses crushed coal for power generation. Here the bed is kept low temperature (below 900°C) and the bed contains calcium-based sorbent material to capture $SO_2$ evolved during combustion. The $NO_x$ emission is low since the combustion temperature in the bed is below the temperature at which thermal $NO_x$ is formed. However, the emission of $N_2O$ emission (a greenhouse gas that has more pronounced effect than $CO_2$ gas emission to the atmosphere) is a disadvantage since it is formed and propagates at low temperatures. Therefore, one must strike a balance to determine the favorable conditions for reducing $N_2O$ emission and enhance sulfur capture.

The criteria pollutants include $NO_x$, $SO_x$, HC, mercury (Hg) and fine particles with particle size smaller than 10 μm (PM10) or even 2.5 μm and under (PM2.5). Staged combustion has been shown to be effective and mature technology to achieve low $NO_x$. However, problems exist on the carbon burnout during fuel rich pyrolysis of the coal that is used for low $NO_x$ emission. Therefore, a better understanding of the physical and chemical transformations of coal, in particular, the change in reactivity of char in the fuel rich high temperature pyrolysis region of the combustion chamber used for $NO_x$ reduction. Selective catalytic reduction technique is a well-established technology for post-combustion cleanup of $NO_x$ from the flue gases. Low $NO_x$ burners that can reduce the $NO_x$ levels by more than 95% are available. The $SO_x$ control is usually obtained by flue gas desulfurization (FGD). The technology most widely used today is wet scrubbing FGD to remove up to about 97% using Ca/S molar ratio of 1–1.5. In dry scrubbing FGD, an alkaline slurry reagent in aqueous form is sprayed into the flue gas. The water is evaporated quickly and the reaction products are removed as dry powder. Dry FGD has the advantages of low cost and less water requirements. However, the disadvantages are lower capture efficiency (about 70–90%), reduced reagent utilization (1.1–1.6%) and exclusion of limestone as reagent option. Other FGD options include ammonia scrubbing to produce ammonium sulfate that can be used as fertilizer and in cases when the sulfur content is very high, the $SO_2$ is oxidized to $SO_3$ and then subsequently to sulfuric acid.

Particulate matter removal technologies from the tail end of the power plant utilize fabric filters (bag houses) for large size particles and electrostatic precipitators for fine particles. This is also a mature technology to achieve 99.9% (or three nines) collection efficiency and results in emission levels of 0.01–0.03 lb/MBtu.

Mercury emissions can be obtained using powered activated carbon. Mercury capture efficiency has been shown to be 50–70% with electrostatic precipitators (ESPs) and up to about 90% with fabric filters. The capture efficiency strongly depends on the form of mercury vapor in the flue gas. Chlorine in the coal or Bromine addition to coal greatly enhances the mercury capture. Thus co-combustion of fuels (for example coal and wastes such elements) is beneficial for enhancing mercury capture efficiency while simultaneously utilizing waste fuels without modifying the combustor.

It is clear that with the increased energy use and greater awareness of the harmful effects of pollutants from combustion, one can only expect to see even more stringent air pollution control from the tail end of power plants. We expect control from larger size plants will occur first than small size plants. Emission of carbon dioxide will continue to be of greater concern to the society as the evidence of close correlation between global $CO_2$ concentration and temperature is very convincing. A schematic diagram of the future fossil fuel power plant of the 2020 is shown in Fig. 20.

## 7 Conclusions

Various methods of power generation, including some recent advances in power generation are presented. With the continued developments, very high efficiency thermodynamic efficiency can be achieved from fossil fuel combustion systems. Costs of coal and other solid fuels are more favorable than clean natural gas. However, with the recent developments in gasification and other

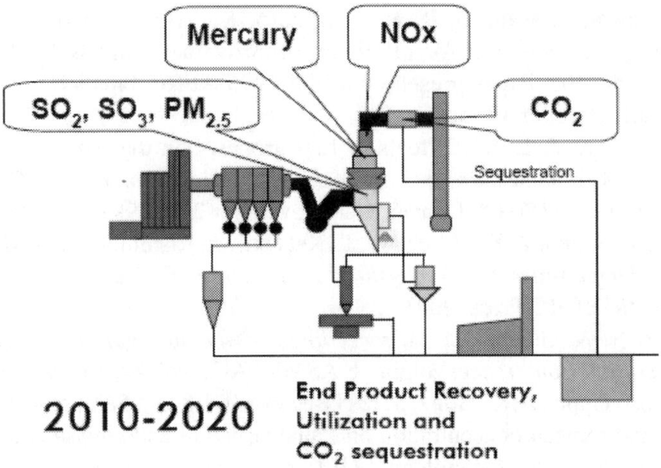


Figure 20: A schematic diagram of the future power plant with advanced pollution control including $CO_2$ capture and sequestration.

developments on integrated power systems, it is reasonable to achieve very high thermodynamic cycle efficiencies, close to 60%, with low pollution emission in the near future. Future large scale energy conversion systems are expected to have $CO_2$ capture technologies and sequestration so that one must seek innovative designs for advanced hybrid systems, which can potentially offer very high efficiency and low pollution emission.

HiTAC has shown a significant potential of HiTAC technology. The HiTAC provides colorless distributed combustion and under certain conditions, green color and other flame colors. The flame size increases under HiTAC conditions. The emission of $NO_x$ is very low at high temperatures and low oxygen concentration air compared to normal combustion air at low temperatures. HiTAC flame gives low noise, high thermal field uniformity (better than 30°C to provide an isothermal reactor) in the furnace. HiTAC has been used in many countries worldwide with significant energy savings demonstrated in field test trails and in many practical systems. Fuel reforming using high temperature steam gasification of wastes, biomass and coals provides a clean conversion of waste fuels and biomass to hydrogen-rich syngas.

## References

[1] Beer, J.M., High efficiency electric power generation: the environmental role. *Progress in Energy and Combustion Science*, **33(2)**, pp. 107–134, 2007.

[2] Schilling, H.D., Cleaner energy from coal with circulating fluid beds. *VGB Kraftwerkstech* (English edn.), **73(8)**, pp. 564–573, 1993.

[3] Booras, G. & Holt, N., Pulverized coal and IGCC plant cost and performance estimates. *Gasification Technologies Conference*, Washington DC, 1994.

[4] Weinberg, F.J., Combustion temperatures: the future? *Nature*, **233**, pp. 239–241, 1971.

[5] Hasegawa, T., Tanaka, R. & Kishimoto, K., High temperature excess-enthalpy combustion for efficiency improvement and $NO_x$ abatement, Paper No. 9E, Paper presented at the 1995 AFRC Japan-USA Meeting, Hawaii, October 1995.

[6] Gupta, A.K. & Li, Z., Effects of fuel property on the structure of highly preheated air flames. *Proceedings of the International Joint Power Generation Conference (IJPGC)*, Denver, CO, November 2–5, 1997.

[7] Tsuji, H., Gupta, A.K., Hasegawa, T., Katsuki, M., Kishimoto, K. & Morita, M., *High Temperature Air Combustion: From Energy Conservation to Pollution Reduction*, CRC Press, 2003.

[8] Boltz, S. & Gupta, A.K., *Proceedings of the International Joint Power Generation Conference*, Volume 1, ASME FACT Vol. 22, pp. 193–205, 1998. See also Gupta, A.K., Boltz, S. & Hasegawa, T. Effect of air preheat temperature and oxygen concentration on flame structure and emission. *Journal of Energy Resources Technology*, **121(3)**, pp. 209–216, 1999.

[9] Tanigawa, T. *et al.*, *Proceedings of the International Joint Power Generation Conference*, Volume 1, ASME FACT Vol. 22, pp. 207–214, 1998.

[10] Taniguchi, H. *et al.*, *Proceedings of the International Joint Power Generation Conference*, Volume 1, ASME FACT Vol. 22, pp. 215–225, 1998.

[11] Gupta, A.K. *et al.*, *Proceedings of the International Joint Power Generation Conference*, Volume 1, ASME EC Vol. 5, pp. 247–258, 1998.

[12] Kitagawa, K. *et al.*, *Proceedings of the International Joint Power Generation Conference*, Volume 1, ASME FACT Vol. 22, pp. 239–242, 1998.

[13] Gupta, A.K., Flame characteristics and challenges with high temperature air combustion. *Proceedings of the 2nd International Seminar on High Temperature Air Combustion*, Stockholm, Sweden, January 17–18, 2000.

[14] Yoshikawa, K., High temperature gassification of coal, biomass, and solid wastes. *Proceedings of the 2nd International. Seminar on High Temperature Air Combustion*, Stockholm, Sweden, January 17–18, 2000.

[15] Shimo, N., Fundamental research of oil combustion with highly preheated air. *Proceedings of the 2nd International Seminar on High Temperature Air Combustion*, Stockholm, Sweden, January 17–18, 2000.

[16] Weber, R., Verlann, A.L., Orsino, S. & Lallemant, N., On emerging furnace design that provides substantial energy savings and drastic reductions in $CO_2$, CO and $NO_x$ emissions, *Journal of the Institute of Energy, UK*, pp. 77–83, September 1999.

[17] Ahadi-Osuki, T., *Heat Flux from Highly Preheated Air Combustion and Swirl Combustion*, MS Thesis, University of Maryland Combustion Laboratory, February 2000.

[18] Hasegawa, T., Environmentally compatible regenerative combustion heating system. *Proceedings of the 2nd International Seminar on High Temperature Air Combustion*, Stockholm, Sweden, January 17–18, 2000.

[19] Gupta, A.K., Lilley, D.G. & Syred, N., *Swirl Flows*, Abacus Press: Tunbridge Wells, Kent, England, 1984.

[20] Hasegawa, T. & Mochida, S., Highly preheated air combustion characteristics and development of a combustion diagnostic on advanced industrial furnace making. *Proceedings of the International Joint Power Generation Conference*, San Francisco, CA, July 25–28, 1999, Volume 1, ASME FACT Vol. 23, pp. 457–466, 1999.

[21] Chang, R.-C. & Chang, W.-C., Research of high temperature air combustion fired heavy oil. *Proceedings of the 2nd International Seminar on High Temperature Air Combustion*, Stockholm, Sweden, January 17–18, 2000.

[22] Suzukawa, Y., Sugiyama, S. & Mori, I., Heat transfer improvement and $NO_x$ reduction in an industrial furnace by regenerative combustion system. *Proceedings of the 1996 IECEC Conference*, pp. 804–809, Paper No. 96360, 1996.

[23] Hasegawa, T. & Tanaka, R., Combustion with high temperature low oxygen air in regenerative burners, Paper presented at the 1997 ASPACC, Osaka, Japan, 1997.

[24] Katsuki, M. & Hasegawa, T., The science and technology of combustion in highly preheated air, *Proceedings of the 27th Symposium (Intl.) on Combustion*, The Combustion Institute, PA, pp. 3135–3146, 1999.

[25] Gupta, A.K. & Hasegawa, T., Air preheat and oxygen concentration effects on the thermal behavior of propane and methane diffusion flames. *Proceedings of the High Temperature Air Combustion Symposium*, Kaohsiung, Taiwan, January 20–22, 1999.

[26] Ishiguro, T., Tsuge, S., Furuhata, T., Kitagawa, K., Arai, N., Hasegawa, T., Tanaka, R. & Gupta, A.K., Homogenization and stabilization during combustion of hydrocarbons with preheated air. *Proceedings of the 27th Symposium (Intl.) on Combustion*, The Combustion Institute, PA, pp. 3205–3213, 1999.

[27] Kitagawa, K., Konishi, N., Arai, N. & Gupta, A.K., Two dimensional distribution of flame fluctuation during highly preheated air combustion. *Proceedings of the ASME International Joint Power Generation Conference (IJPGC)*, Baltimore, MD, August 23–26, 1998, ASME FACT Vol. 22, pp. 239–242.

[28] Gupta, A.K., High temperature air combustion: experiences from the USA-Japan Joint Energy Project, Invited keynote lecture at the 4th High Temperature Air Combustion and Gasification Symposium, Rome, Italy, November 27–30, 2001.

[29] Gupta, A.K., Flame length and ignition delay during the combustion of acetylene in high temperature air. Invited paper, *Proceedings of the 5th High Temperature Air Combustion and Gasification (5th HTACG)*, Yokohama, Japan, October 28–31, 2002.

[30] Konishi, N., Kitagawa, K., Arai, N. & Gupta, A.K., Two-dimensional spectroscopic analysis of spontaneous emission from a flame using highly preheated air combustion. *Journal of Propulsion and Power*, **18(1)**, pp. 199–204, 2002.

[31] Hasegawa, T., Mochida, S. & Gupta, A.K., Development of advanced industrial furnace using highly preheated combustion air. *Journal of Propulsion and Power*, **18(2)**, pp. 233–239, 2002.

[32] Kitagawa, K., Konishi, N., Arai, N. & Gupta, A.K., Temporally resolved 2-D spectroscopic study on the effect of highly preheated and low oxygen concentration air on combustion. *Journal of Engineering for Gas Turbines and Power*, **125**, pp. 326–331, 2003.

[33] Hasegawa, T., *Proceedings of the International Joint Power Generation Conference*, Volume 1, ASME EC Vol. 5, pp. 259–266, 1997.

[34] Gupta, A.K., Thermal characteristics of gaseous fuel flames using high temperature air. *Journal of Engineering for Gas Turbines and Power*, **126(1)**, pp. 9–19, 2004.

[35] Mortberg, M., Blasiak, W. & Gupta, A.K., Experimental investigation of flow phenomena of a single fuel jet in cross-flow during highly preheated air combustion conditions. *Journal of Engineering for Gas Turbines and Power,* **129(2)**, pp. 556–564, 2007.

[36] Mortberg, M., Blasiak, W. & Gupta, A.K., Combustion of low calorific value fuels in high temperature and oxygen deficient environment. *Combustion Science and Technology*, **178**, pp. 1345–1372, 2006.

[37] Mochida, S., Araake, T., Hasegawa, T. & Gupta, A.K., Invisible-HiTAC-flame control for improving steam reformer heating performance using flame ionization monitoring technique, *AFRC/JFRC Joint Conference*, Maui, October 10–14, 2004.

[38] Cichonski, W. & Gupta, A.K., Hydrogen generation from wastes using ultra-high temperature steam gasification. *Proceedings 6th International Conference on High Temperature Air combustion and Steam Gasification (HTACG)*, Essen, Germany, October 17–19, 2006.

[39] Abe, T., Mochida, S., Yasuda, T. & Gupta, A.K., Advanced gasification technology for biomass and plastics using super high temperature steam. *AFRC/JFRC International Symposium*, Waikoloa, Hawaii, October 22–24, 2007.

[40] Ahmed, I., Noosai, N. & Gupta, A.K., Parameters influencing clean syngas production from biomass, solid wastes, and coal during pyrolysis and steam gasification. *7th International High Temperature Air Combustion and Gasification Conference (7th HTACG)*, Phuket, Thailand, January 13–16, 2008.

# CHAPTER 3

# Virtual engineering and the design of power systems

D.S. McCorkle & K.M. Bryden
*Department of Mechanical Engineering,*
*Iowa State University, Ames, IA, USA.*

## Abstract

The art and science of power system design is rapidly changing. Future power systems will have significantly lower emissions, operate at higher pressures and temperatures, utilize new fuels and new technologies, and be far more complex than current power plants. Power plant sensors and controls will be smaller and more capable, changing how power systems are operated and maintained. To achieve these goals, a new generation of power system design tools is needed. These tools will integrate all the models needed to describe and examine the performance of a power plant. This integration will be much more comprehensive than the current systems integration supported by process modeling software, in which simplified component descriptions are integrated into a complex system-level model of the power plant. Rather, these new engineering design tools will integrate micro-scale phenomena, component-level phenomena, and system-level demands to create an integrated hierarchy of models, which will simultaneously model all aspects of power system performance, operation, and maintenance. This integrated modeling environment will then be used to examine the impact of changes at the micro level on efficiency, emissions, and cost. These broadly integrated models will enable integrated optimization of power systems and will explicitly bring human judgment into the optimization and design of power systems during both design and operation. This chapter will examine how these virtual engineering design environments will be created and utilized to build a new generation of power plants. Examples of how virtual engineering is currently being used in the power industry will be given. In addition, the US Department of Energy's use of virtual engineering in the development of near-zero-emission power plants will be discussed.

# 1 Virtual engineering

Virtual engineering is an emerging technology that is growing in popularity. Virtual engineering seeks to create a virtual environment in which all aspects of engineering decision making can be addressed [1, 2]. The goal of virtual engineering is to enable a design team to alter the shape, size, or other characteristics of components within a power plant and immediately see the impact of the changes on systems throughout the plant. For example, an engineer seeking to modify a tangentially fired, pulverized coal furnace to co-fire coal and switchgrass could alter the burner configuration and immediately see the impact of the change on the performance of the furnace. In another example, an engineer could investigate the performance of a heat exchanger by altering the geometry, thickness, or spacing of any or all of the components and would immediately see the changes in heat transfer and fluid flow on both sides of the heat exchanger. Additionally, because the virtual engineering-based power plant is presented in a natural and accessible visual format, the design and changes can be readily shared with other plant stakeholders. This can provide a design platform that allows a number of concerns including those of construction, operations, maintenance, and engineering to be incorporated and tested before commitment to a physical plant. This will improve plant performance, reduce cost, and enable a wide range of new technologies to be readily incorporated into power plants.

The four key aspects of a virtual engineering environment are (1) the integration of models, data, and other engineering information to create a complete model of the system or component; (2) visualization of the resulting systems and information; (3) interactive analysis and design of the system, e.g. the ability to easily ask "what if" types of questions; and (4) high-level decision making tools.

- *Integration of models, data, and other engineering information* – All real systems are composed of a wide variety of models. Even a single component will need a number of models to describe the operation and performance. These models can range from simple integral models and algebraic expressions to detailed models of micro-scale phenomena. The first goal of virtual engineering is to couple these models together to create a single unified model that describes the behavior of a component. In addition, a power plant consists of a number of components. Each of these unified model sets describing a component's behavior needs to be assembled to create a complete plant. Currently, process simulation programs such as Aspen™ enable the user to link simplified models of components to create complex models of processing plants. The goal of virtual engineering is to take this one step further. First, a set of virtual components is created, each of which is composed of a suite of models. These virtual components are then linked to create a virtual system model that can understand how changes in one component impact the overall performance of the power plant. For example, a virtual pump may be composed of several models at varying levels of fidelity, including a computational fluid dynamics (CFD) model, a finite element analysis (FEA) model, a cavitation model, and a set of pump curves. This virtual pump would then be coupled within the system to other virtual components, each of which is represented by an integrated set of models, to create a "complete" virtual system. The word "complete" does not imply that every detail is addressed by the virtual systems, though they could be, but rather that the virtual system can completely address the engineering questions that are being asked.

- *Visualization of the resulting systems and information* – Engineering experience and judgment are critical to the decision making process. Modeling results must be checked against experience to ensure that they are credible and usable. In addition, engineering judgment is needed to decide between various options with similar benefits and to include a variety of intangibles in the engineering process. While computer modeling can provide a variety of options and can help the engineer understand the consequence of various decisions, computational modeling cannot bring this needed engineering experience and judgment into the decision making process without human interaction. In the virtual engineering environment, the goal is to present the engineering system and information in a familiar, natural, and readily understandable interface that enables the engineer to quickly understand, explore, and design the system. Coupled with an appropriate expert (e.g. a design engineer, a plant engineer, or a construction manager), this can reduce design time and cost while resulting in better solutions.
- *Interactive analysis and engineering* – Today, nearly all aspects of power plant simulation require extensive offline setup, calculation, and iteration of a variety of separate models. The time required for each iteration for a single component can range from one day to several weeks. The results of the separate component analyses must then be integrated, often within the engineer's mind, to create an overall understanding of the system's response to various changes. This is a time consuming and inexact process. Tools are needed that enable the engineer to establish a dynamic thinking process to permit exploration of the "what if" questions that are essential to engineering. In nearly all circumstances, an engineering answer now has a much greater value than an answer tomorrow, next week, or next month. Although many high-fidelity engineering analysis techniques have been developed, they are often not routinely used as a fundamental part of the engineering design, operations, control, and maintenance process because the time required to set up, compute, and understand the result, then repeat the process until an adequate answer is obtained, significantly exceeds the time available. This includes techniques such as CFD, FEA, and optimization of complex systems. Instead, these engineering tools are used to provide limited insight into the problem, to sharpen an answer, or to understand what went wrong after a bad design and how to improve the results next time. This is particularly true of CFD analysis. While faster, less expensive processors will provide faster solutions, we also need better algorithms that enable the engineer to understand the error in the analysis, automatically create needed reduced order models, and choose the right level of fidelity and accuracy to answer the question being asked.

- *High level engineering decision support tools* – Engineering is a process of making decisions for complex and uncertain systems. Therefore, tools that start with physics-based models and help the engineer understand the options and make the best decisions are essential to the virtual engineering process. The most commonly recognized decision support tools are optimization tools. Optimization tools require the user to provide a set of constraints in exchange for a set of best (or, in some cases, better) solutions that address a desired objective or set of objectives. Today, this process is generally hands-free in that the user specifies the needed conditions and preferences and then gets the results with human intervention, *a priori* interaction. Following the optimization process, the user specifies a new set of conditions based on the newly acquired information and repeats the process, *a posteriori* interaction. Once an optimized solution is found, it will often provide a starting place for further design or exploration. Because of this, tools are needed that support a paradigm of progressive interactions in which the engineering team works together in real time with the optimization process to direct the search for the needed design. Progressive interaction is a human-guided preference articulation method where the designers' interaction continuously controls the engineering optimization through visualization, modification, and controlled re-optimization. In progressive interaction, designer preference articulation is done continuously during the optimization process [3]. In addition, a variety of tools that extend beyond optimization are needed. These include agent-based design tools, inverse engineering tools, pattern recognition tools, and other tools. Agent-based design tools send a group of computational agents out to explore the design space and bring back specific needed observations. Inverse engineering tools start with the needed result and then deduce the steps required to achieve it. Pattern recognition tools find interesting and important patterns and bring them to the engineer's attention.

The most important aspect of a virtual engineering environment is that all these components – integration of models, visualization, interactive analysis, and decision making tools – are integrated into a single seamless interface. In short, a virtual engineering environment is one that provides a user-centered, first-person perspective, enabling the user to interact with the engineered system naturally and providing him or her with a wide range of accessible tools. This requires an engineering model that includes the geometry, physics, and any quantitative or qualitative data from the real system. The user should be able to walk through the operating system and observe how it works and how it responds to changes in design, operation, or any other engineering modification. Interaction within the virtual environment should provide an easily understood interface, appropriate to the user's technical background and expertise that enables the user to explore and discover the unexpected but critical details about the behavior of the system. Similarly, engineering tools and software should fit naturally into the environment and allow the user to maintain her or his focus on the engineering problem at hand. A key aim of virtual engineering is to engage the human capacity for complex evaluation [4].

As shown in Fig. 1, CAD and geometric modeling are the foundation of a virtual engineering system that also includes tools that facilitate spatial configuration and planning, manufacturing simulation, factory simulation, high-fidelity numerical models, and other engineering tools that require geometric information. Then, multidisciplinary analysis and simulation tools are integrated at the second level, which sustains the interaction capability for the system. The tools at the third level facilitate engineers' decision making and collaboration. Product data management tools allow designers to easily access and share data. The distributed product realization tools support geographically dispersed engineers working collaboratively. Decision making and optimization tools provide suggestions to design changes. At any of the three levels, designers can visualize the product model and data in the virtual environment whenever necessary. In almost all current visualization applications, data only flow from the engineering tools to the visual environment. In a virtual engineering system, designers can control the geometric and engineering models in a virtual environment, shown as the double arrowed lines in Fig. 1. In addition to the two-way communication flow between engineers and the engineering tools, each capability noted on the left hand side of the diagram is available throughout the life of the product, as noted by the continuous box on the right hand side of the diagram.

Several research groups have worked to create virtual engineering environments for specific types of problems [5–8]. One of the earliest power plant applications was Boilermaker [5]. Boilermaker couples a computational model of an industrial furnace with a virtual environment to enable the user to make changes to the furnace nozzle configuration and then see the resulting furnace performance on the fly. This enabled engineers to explore various furnace configurations on the fly and consider a number of variables to create the optimum design for a particular furnace. Fuel Tech NV funded a portion of the Boilermaker project and uses virtual engineering software in many of their engineering decision making processes today. Another example of coupling engineering analysis and virtual environments is DN-Edit [7], which allows a Non-Uniform Rational B-Splines (NURBS)-based

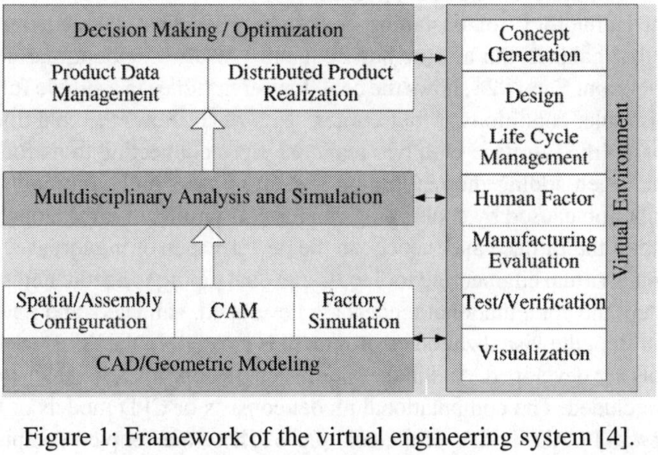


Figure 1: Framework of the virtual engineering system [4].

surface geometry to be altered interactively. Virtual cursors enable interaction with geometry surfaces, allow surface points to be displaced and material to be added or removed, and provide exact control of surface normals at specific points. The resulting surfaces are NURBS-based and can be exported to various CAD or analysis programs. Applications coupling virtual environments and engineering tools have been developed for assembly [6], hose routing [9], and design of three-dimensional mechanisms [7].

## 2 A virtual engineering application

As a coal-air mixture flows through the coal transport piping from the mills to the furnace, it travels through a number of twists and turns in the piping. Consequently, the coal particles that were originally uniformly distributed across the flow cross section group together and flow as a single stream of particles. Viewed from the outside, this group of coal particles traveling together looks like a rope. The primary purpose of the coal transport system is to balance the coal flow from a few coal mills to a large number of individual burners. Riffle type bifurcators are commonly used to split the piping system. Without a coal rope in front of a distributor, the particle distribution in the inlet cross section of a distributor is uniform; the mass flow is divided relatively uniformly into both outlets [10]. However, if coal roping occurs in front of the distributor, particle distribution is not uniform over the pipe cross section and the coal will no longer be evenly divided between the two outlets. Thus, the fuel inputs to the burner are not balanced, so combustion efficiency decreases and emissions increase. Coal roping also affects the accuracy of traditional coal sampling probes that rely on a homogeneous flow. The failure of these probes to obtain accurate measurements adds difficulty to the flow control system. Therefore, despite the use of matched outlet pipes and flow splitting devices, uneven distribution of pulverized coal inevitably occurs. The most common way to deal with coal roping is to install orifices in the coal pipes to balance airflow and to create three-dimensional eddies that can rapidly break up the coal rope. However, the coal can readily reform a rope downstream of an orifice.

Current technology for designing coal piping systems is based on empirical methods that balance the airflow but frequently result in suboptimal designs for coal distribution. Figure 2a shows the coal flow streamlines in a simple furnace with the uniform inlet conditions at four corners. Figure 2b shows the coal flow streamline in a simplified furnace with two real coal pipes connecting to its four corners. As shown, when adding the real piping system to the simulation model, unequal coal distribution caused by coal roping can have a significant impact on coal distribution within the furnace and, hence, on the performance of the furnace.

Building a virtual engineering tool to design coal piping is a multistep process. In the first step, the computational models are developed, validated, and integrated. In the second step, the visualization environment is created. In the third step, the interaction tools are developed. And in the fourth step, the optimization/decision making tools are included. The computational model consists of CFD models of the multiphase flow and CAD models of the coal pipes. These are coupled computationally

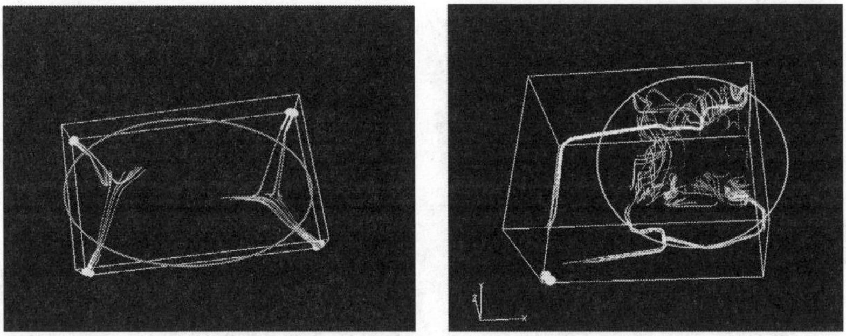

(a) (b)

Figure 2: Coal density in a cold furnace simulation (a) with uniform inlet conditions and (b) with non-uniform inlet conditions caused by coal roping.

such that changes in the CAD model are automatically updated in the CFD model. In this way, the user can change the geometry of the coal pipes, orifices, or bifurcator and rerun the model without extensive redrawing within the CFD package. In this case, the multiphase flow model of the coal-air mixture is modeled using the commercial CFD package Fluent$^{TM}$. The results were validated using available experimental data, both numerical and qualitative. The specific details of the model development and validation are given in [11]. The computational results are processed on a computing cluster that is linked to the interaction and visualization environment using Common Object Request Broker Architecture [12].

The visual environment and user interactions are created using VE-Suite [13]. VE-Suite is an open source virtual engineering software package being developed at Iowa State University. VE-Suite is discussed in more depth later in this chapter. The visual environment includes the piping being analyzed (Fig. 3), the input parameters (Fig. 4), and the output parameters (Fig. 5). In addition to showing the coal pipe geometry, the visualization environment can simultaneously show three-dimensional air flow, coal density, and turbulence, as well as a number of other variables including three-dimensional isosurfaces, cutting planes, streamlines (where appropriate) and other common CFD visualization tools. The results can be shown and interacted with in a variety of environments, including a standard desktop or laptop environment, large-scale single wall stereo or non-stereo environments, tiled displays, and large-scale multiwall displays (e.g. CAVES$^{TM}$).

A variety of optimization methods are available. These methods can be broadly classified into four main areas: optimization by intuition, optimization by trial-and-error modeling, optimization by numerical algorithm, and optimization by evolution [14]. Optimization by intuition is commonly used as a starting place in designing complex systems. Optimization by trial and error is appropriate when the first feasible design is not the optimal one. Before the advent of fast computers, this was the predominant technique. Optimization by numerical algorithm includes such techniques as gradient-based search. Optimization by evolution mimics the process of evolution by comparing designs, eliminating the less effective designs,

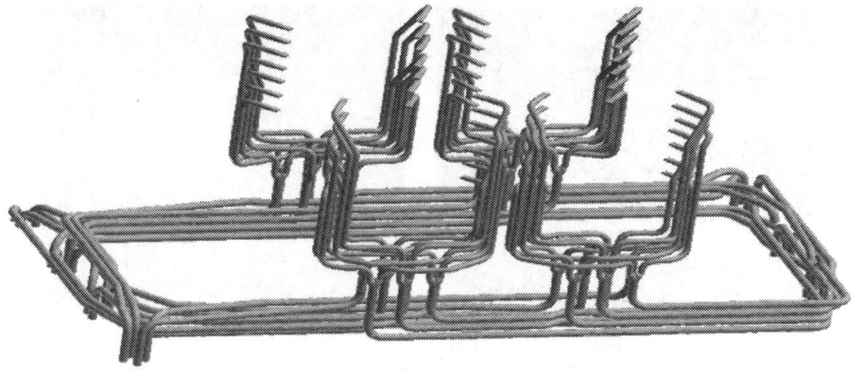

Figure 3: Coal piping being analyzed.

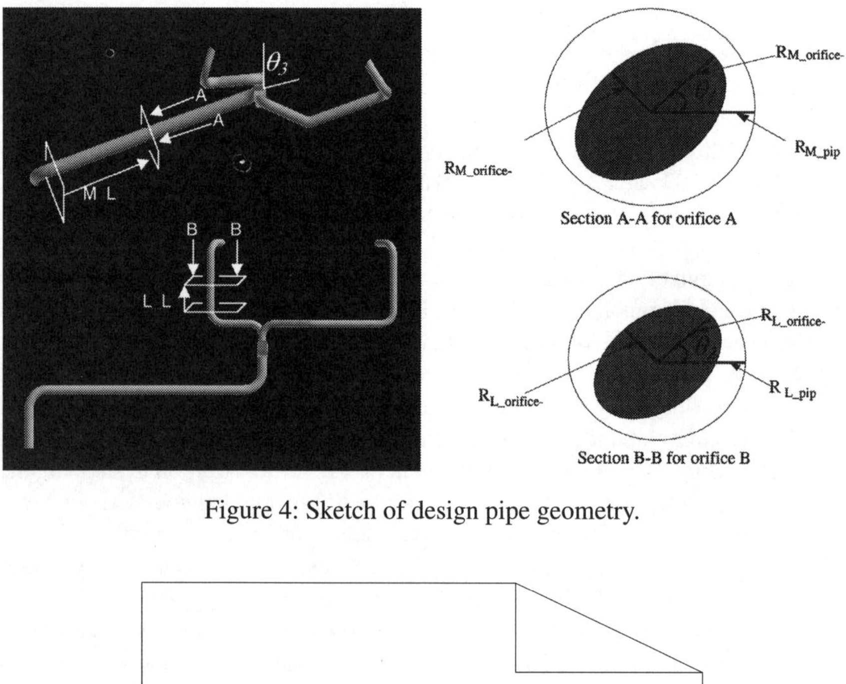


Figure 4: Sketch of design pipe geometry.

| | | |
|---|---|---|
| MAIN_ORIFICE_ANGLE | 0 | 90 |
| MAIN_ORIFICE_MAJOR_RADIUS | 0.85 | 1.0 |
| MAIN_ORIFICE_LOCATION | 2.0 | 43.0 |
| LEFT_ORIFICE_ANGLE | 0 | 90 |
| LEFT_ORIFICE_MAJOR_RADIUS | 0.74 | 0.8 |
| LEFT_ORIFICE_LOCATION | 2.0 | 10.0 |
| BIFURCATOR_ANGLE | -45.0 | 45.0 |

Figure 5: Constraints for the design variables.

and "breeding" the remaining designs to create better ones. This is very similar to the engineering design process and works well when combined with user knowledge and preference in a progressive interaction mode. Therefore, evolutionary optimization is used here.

Coupling the coal piping shape optimization/design tools with the integrated CFD/CAD models requires substantial computational expense. In the case of a single-phase fluid flow shape optimization, finding the optimal solution using a CFD model may take hours, days, or even weeks. The CFD model for two-phase flow is more complex than a single-phase model. The numerical model used for two-phase flow requires that the continuous flow field be calculated first. After this continuous flow field converges, the particle phase is injected into the flow field and the solution is iterated until the interaction between the two phases is resolved. Because of this process of iteration, the computational time required is significantly longer than the time required for a single-phase flow calculation. Therefore, integrating a two-phase CFD solver, which is required for coal pipe system design, into an optimization searching process becomes extremely time-consuming. Thus, computational cost for new design systems must be reduced [15]. There are three ways to reduce the computational cost:

1. *Reduce the number of calls to the CFD solver* – This requires that the optimization method be computed efficiently so that better designs can be found with fewer calls to the CFD solver. Currently, this is the primary focus of increasing the optimization's performance amongst the design community. A substantial amount of research has already been carried out on methods for dealing with this issue.
2. *Speed up the convergence rate when calculating the CFD model* – As in any design process, the faster the CFD code the better, because more design iterations performed within a given timescale increase the designer's ability to achieve the design target within real world time constraints. This method, which focuses on reducing the computational cost by speeding up the convergence rate of the CFD model, is commonly neglected by the design community, mainly because of the long history of separation between the design community and the CFD community.
3. *Introduce human experience into the optimization process to narrow the search space to a manageable size* – On the whole, the basis of the design process is trial and error, and the success of the final design depends largely on the designer's knowledge and intuition. By adding progressive human interaction to the design and optimization process, high-fidelity models like CFD can use the full extent of their capability when coupled with numerical optimization methods.

This discussion will focus primarily on the third method of reducing computational time: human interaction within the optimization process.

The shape of the pipe to be designed is represented by a number of variables such as degree of bends, length of pipes, geometrical definition of orifices, and so on. The set of parameters can vary according to the desired degrees of freedom. The pipe used to demonstrate the process has eight bends, two orifices, and one bifurcator in the system (Fig. 4). The design parameters were chosen based on observations of the

CFD results of the baseline model and the desired outcome. Preliminary results indicated that an elliptical orifice can increase the mixing mechanism and break up coal ropes more quickly. In this case, it was assumed that the geometry of the piping was fixed (as in a redesign of an existing pipe). The location, diameter, and angles of two elliptical orifices, as well as the angle of the bifurcator, were allowed to be interactively changed. Orifice A was placed in the main pipe upstream from the bifurcator to prevent unbalanced flow rates among the four pipes extending from the same mill. Orifice B was placed in the left branch of the pipe after the bifurcator to balance the flow resistance between these two pipe branches downstream to the bifurcator. In current power plant design, the locations of these two orifices in the pipes are undefined. They are mainly installed in the pipe wherever convenient for installation and replacement. However, our preliminary results showed that when the coal rope exists in the pipe, the location of the orifices is important to maximize the dispersion effect that the orifice inserts into the flow field. Assuming accessibility for installation and maintenance is granted wherever the orifices are located, orifice A is put in the long horizontal pipe after the horizontal-to-vertical bend and orifice B is put in the long vertical pipe after the horizontal-to-vertical bend. Both orifices can be moved along the length of the pipe. Also, because our goal is to balance the coal flow rate, we will maintain a consistent pressure drop. Therefore, the opening of the orifice that determines the pressure drop is constant throughout the design process. Based on this, the orifice's radius can be modified as long as the orifice's opening remains constant. The best angular position of the rope was found to depend strongly on the pipe geometry, especially the length of pipe between two bends. Therefore, although it is omitted in practice, the bifurcator's orientation in this study is also taken into consideration. The assumption is that if the centerlines of the bifurcator are in line with the center of the rope, coal flow may be more uniformly distributed.

An optimization problem in the virtual engineering environment can now be set up using the following preprocessing steps:

- *Define the design variables* – This is mainly done in the problem definition step.
- *Define the constraint file* – Seven inequality constraints are used in the optimization to obtain realistic configurations. Fig. 5 shows the complete constraint file for this model. In addition to the inequality constraints mentioned above, two side constraints are used to maintain the consistency of the two orifice openings with the baseline design. Therefore, only one radius of the elliptical orifice is the independent variable. It is worth mentioning that users can manually input this information through the graphical user interface (GUI) the system provides.
- *Define fitness evaluation functions* – The design of a coal transport system for a coal-fired power plant needs to address two main objectives. First, the piping system should deliver equal mass flow rates (gas and solid) at the main outlets (left and right). The objective function measuring the coal flow balance can be expressed as:

$$F(\vec{x}) = \sqrt{\frac{1}{2}\left((Q_L - \bar{Q})^2 + (Q_R - \bar{Q})^2\right)} \quad (1)$$

where $\vec{x}$ is the design variable vector and $Q_L, Q_R, Q$ denote the coal mass flow rate associated with the left and right outlet and the average outflow rate respectively. Second, the particle distribution at the cross section immediately before the bifurcator should be as uniform as possible. This is an important factor in extending the distributor's life and meeting the requirement that the flow splitter split both the coal and air flow evenly. The object function measuring the quality of the particle distribution profile is determined by the standard deviation of the particle concentration in the cross section right before the bifurcator *MI*,

$$MI = \frac{1}{C_p}\left[\frac{1}{n-1}\sum_{k=1}^{n}\left(C_m(k) - C_p\right)^2\right]^{1/2} \tag{2}$$

where $C_p$ is the mean particle concentration in the pipe's cross section, $C_m(k)$ is the local particle concentration measured at different locations on the surface, and $n$ is the total number of cells on the surface. Because the correlation between these two objective functions is unclear, the weighted-sum approach is chosen to linearly combine these two objectives into one fitness evaluation function. In this way, the optimization problem can be simplified as a single objective rather than a multi-objective optimization problem.

The design optimization history can be seen in Fig. 6, which monitored the development of the objective function. Starting from a randomly generated initial parameter set with 32 individuals, with periodic help from the user, the fitness converged very quickly. As indicated, no significant changes take place beyond the twelfth design iteration. Therefore, we can consider that the optimizer obtains the optimum after 12 iterations (generations). In a matter of 12 design iterations (about 5 days), the coal mass flow unbalance decreased from 15% (baseline model) to 1%.

Figure 7 shows a configuration summary of the design variables for the optimum design. Figure 8 shows the geometry of the optimal pipe. Because all seven design variables were changed, it is difficult to define which one is the dominant

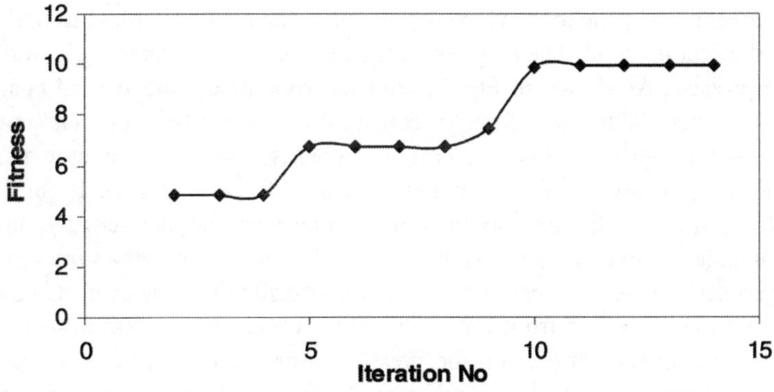


Figure 6: Iteration history for fitness.

| | |
|---|---|
| MAIN_ORIFICE_ANGLE | 90 |
| MAIN_ORIFICE_MAJOR_RADIUS | 0.9 |
| MAIN_ORIFICE_LOCATION | 43 |
| LEFT_ORIFICE_ANGLE | 33 |
| LEFT_ORIFICE_MAJOR_RADIUS | 0.8 |
| LEFT_ORIFICE_LOCATION | 2.1 |
| BIFURCATOR_ANGLE | 45 |

Figure 7: Summary of design variables.

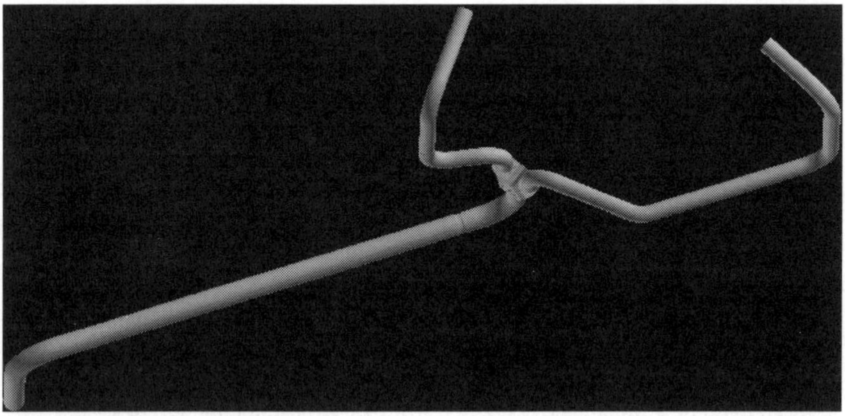

Figure 8: Geometry of the optimal pipe.

parameter. The main changes in orifice A (from circular shape to elliptical shape, from near the center of the long horizontal pipe to its end right before the second bend) affect the way the rope, which is generated after the first vertical-to-horizontal, enters the second bend. The engineer can view the results interactively during the design process. As shown in Fig. 9, after the rope enters the second bend, it is slightly twisted. With the bifurcator rotated 45°, this twisted rope can enter the bifurcator through its centerline. This will greatly help the bifurcator distribute particle flow evenly. To further understand the effects of orifice B (in the left branch) on the coal distribution, the optimum case can be interactively and visually compared with a similar design (see Fig. 10 for its geometry summary). The only geometry difference between these two is orifice B. However, its coal flow rate unbalance increased from 1% to 5%. This implies that the location of the orifice has a significant impact on the pressure drop, mainly because the pressure fluctuation happened in the gas–solid flow. Unlike single-phase flow, the pressure is dynamically changed (pattern unknown so far) instead of linearly increased with

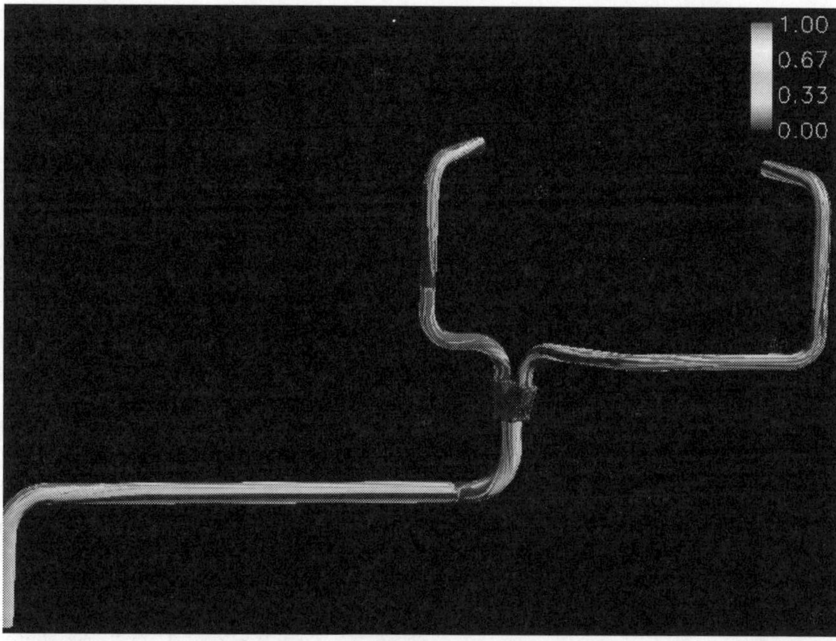


Figure 9: Particle distribution in the optimum pipe.

| | |
|---|---|
| MAIN_ORIFICE_ANGLE | 90 |
| MAIN_ORIFICE_MAJOR_RADIUS | 0.9 |
| MAIN_ORIFICE_LOCATION | 43 |
| LEFT_ORIFICE_ANGLE | 65 |
| LEFT_ORIFICE_MAJOR_RADIUS | 0.77 |
| LEFT_ORIFICE_LOCATION | 10 |
| BIFURCATOR_ANGLE | 45 |

(a)

| | |
|---|---|
| MAIN_ORIFICE_ANGLE | 90 |
| MAIN_ORIFICE_MAJOR_RADIUS | 0.9 |
| MAIN_ORIFICE_LOCATION | 43 |
| LEFT_ORIFICE_ANGLE | 33 |
| LEFT_ORIFICE_MAJOR_RADIUS | 0.8 |
| LEFT_ORIFICE_LOCATION | 2.1 |
| BIFURCATOR_ANGLE | 45 |

(b)

Figure 10: Summary of design variables for (a) case with same parameters for orifice A and bifurcator as the optimum case and (b) optimum design.

the pipe length for gas–solid flow. Therefore, the location of the orifice is a very important variable to determine if the flow resistance is the same for the two branches, which will in turn affect the flow rate balance. This shows the importance of using a high-fidelity CFD solver within a virtual engineering environment for complex problems. Other specific applications of virtual engineering to power plant components include the design of $NO_x$ reduction systems [16] and fire protection for nuclear power plants [17].

## 3 Current development efforts

There are currently two development efforts focused on the broader issues of virtual engineering for power plants. These are:

- The Integrated Environmental Control Model (IECM) is being developed at Carnegie Mellon University and interactively calculates the performance, emissions, and cost of alternative power plant configurations of interest for preliminary design and scoping studies [18]. Plant options currently available in the IECM include a pulverized coal plant, a coal-based integrated gasification combined cycle plant, an oxyfuel combustion plant, and a natural gas-fired combined cycle plant. This demonstrates a broader virtual engineering tool that can be used during the initial development of a power plant concept. IECM has recently been coupled to a virtual environment and new detailed models of components can be imported and displayed within the virtual engineering environment.
- Advanced Process Engineering Co-Simulator (APECS) is currently under development at the National Energy Technology Laboratory. APECS links various solvers using the CAPE-Open standard for linking processing models as the integration method. APECS has been demonstrated for power plant design using the process simulation software Aspen™ and Fluent™ as the CFD package.

### 3.1 Integrated Environmental Control Model

IECM (Fig. 11) is a process simulator that can provide the key power plant operating parameters including performance, emissions, and cost during the policy level of the preliminary design of new power plants. IECM's strength is in linking high-level decisions on plant configuration with economic cost models and uncertainties in performance. With these aspects of plant operation and design explicitly characterized, the overall risks and payoffs, including the environmental payoff, of an advanced system design can be quickly determined.

IECM currently includes the necessary virtual models to create a pulverized coal plant, a coal-based integrated gasification combined cycle plant, an oxyfuel combustion plant, or a natural gas-fired combined cycle plant. Environmental control options include virtual components that allow designers to investigate power plants with near-zero emissions of critical air pollutants, air toxins, and greenhouse gases. Model parameters and virtual plant components can be adjusted to represent either current technology or advanced, high-performance, low-emission designs. IECM has recently been integrated into a broader virtual engineering framework using VE-Suite. This enables the engineer to bring the detailed analyses of various components (e.g. CFD analysis of a gasifier) into IECM, extending the types of components that can be examined. In addition, the user can display the results in a plant walkthrough format, enabling the designer to readily understand the space requirements and floor plan of a proposed power plant.

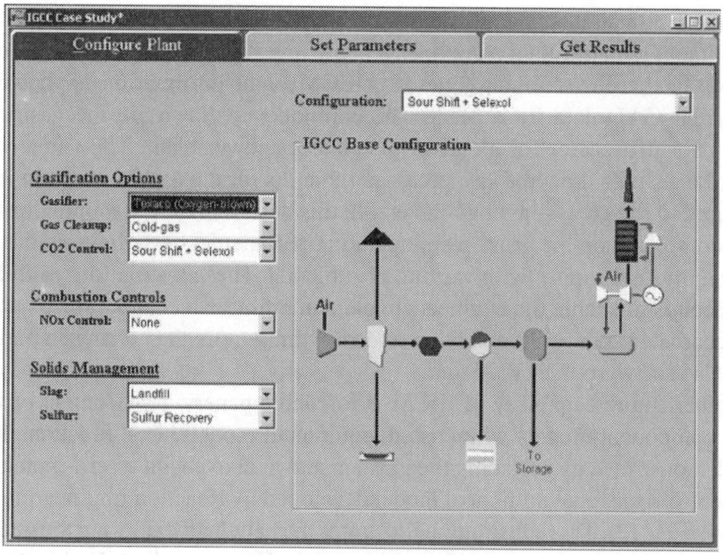


Figure 11: IECM interface.

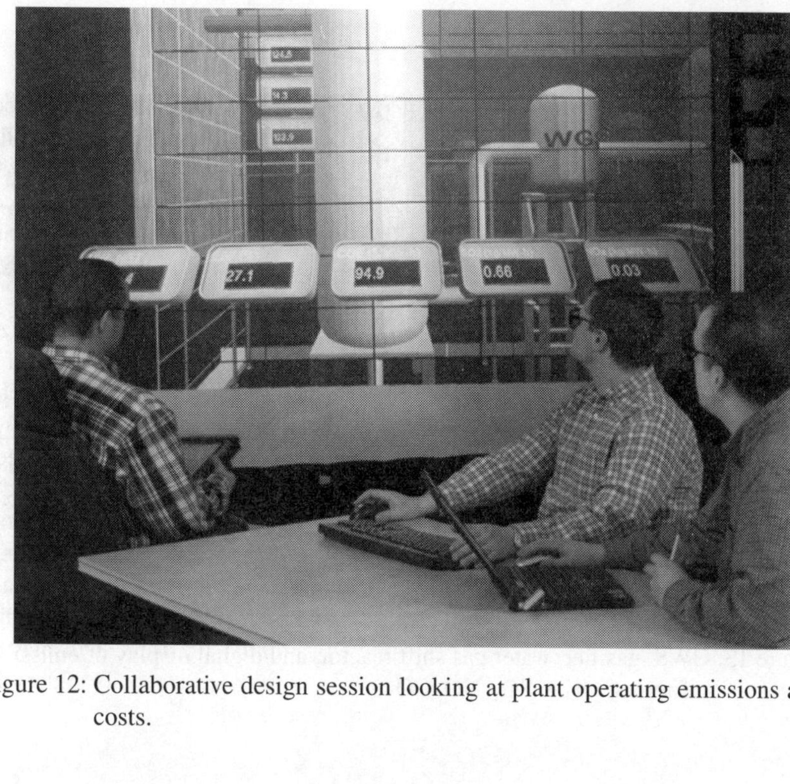


Figure 12: Collaborative design session looking at plant operating emissions and costs.

Figure 12 illustrates how overlaying abstract design information in an intuitive environment with the plant components being investigated can improve the design process. In the environment (Fig. 12), IECM plant information is shown with conceptual CAD data for a plant. The components shown are the gasifier and water gas shift reactor for an advanced concept power plant. The gauges to the left of the gasifier show the key stream parameters for the gasifier that can then be compared to the gross plant parameters. In this design review, a scoping review is being completed on the gross plant operating parameters by examining the effect of changing fuel supply with varying coal blends. These simulations with IECM take seconds to enable the engineers to gain an intuitive feel about the plant operating characteristics without having to do anything more than change a dialog on the IECM software with VE-Suite.

Another future capability of IECM is to enable generic replacement of IECM native component models with custom equipment models. This functionality has been demonstrated by replacing the gasifier shown above with a zero-dimensional and three-dimensional numerical model developed by Reaction Engineering International (Fig. 13). This substitution illustrates the benefit of exploring the increased fidelity of information on specific sub-systems of a power plant. In this example, three-dimensional stream information is available to the rest of the IECM simulation. For the water gas shift reactor, this means that the detailed model helped improve the shift reactor results and gave improved overall plant operating parameters for emissions and cost.

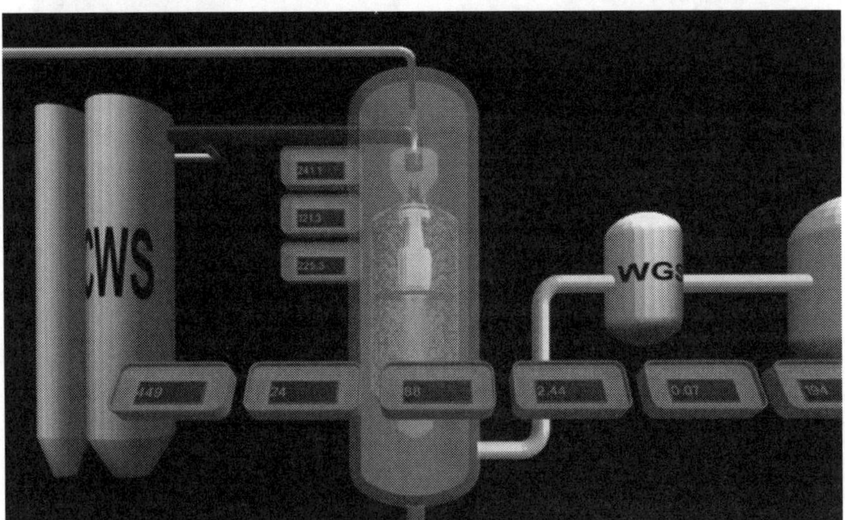

Figure 13: CWS, gasifier, water gas shift reactor, and digital display of emissions information (i.e. $SO_x$, NO, $CO_2$).

### 3.2 Advanced Process Engineering Co-Simulator

New power plants are large capital investments that will be in operation for sixty or more years and require periodic costly maintenance. Small changes in operation can impact efficiency and significantly increase fuel cost. As a result, power companies and those who fund the construction of new power plants want to ensure success and have little interest in unproven and untried technologies even if they promise reduced emissions and improved efficiency. Because of this, new types of power plants have traditionally been built and designed in a series of lab scale experiments, prototypes, demonstration-scale and full-scale power plants using five to ten physical models. To reduce the cost and speed up the introduction of a new fleet of near-zero emission power plants, the US Department of Energy is working to create high-fidelity computer simulations to design and evaluate virtual power plants. Key to the development of these virtual power plants is the development of the APECS. APECS (Fig. 14) couples CFD, reduced order models, and custom engineering models with any process simulator supporting the CAPE-Open standard [19–21]. This significantly extends the capability of process

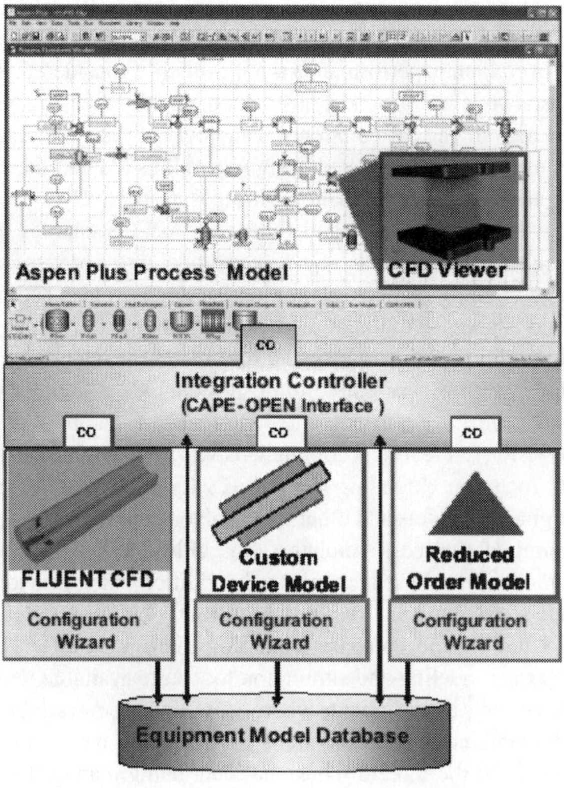


Figure 14: Software tools within APECS.

simulation software by enabling the designer to examine the impact of fine grain component behavior on the overall process.

Process simulators perform material and energy balances, thermodynamic calculations, and chemical reaction computations for the entire plant. Such tools are used throughout the lifecycle of the plant for designing the overall plant configuration, evaluating plant economics, analyzing safety and environmental issues, and calculating plant operating efficiency and performance under start up, shut down, and upset conditions. These tools offer libraries of simplified equipment models, which require input data that are not easily measured and that may change with operating conditions (e.g. pressure drop, heat transfer coefficient, and reactant conversion). Therefore, a plant design based on simplified equipment models may be suboptimal or may violate constraints imposed by a certain equipment item (e.g. local reactor catalyst temperature should not exceed the sintering temperature). This modeling process is now widely used by the process industries and has changed the way that engineers design and optimize plants.

Currently, two main levels of model abstraction are used in the power plant design process. Process system models are used to define the overall plant performance and detailed models of components or parts of components are used to gain additional insight and direct the design of the components. These models are not linked together. Consequently, the impact of component-specific behavior (e.g. three-dimensional flow issues) on overall plant performance is not explicitly considered. APECS explicitly couples these two levels of model abstraction, enabling the engineer designer to consider both design of various components and overall system performance simultaneously. This has been demonstrated using Aspen Plus™ and Fluent™ as the process simulation and component modeling packages, respectively.

When used with Aspen Plus™ and Fluent™, APECS enables an engineer to graphically integrate a high-fidelity equipment model into a process simulation in a matter of minutes. This graphical integration, as opposed to a coding integration, reduces time and enables equipment models to be coupled to Aspen Plus™ on an as-needed basis without having to consider the cost of the integration. With an efficient integration workflow accessible to design engineers, another component of the comprehensive virtual environment can be realized. The final goals of APECS will further the vision of creating a comprehensive virtual engineering environment. Current efforts focus on extending the APECS toolkit to support reduced order models of equipments models and other CFD solvers, enable integration with other CAPE-Open compliant process simulators (e.g. gPROMS), and enable data persistence through modern database technology. In addition to the previously described components under development, research is currently being completed on integrating the APECS toolkit and database technology into VE-Suite. This integration effort will bring the modeling and simulation tool suite available with APECS into the virtual engineering environment, where it will also be available to other VE-Open-compliant toolkits. In addition, all of VE-Suite's visualization capability can be used to interrogate the three-dimensional data being managed by APECS. For example, the process simulation data from Aspen Plus™ and Fluent™ could be integrated with large-scale environment simulation to understand the potential environmental impact of the power plant under design on the surrounding area throughout the plant's life. This type of integration is currently a human-intensive integration process and is thus less likely to be used in the decision making process for the plant. With the future toolset of APECS, Aspen Plus$^{TM}$, Fluent$^{TM}$, and VE-Suite all transferring information, this type of modeling and simulation effort can become a reality. Each of these APECS extensions will enable more physically based information to be at the fingertips of power plant design teams up to policy makers.

## 4 Building a virtual engineering application

Development of a virtual engineering environment from scratch is currently a significant challenge. A better way to approach the development of an application is to utilize VE-Suite [13, 22, 23]. VE-Suite, an object-oriented, open source software package developed for virtual engineering of complex engineering systems (e.g. power plants) design and analysis, is currently the only generalized virtual engineering software package. This section provides a brief overview of VE-Suite and the process of developing a virtual engineering application.

The architecture of VE-Suite is composed of three core engines driving the entire product design and analysis process: the computational engine (VE-CE), the graphical engine (VE-Xplorer), and the GUI (VE-Conductor) (Fig. 15). VE-CE

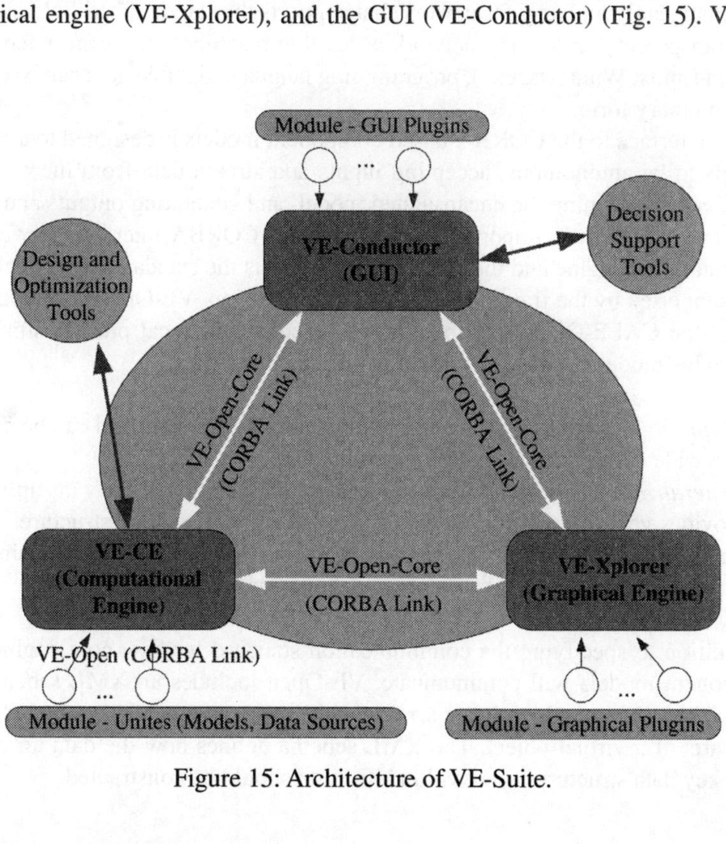


Figure 15: Architecture of VE-Suite.

handles the scheduling and synchronization of the data sources (model and information) being used within a product design and analysis process. VE-Xplorer displays all necessary three-dimensional data. VE-Conductor provides the user interface to control the software tools, engineering models, and graphical display. Similar to CAPE-Open [24], VE-Open [13] is a communication standard that supports the integration and communication between these core engines, as well as the integration with the specific modules developed by the user.

## 4.1 System integration mechanism: VE-Open

VE-Open is the mechanism that supports the information transfer, coordinates the operation of these engines, and enables models to be "plugged" into the framework in a straightforward manner. Component models are mathematical representations of individual virtual objects that are used by the framework to construct an overall simulation. The key to making the simulation framework extensible is to provide a mechanism by which component models can easily be integrated without extensive software development. To address this need, the component architecture design has been adopted. CORBA is used along with a standard model interface definition, which is implemented as an interface definition language (IDL), to create componentized computational models. These models can be used interchangeably with any framework that supports the standardized IDL, are location transparent (run on any network accessible machine), platform independent (run on Linux, Windows, etc.), programming language flexible, and can be distributed in binary form.

The interface to the CORBA-based component models is designed to allow the models to be autonomous, accepting inputs and stream data from the computational engine, running the encapsulated model, and generating outputs and modified stream data. It is important to note that the CORBA interface between the computational engine and the component models is the standardized model interface supported by the framework for model integration. VE-Open is analogous to that of the CAPE-Open [24] specification used by chemical process simulation tools. This model interface has a number of characteristics:

- *Simplicity:* The functions that are implemented are general and can be adapted to a wide variety of simulation environments.
- *Generalization:* The new interface removes the specificity of any discipline and provides generic structure for data types and software engine structure.
- *Enhanced data passing:* The new interface provides facilities for passing data beyond the level of simple scalars to downstream models.

In addition to specifying the communication standard for how core engines and component models will communicate, VE-Open includes an XML schema that defines how commands and data arrays can be constructed and passed to the various parts of a virtual object. The XML schema defines how the data arrays and other key data structures used within VE-Suite should be constructed.

## 4.2 VE-Conductor, VE-CE, and VE-Xplorer

VE-Suite is general in nature and the three key components can run separately on a geographically diverse set of heterogeneous computer platforms. This separation is convenient because VE-CE can run on the same machine as the application (computational unit), and VE-Conductor, which presents a GUI to the user, can execute remotely on a separate machine. For example, VE-CE can run on a Linux cluster, VE-Xplorer can run on a graphics rendering cluster, and VE-Conductor can run on a tablet PC. Therefore, the framework components can be distributed across computational resources to make the most efficient use of these resources. This architecture is also advantageous because, unlike VE-Conductor, VE-CE must exist throughout the application's lifetime. VE-Conductor is transitory and can connect to the server many times throughout the server's (application's) existence. Because the client may use visualizations for data interpretation, the end user may choose to run the client on a high-performance graphics system. Also, the three core components of VE-Suite can function as complete stand-alone applications, provided the user prepares the necessary input files. The ability to integrate models and query for data seamlessly within VE-Suite is one aspect of the flexibility of the framework. Since the data structure being passed within VE-Suite is XML, it is possible for each of the core engines within VE-Suite to benefit from its self-describing nature. This ability is also essential in creating an interactive design environment, which enables engineers to swap components in and out of the virtual design space and to have the rest of the product realization environment update automatically. Since the results are not sent to a specific location, any module within the VE-Suite framework can query the models for results data. This also makes it possible to access any information from any model in the VE-Suite framework. The two-dimensional UI environment and the virtual environment can then create custom graphical data representations based on what the engineer requires.

VE-Conductor exists independently from VE-CE and VE-Xplorer. The communication link between VE-Conductor and the rest of VE-Suite is bidirectional, handling items such as model parameters passed to VE-CE and receiving items such as execution status and results from VE-CE. VE-Conductor can also connect to the graphical environment and control graphical representations in VE-Xplorer. VE-Conductor can dynamically discover, identify, and load user interface elements for new component models. This capability minimizes the level of difficulty involved in integrating new component models.

VE-CE can construct, coordinate, schedule, and monitor software and engineering models in the product design–analysis process. An integral part of this process is reviewing the graphical design process by interacting with all aspects of the graphical data during the process. This requires VE-CE to be capable of transforming simple scalar values to three-dimensional data to the rest of VE-Suite. In the graphical design process, an engineer must be able to move and adjust product parameters, much like he or she would do when using CAD software. Hence, in the virtual space, VE-Xplorer must have the flexibility to transfer geometrical data from the three-dimensional virtual design space to the two-dimensional user interface and to VE-CE for use in updating the underlying numerical model. VE-Suite supports intensive interaction features between activities because in the product design and analysis process, the data must be handled interactively to give engineers the necessary information to make decisions.

VE-Xplorer handles the creation of the virtual environment and can run with any type of virtual environment, from a regular two-dimensional screen to a six-walled immersive virtual space. Due to the generality of the visualization requirements, the VE-Xplorer core provides a complete visualization toolset so that users can navigate and control the scene. Thus, although most VE-Suite users are not necessarily expert software developers, they need not worry about the complexities of graphical details and virtual reality programming and can instead spend time on their applications.

## 4.3 Detailed application development

When building a virtual engineering-based application, the implementation can be broken down into three tasks: user interfaces, the computational unit, and the visualization plug-in. These tasks comprise the VE-Suite API. In the descriptions below, the coal pipe problem described previously will be used as an example of how to develop a virtual engineering application.

### 4.3.1 User interface development

For example, in the virtual coal pipe design tool discussed earlier, the variables being utilized must be considered. The orifice angle, orifice major radius, orifice location, elbow angle, and elbow radius were utilized as the design parameters to investigate the coal pipes. A custom user interface can then be developed to enable the user to change the coal pipe design intuitively (Fig. 16). Through this interface, the user is able to create the system configuration, set model inputs, start and stop simulation execution, and view simulation results. The plug-in is then compiled into a Dynamic Link Library in Windows or a shared library in Linux/Unix. In general, these plug-ins are more likely to meet user-specific requirements because the user builds the interface, so it is tailored to the user's needs and can easily be modified if the user's focus of interest changes. VE-Suite is able to dynamically discover, identify, and load the user's GUI from these shared libraries. For the coal pipes, the user interface is specifically designed to optimize coal pipes. As noted previously, the application being developed can utilize all the tools already contained within VE-Conductor.

### 4.3.2 Computational unit

The computational unit for the coal pipe application wraps a Fluent$^{TM}$ computational model. This unit is designed to take the variables described above for the GUI plug-in and create and modify the Fluent$^{TM}$ numerical grid and rerun the model. In general, VE-CE computational units are capable of running a simulation containing a multitude of different types of models, each accepting and generating

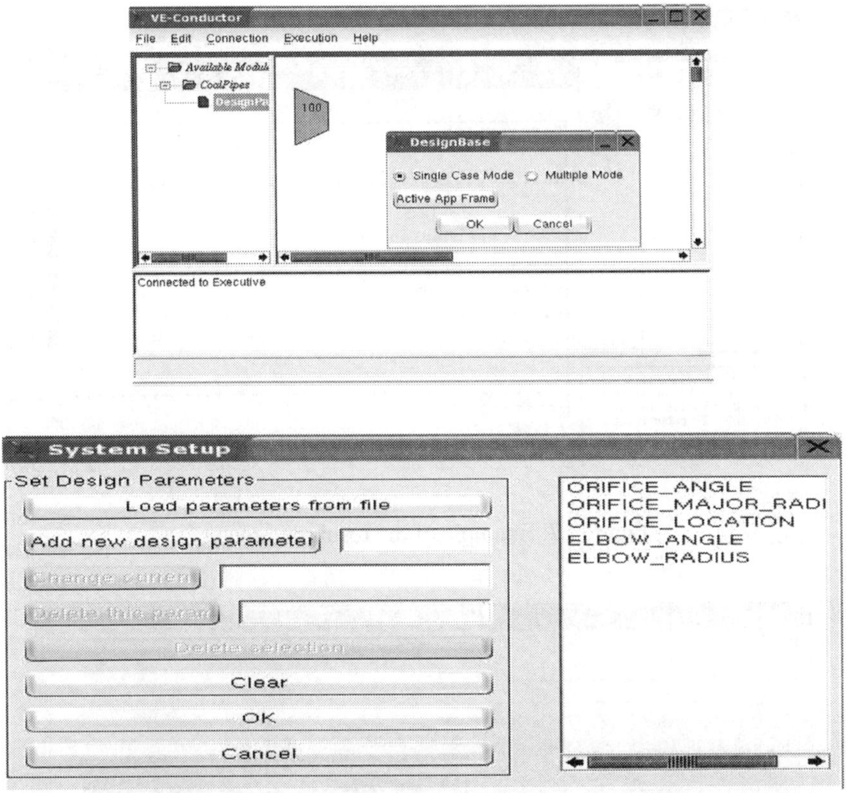


Figure 16: VE-Conductor plug-in for coal pipe design.

a myriad of data types. Since users from different fields have a wide variety of needs, the analysis tools and their use require a detailed understanding of the problem. Therefore, it is the user's responsibility to develop their application objects (computation unit). Changes to the GUI or the computational unit will not affect the other unless there is a change in system configuration. Once a client–server connection is made, the GUI is able to send parameters and commands to the unit, and the unit is able to send results, messages, updates, and communications back to GUI in real time. Because of this, existing commercial, in-house, and open source analysis packages can be used almost directly with VE-Suite. The analysis packages can vary from Microsoft Excel$^{TM}$ spreadsheets to process models to CFD models. The analyst only needs to provide a few routines to declare the communication variables. Once the computational unit has completed a run, the results from the run are available to the rest of the VE-Suite framework (Fig. 17). In the case of the coal pipes, the three-dimensional simulation data are available for review in VE-Xplorer through the graphical plug-in (described below) and overall design parameters are available in VE-Conductor (Fig. 18).

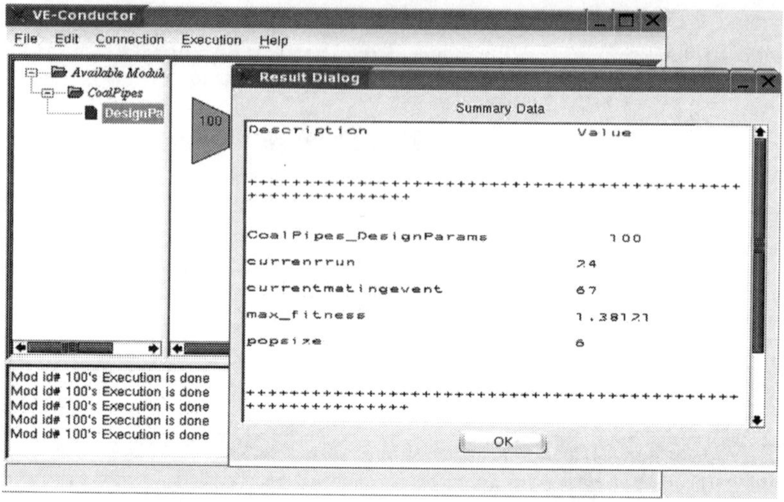


Figure 17: Example of the result parameter.

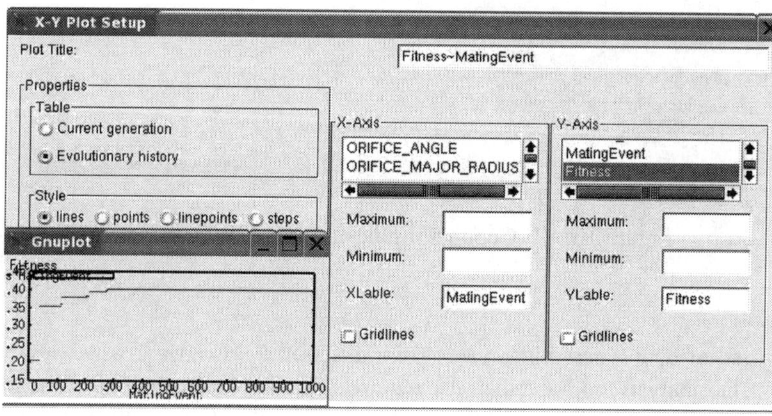


Figure 18: Example of a result plot.

### 4.3.3 Graphical plug-in

A key aim of virtual engineering is to fully engage the human capacity for problem solving by creating a realistic experience for the user so that he or she can focus entirely on the engineering problem. The advantage is that previously indescribable complexities can be understood and the full range of engineering solutions can be explored. VE-Xplorer provides the core visualization functionality for the virtual engineering aspect of the framework. VE-Xplorer can load geometry files, three-dimensional simulation data, and experimental data of almost every format into a scene. To enable graphical display of the coal pipe data in VE-Xplorer, a graphical plug-in is created that is designed to retrieve and store all the designs generated by the user. The graphical plug-in does not have the capability to graphically display the Fluent™ data but utilizes the capability of VE-Xplorer to handle this feature set. The graphical plug-in in this case adds the capability to catalogue and remotely retrieve the Fluent™ data from the computational unit described above. In addition, the graphical plug-in takes the user's input from VE-Conductor and changes the graphical CAD data so that the user can verify the design before sending the job to the computational unit to run a computationally expensive model.

## 5 Summary

Virtual engineering is an emerging technology that can provide a high-level integration of models to create a "game-like" engineering decision environment for power plant design. The strength of a virtual engineering environment is the explicit inclusion of multiple options and the integration of human experience and values with computational modeling to create a progressive interaction approach to engineering design. This integrated modeling environment will then be used to examine the impact of changes at the micro level on efficiency, emissions, and cost. These broadly integrated models will enable integrated optimization of power systems and will explicitly bring human judgment into the optimization and design of power systems during both design and operation.

## Acknowledgments

We would like to thank Dr. Gengxun Huang, Dr. Steve Zitney, Professor Ed Rubin, and Dr. Mike Bockelie for their assistance in developing this chapter.

## References

[1] Huang, G., Bryden, K.M. & McCorkle, D.S., Interactive design using CFD and virtual engineering. *Proc. of the 10th AIAA/ISSMO Multidisciplinary Analysis and Optimization Conference*, AIAA: Herndon, VA, AIAA-2004-4364, 2004.

[2] Bockelie, M., Swensen, D.A., Denison, M.K., Maguire, M., Yang, C., Chen, Z., Sadler, B., Senior, C.L. & Sarofim, A.F., A computational workbench environment for virtual power plant simulation, Contract DE-FC26-00NT41047, Final Report, December 2004.

[3] Branke, J., Kaußler, T. & Schmeck, H., Guidance in evolutionary multi-objective optimization. *Advances in Engineering Software*, **32(6)**, pp. 499–507, 2001.

[4] Xiao, A. & Bryden, K.M., Virtual engineering: a vision of the next-generation product realization using virtual reality technologies. *Proceedings of ASME Design Engineering Technical Conference and Computers in Engineering Conference*, ASME: New York, DETC2004/CIE-57698, 2004.

[5] Diachin, D., Freitag, L., Heath, D., Herzog, J., Michels, W. & Plassmann, P., Collaborative virtual environments used in the design of pollution control systems. *International Journal of Supercomputer Applications and High Performance Computing*, **10(2–3)**, pp. 223–235, 1996.
[6] Ryken, M.J. & Vance, J.M., Applying virtual reality techniques to the interactive stress analysis of a tractor lift arm. *Finite Elements in Analysis and Design*, **35(2)**, pp. 141–155, 2000.
[7] Kihonge, J.N., Vance, J.M. & Larochelle, P.M., Spatial mechanism design in virtual reality with networking. *Journal of Mechanical Design*, **124(3)**, pp. 435–440, 2002.
[8] Perles, B. & Vance, J.M., Interactive virtual tools for manipulating NURBS surfaces in a virtual environment. *Journal of Mechanical Design*, **124(2)**, pp. 158–163, 2002.
[9] Chipperfield, K.A. & Vance, J.M., Modeling of hydraulic hose paths. *Proc. of 2002 ASME Design Engineering Technical Conference and Computers in Engineering Conference*, ASME: New York, DETC2002/DAC34152, 2002.
[10] Schneider, H., Frank, T., Pachler, D.K. & Bernert, K., A numerical study of the gas-particle flow in pipework and flow splitting devices of coal-fired power plant. *Proceedings of the 10th Workshop on Two-Phase Flow Predictions*, pp. 227–236, 2002.
[11] Huang, G., *An Interactive Design Environment for Coal Piping Systems*, PhD Thesis, Iowa State University, Ames, IA, 2006.
[12] CORBA, http://www.omg.org, accessed July 12, 2007.
[13] VE-Suite, http://www.vesuite.org, accessed July 12, 2007.
[14] Siddall, J.N., Frontiers in optimum design. *Transactions of ASME Journal of Mechanical Design*, **101(4)**, pp. 674–681, 1979.
[15] McCorkle, D.S., Bryden, K.M. & Carmichael, C.G., A new methodology for evolutionary optimization of energy systems. *Computer Methods in Applied Mechanics and Engineering*, **192(44–46)**, pp. 5021–5036, 2003.
[16] McCorkle, D.S., Bryden, K.M. & Swensen, D.A., Using virtual engineering tools to reduce $NO_x$ emissions. *Proceedings of the ASME Power 2004 Conference*, ASME: New York, pp. 441–446, 2004.
[17] Muth Jr., D., Bryden, K.M. & Buell, D., Using virtual engineering tools to assess fire scenario hazards. *Proceedings of the American Nuclear Society Winter Meeting*, American Nuclear Society: La Grange Park, IL, pp. 305–306, 2005.
[18] IECM, http://www.iecm-online.com/, accessed July 11, 2007.
[19] Osawe, M.O., Felix, P., Syamlal, M., Lapshin, I., Cleetus, K.J. & Zitney, S.E., An integrated process simulation and CFD environment using the CAPE-OPEN interface specifications. *Proceedings of the AIChE 2002 Annual Meeting*, Paper No. 250c, 2002.
[20] Zitney, S.E., Hockley, R., Scheele, T. & Syamlal, M., Using fluid flow models in a process simulator for detailed plant design. *Proceedings of SIMS 2001, the 42nd Conference on Simulation and Modeling*, 2001.

[21] O'Brien, T.J., Zitney, S.E., Rogers, W.A. & Cugini, A., Simulation tools for the design of Vision 21 energy plants. *Proceedings of the 226th ACS National Meeting*, New York, NY, 2003.
[22] Bryden, K.M. & McCorkle, D.S., VE-Suite: A foundation for building virtual engineering models of high performance, low emission power plants. *Proceedings of the 29th International Technical Conference on Coal Utilization & Fuel Systems*, pp. 91–100, 2004.
[23] McCorkle, D.S., Bryden, K.M. & Kirstukas, S.J., Building a foundation for power plant virtual engineering. *Proceedings of the 28th International Technical Conference on Coal Utilization and Fuel Systems*, pp. 118–127, 2003.
[24] CAPE-OPEN, http://www.colan.org/, accessed July 12, 2007.

# CHAPTER 4

# Steam power plants

E. Khalil
*Department of Mechanical Power Engineering,*
*Cairo University, Cairo, Egypt.*

## Abstract

The efficient utilization of fossil energy in power generation together with low pollution in conventional thermal power plants is a topic that is gaining interest internationally. The energy availability and sustainability scenario is an area of growing interest and demand in many countries worldwide due to the greater desire to enhance standards of living, increase productivity and preserve a clean environment. Efficient energy use is favorable for better productivity, product quality, costs, and quality of human life but the use of energy adversely impacts our environment. The fundamental concepts of power generation had been refined to enhance the power generation efficiency through the use of modern techniques of waste heat recovery and co-generation. This chapter summarizes the basic power cycles in steam power plants and outlines the various methods of improvement. The main goal of efficient power generation is, among others, to rationalize the use of fossil fuels and enhance the combustion efficiencies. This is outlined in this chapter through a review of the various combustion modeling techniques for furnace flames under steady and time dependent configurations. The ability of numerical computations to predict the boiler furnace thermal behavior is an ultimate goal. The heat transfer to furnace walls through thermal radiation is reviewed briefly to demonstrate the present capabilities. Boiler furnace walls are subject to the major problems of fouling that result in deterioration of the performance and drastic reduction of the heat transfer characteristics. This work briefly highlights the fouling problem in power plant water walls and proposes a monitoring, inspection, and maintenance schedule. The information provides a quick guide on the commonly faced operation problems and methods to enhance energy conversion efficiency.

# 1 Introduction

Centralized power generation became possible when it was recognized that alternating current power lines can transport electricity at low costs across great distances by taking advantage of the ability to raise and lower the voltage using power transformers. Since 1881, electricity has been generated for the purpose of powering human technologies from various sources of energy. The first power plants were run on water power or coal, and today we rely mainly on coal, nuclear, natural gas, hydroelectric, and petroleum with a small amount from solar energy, tidal harnesses, wind generators, and geothermal sources. Rotating turbines attached to electrical generators produce most commercially available electricity. Turbines are driven by a fluid, which acts as an intermediate energy carrier. The fluids typically used are:

- *Steam in steam turbines* – Water is boiled by nuclear fission or the burning of fossil fuels (coal, natural gas, or petroleum). Some newer plants use the sun as the heat source: solar parabolic troughs and solar power towers concentrate sunlight to heat a heat transfer fluid, which is then used to produce steam. Another renewable source of heat used to drive a turbine is geothermal power. Either steam under pressure emerges from the ground and drives a turbine or hot water evaporates a low-boiling liquid to create vapor to drive a turbine.
- *Water in hydraulic turbines* – Turbine blades are acted upon by flowing water, produced by hydroelectric dams or tidal forces.
- *Wind* – Most wind turbines generate electricity from naturally occurring wind. Solar updraft towers use wind that is artificially produced inside the chimney by heating it with sunlight.
- *Hot gases in gas turbines* – Turbines are driven directly by gases produced by the combustion of natural gas or oil.

*Combined cycle gas turbine plants* are driven by both steam and gas. They generate power by burning natural gas in a gas turbine and use residual heat to generate additional electricity from steam. These plants offer efficiencies of up to 60%.

Various other technologies have been studied and developed for power generation. Solid-state generation (without moving parts) is of particular interest in portable applications. This area is largely dominated by *thermoelectric* (TE) devices, though *thermionic* (TI) and *thermo-photovoltaic* (TPV) systems have been developed as well. Typically, TE devices are used at lower temperatures than TI and TPV systems. *Piezoelectric devices* are used for power generation from mechanical strain, particularly in power harvesting. *Betavoltaics* are another type of solid-state power generator, which produces electricity from radioactive decay. *Fluid-based magnetohydrodynamic* power generation has been studied as a method for extracting electrical power from nuclear reactors and also from more conventional fuel combustion systems. Electrochemical electricity generation is also important in portable and mobile applications. Currently, most electrochemical power comes from closed electrochemical cells ("batteries"), which are arguably utilized more as storage systems than generation systems, but open electrochemical systems, known as fuel cells, have been undergoing a great deal of research and development in the last few years. Fuel cells can be used to extract power either from natural fuels or from synthesized fuels (mainly electrolytic hydrogen) and so can be viewed as either generation systems or storage systems depending on their use.

Factors and parameters affecting the selection of steam power plant capacities and sites [1]:

1. *Types of* service (base load or peak load).
2. *Location* (relative to water and fuel).
3. *Space available* (each power plant has a certain area/unit energy produced).
4. *Reliability:* Steam turbine life is extremely long. There are steam turbines that have been in service for over 50 years. Overhaul intervals are measured in years. When properly operated and maintained (including proper control of boiler water chemistry), steam turbines are extremely reliable. They require controlled thermal transients as the massive casing heats up slowly and differential expansion of the parts must be minimized.
5. *Environment:* Emissions associated with a steam turbine are dependent on the source of the steam. Steam turbines can be used with a boiler firing any one or a combination of a large variety of fuel sources, or they can be used with a gas turbine in a combined cycle configuration. Boiler emissions vary depending on fuel type and environmental conditions. Boiler emissions include nitrogen oxide ($NO_x$), sulfur oxides ($SO_x$), particulate matter (PM), carbon monoxide (CO), and carbon dioxide ($CO_2$). Recently, $NO_x$ control has been the primary focus of emission control research and development in boilers. The following provides a description of the most prominent emission control approaches [2]. Combustion control techniques are less costly than post-combustion control methods and are often used on industrial boilers for $NO_x$ control. Control of combustion temperature has been the principal focus of combustion process control in boilers. Combustion control requires tradeoffs – high temperatures favor complete burn-up of the fuel and low residual hydrocarbons and CO, but promote $NO_x$ formation. Very lean combustion dilutes the combustion process and reduces combustion temperatures and $NO_x$ formation. However, if the mixture is too lean, incomplete combustion occurs, increasing CO emissions [3].
6. *Cost* (capital cost and maintenance cost): There is broad consensus among scientists that we are not close to running out of fossil fuels. Despite this abundance, political considerations over the security of supplies, environmental concerns related to global warming and sustainability might move the world's energy consumption away from fossil fuels. A government-led move away from fossil fuels would most likely create economic pressure through carbon emissions trading and green taxation. Some countries are taking action as a result of the Kyoto Protocol, and further steps in this direction are proposed. For example, the European Commission has proposed that the energy policy of the European Union should set a binding target of increasing the level of renewable energy in the EU's overall mix from <7% today to 20% by 2020.

# 2 Energy scenarios

## 2.1 Crude oil production

World crude oil production during August 2005 was 74 million barrels per day, up 0.2 million barrels per day from the level in the previous month. OPEC (Organization of the Petroleum Exporting Countries) production during August 2005 averaged 31 million barrels per day, down 0.1 million barrels per day from the level in the previous month. During August 2005, production increased in the United Arab Emirates by 50,000 barrels per day; Algeria by 20,000 barrels per day; Libya by 15,000 barrels per day; and Iran by 10,000 barrels per day. Production decreased in Nigeria and Iraq each by 100,000 barrels per day and remained unchanged in Saudi Arabia, Venezuela, Kuwait, Indonesia, and Qatar. Among the non-OPEC nations, production during August 2005 increased in Mexico by 332,000 barrels per day; Russia by 150,000 barrels per day; the United States by 29,000 barrels per day; and China by 14,000 barrels per day. Production decreased in the United Kingdom by 215,000 barrels per day; Canada by 92,000 barrels per day; Norway by 72,000 barrels per day; and Egypt by 3000 barrels per day [4].

## 2.2 Petroleum consumption

In July 2005, consumption in all OECD (Organization for Economic Cooperation and Development) countries was 49 million barrels per day, 1% (percentage changes are based on unrounded data) lower than the July 2004 rate. Comparing July rates in 2005 and 2004, consumption was higher in 2005 in Canada (+1%) and South Korea (<+1%). The July 2005 consumption rate was lower in Italy (–7%); Germany (–4%); France and Japan (each –2%); the United Kingdom (–1%); and the United States (less than –1%), compared with the rate one year earlier.

## 2.3 Petroleum stocks

For all OECD countries, petroleum stocks at the end of July 2005 totaled 4.2 billion barrels, 4% (percentage changes are based on unrounded data) higher than the ending stock level in July 2004. Stock levels were higher in July 2005 in the United States (+6%); France (+4%); Germany (+3%); and Canada and Japan (each +2%). Stock levels were lower in the United Kingdom (–6%) and South Korea and Italy (each –2%), compared with levels one year earlier.

# 3 Steam power plants cycles

## 3.1 Basic cycle description

Rankine cycles describe the operation of steam heat engines commonly found in power generation plants as schematically shown here in Fig. 1. In such vapor

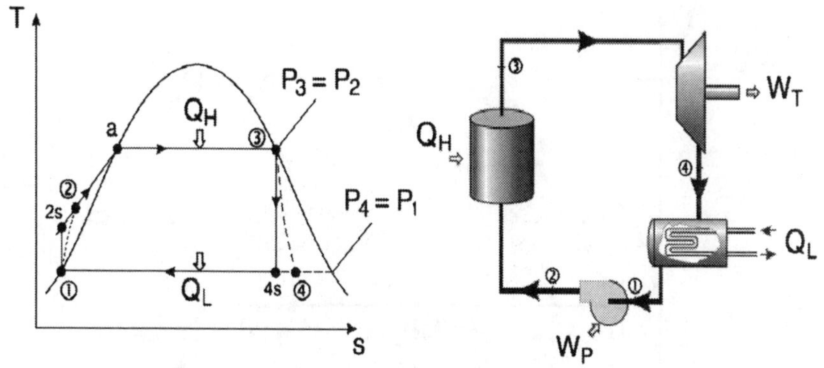


Figure 1: Simple Rankine cycle.

power plants, power is generated by alternately vaporizing and condensing a working fluid (in many cases water, although refrigerants such as ammonia may also be used) [1].

There are four processes in the Rankine cycle, each changing the state of the working fluid. These states are identified by number in Fig. 1.

- *Process 1-2$_s$:* First, the working fluid is pumped (ideally isentropically) from low to high pressure by a pump. Pumping requires a power input (for example mechanical or electrical).
- *Process 2$_s$-3:* The high pressure liquid enters a boiler where it is heated at constant pressure by an external heat source to become a saturated vapor. Common heat sources for power plant systems are coal, natural gas, or nuclear power.
- *Process 3-4$_s$:* The saturated vapor expands through a turbine to generate power output. Ideally, this expansion is isentropic. This decreases the temperature and pressure of the vapor.
- *Process 4$_s$-1:* The vapor then enters a condenser where it is cooled to become a saturated liquid. This liquid then re-enters the pump and the cycle repeats.

## 3.2 Actual Rankine cycle

In actual situations, both the water pumps and the steam Turbines do not operate isentropically and losses result in more power demand for pumping and less power actually generated by steam to blades [1]. The actual Rankine cycle is shown in Fig. 2.

Such losses are clearly shown in the following comparisons. That is,

$$h_3 - h_4 < h_3 - h_{4s} \quad \text{and} \quad h_2 - h_1 > h_{2s} - h_1$$

The performance of an actual turbine or pump is usually expressed in terms of isentropic efficiency. The isentropic efficiency of a turbine ($\eta_T$) is defined as the ratio of "work delivered by the actual turbine" to "work delivered by an isentropic turbine."

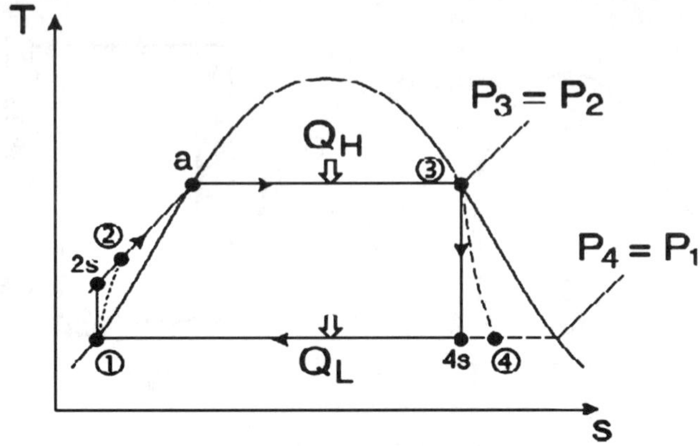


Figure 2: Practical Rankine cycle.

$$\eta_T = \frac{h_3 - h_4}{h_3 - h_{4s}} \tag{1}$$

The isentropic efficiency of a pump ($\eta_P$) is defined as the ratio of "work required by an isentropic pump" to "work required by the actual pump."

$$\eta_P = \frac{h_{2s} - h_1}{h_2 - h_1} \tag{2}$$

### 3.3 Efficiency improvements in power plants

It is well known that the cycle efficiency is generally proportional to

$$\eta \propto 1 - \frac{T_L}{T_H} \tag{3}$$

The question that emerges is how to improve the cycle efficiency; naturally through lowering the heat sink temperature $T_L$ and/or raising the heat source temperature $T_H$.

#### 3.3.1 Lowering the condenser pressure

Setting the steam condenser pressure is generally restricted by the temperature of the available condenser water (lake, river, etc.) typically around 25°C or condenser saturation pressure of around $P_{sat} \cong 3.2$ kPa.

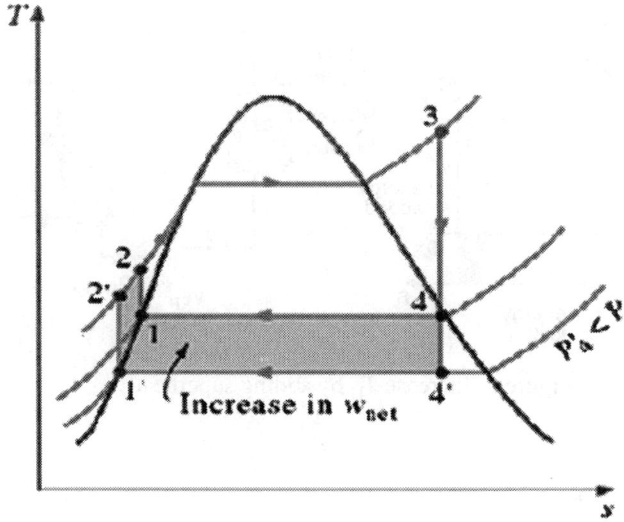


Figure 3: Lowering condenser temperature $T_L$ through decreasing condenser operating pressure [1].

The pressure at the exit of the turbine can be less than atmospheric pressure with a condenser and the closed loop of the condenser permits the use of high water quality on the steam cycle side. However, lowering condenser pressure is not unlimited as it depends on the design condenser temperature and the limits of lower steam quality at turbine exit. In Fig. 3, the shaded area represents the gain in network of the system due to lowering the condenser pressure.

### 3.3.2 Superheating the steam to high temperatures

The average temperature at which heat is supplied in the boiler can be increased by superheating the steam. Dry saturated steam from the boiler is passed through a second bank of smaller bore tubes within the boiler until the steam reaches the required temperature. The value of $T_H$, the mean temperature at which heat is added, increases, while $T_L$ remains constant. Therefore, the efficiency increases.

The quality of the turbine exhaust termed, $x$, increases, the value of steam dryness fraction at turbine exit should not be lower than about 0.9 to prevent water droplets effects on blading efficiency, as outlined in Fig. 4. With sufficient superheating, the turbine exhaust conditions may well fall in the superheated region.

### 3.3.3 Increasing the boiler pressure

Increasing the operating pressure of the boiler automatically raises the temperature at which boiling takes place. This consequently raises the average temperature

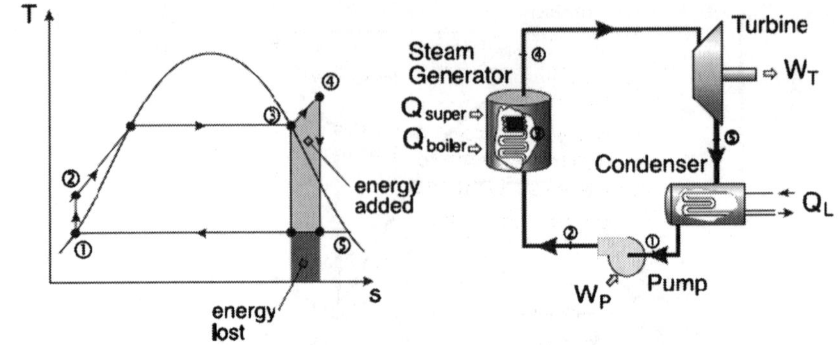


Figure 4: Increase $T_H$ by adding superheat.

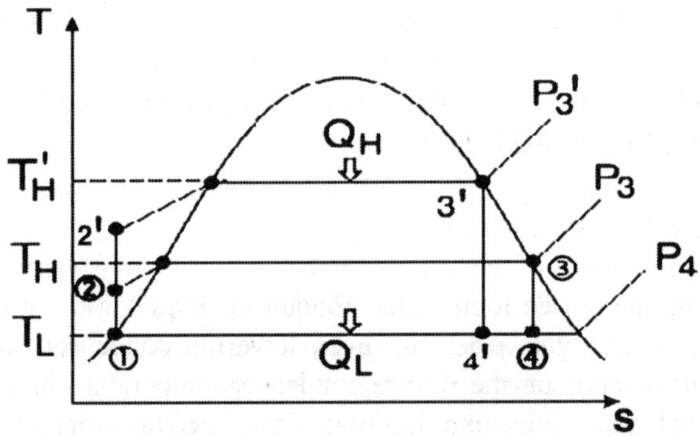


Figure 5: Effect of increasing the boiler pressure.

at which heat is added to the steam and thus raises the thermal efficiency of the cycle as indicated in Fig. 5.

An increase in boiler pressure results in a higher $T_H$ for the same $T_L$, therefore higher cycle thermal efficiency $\eta$. But state 4′ indicates a lower steam quality than state 4, then more wet steam at the turbine exhaust is expected; this may result in cavitation of the turbine blades at the last low pressure stages. Consequently, the efficiency is decreased and the cost of maintenance increases. It is recommended to keep the steam quality higher than 90% at the turbine exhaust section.

### 3.3.4 Rankine cycle with reheat

1. The moisture content at the exhaust of the turbine should be no greater than 10% – this can result in physical erosion of the turbine blades.

2. As higher boiler pressures are required for high efficiency, this leads to a higher moisture content ratio in the low pressure turbine expansion [1].
3. To improve the turbine exhaust steam conditions, the steam can be reheated between two turbine expansion stages or steps as indicated in Fig. 6. The following points emerge:

   - The temperature of the steam entering the turbine is limited by metallurgical constraints.
   - Modern boilers can handle up to 30 MPa and a maximum temperature of $t_{max} \approx 650°C$.
   - Materials, such as ceramic blades, can handle temperatures up to 750°C.

*Advantages of using Rankine cycle with reheat*: This arrangement provides high steam quality or even slightly superheated vapor at turbine exit. Therefore, for a given $T_H$ the Rankine cycle, efficiency increases without reducing the steam quality at turbine exit.

*Rankine cycle with regeneration:* To increase the cycle efficiency, to near the Carnot cycle efficiency, added heat $Q_H$ should be at as high temperature $T_H$ as it possibly can. Also, heat should be rejected, $Q_L$, at the lowest possible $T_L$. In such configuration, the Rankine cycle is provided with *feed water heaters (FWHs)* to heat the high-pressure sub-cooled water at the exit of the pump to the saturation temperature. As shown in Fig. 7, most of the heat addition ($Q_H$) is performed at high temperature.

*Feed water heaters:* There are two different types of FWHs commonly used in power plants: *open FWH*, where the two streams of high temperature steam and low temperature water mix in an open heater at constant pressure; *closed FWH*, where a heat exchanger is used to transfer heat between the two streams but the two streams do *not* mix. The two streams can be naturally maintained at different pressures.

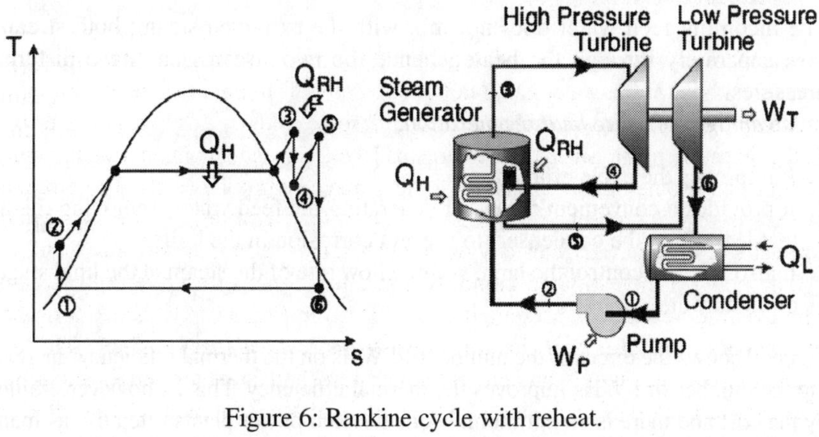


Figure 6: Rankine cycle with reheat.

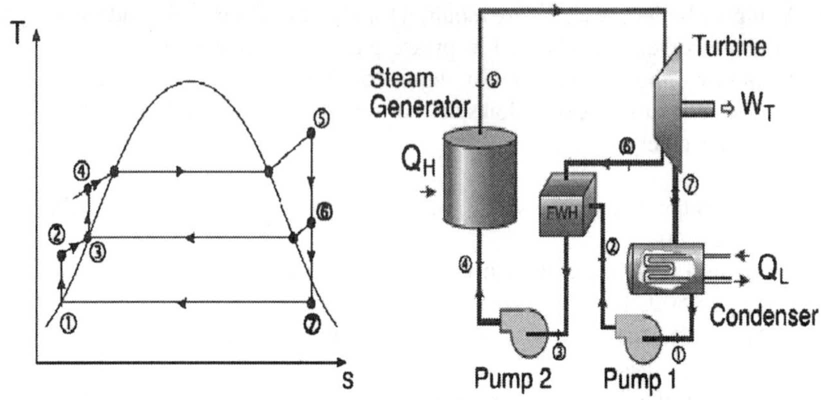


Figure 7: Rankine cycle with regeneration [1].

1. *Open FWH:* In this arrangement, the working fluid passes isentropically through the turbine stages and pumps. Steam enters the first stage turbine at state 1 and expands to state 2 – where a fraction of the total flow is bled off into an open FWH at $P_2$. The rest of the steam expands into the second stage turbine at state point 3. – This portion of the fluid is condensed and pumped as a saturated liquid to the FWH at $P_2$. A single mixed stream exits the FWH at state point 6. The mass flow rates through each of the components are typically calculated by performing a mass balance over the turbine. A heat balance is also performed to calculate the various enthalpies at various states.
2. *Closed FWH:* Such configuration can be practically realized in two alternatives typically:
   - Pump the condensate back to the high-pressure line (Fig. 8a).
   - A steam trap is inserted in the condensed steam line that allows only liquid to pass (Fig. 8b).

The incoming feed water does not mix with the extracted steam; both streams flow separately through the heater, hence the two streams can have different pressures.

*Advantages of using heat regeneration:*

1. It improves the cycle efficiency.
2. It provides a convenient means of deaerating the feed water (removing the air that leaks in at the condenser) to prevent corrosion in the boiler.
3. It also helps to control the large volume flow rate of the steam at the final stages of the turbine.

Figure 9 shows the effect of the number of FWHs on the thermal efficiency; increasing the number of FWHs improves the thermal efficiency. This is, however, limited by the cost and there is an optimum. Modem steam power plants often use as many as eight FWHs. Some of these may be open FWHs, which allows for deaerating of

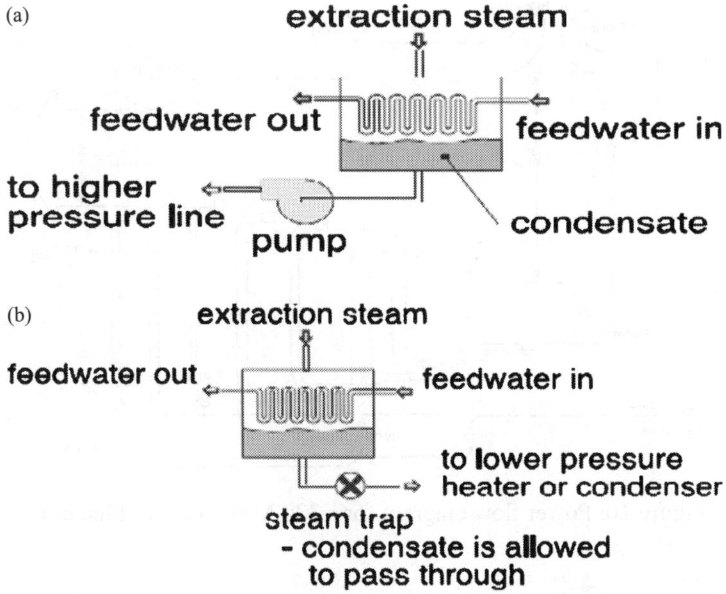


Figure 8: (a) Forward type feed water heaters; (b) cascaded type feed water heaters.

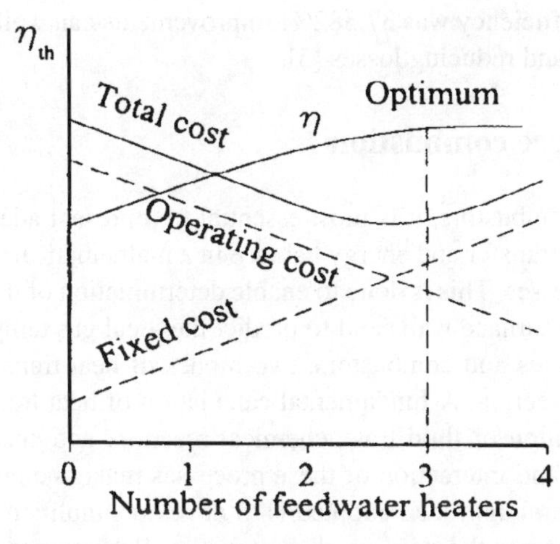


Figure 9: Effect of number of FWHs on the thermal efficiency.

the feed water to prevent boiler corrosion and some are also of the closed type. The optimum number of FWHs is usually determined from economic considerations [1].

An example of the energy balance (Sankey diagram) of a multistage steam turbine unit in operation in one of the power plants in Egypt is shown in Fig. 10. The diagram identifies the input from the boiler and losses at the various components.

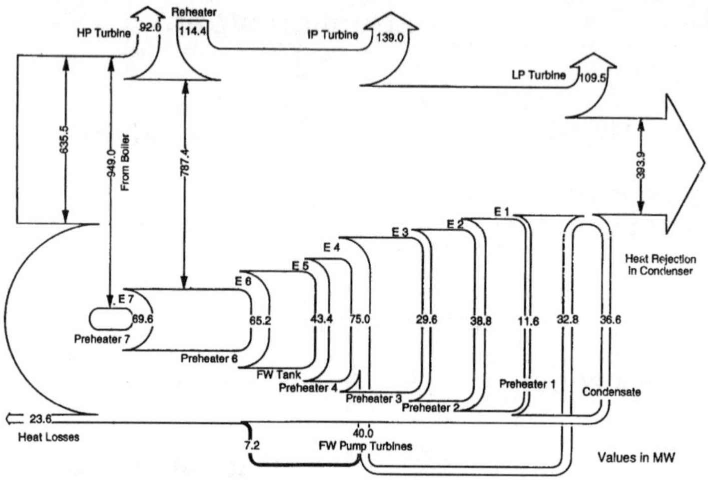


Figure 10: Power flow diagram for a 330 MW steam turbine unit.

The output power is indicated as the sum of the high-, intermediate- and low-pressure stages. Heat rejected in the condenser was shown to be 393.9 MW [5].

Table 1 lists the performance of a 330 MW steam turbine working in a Steam Power Plant in Cairo, Egypt. The performance indices are those of July 2007. The average thermal efficiency was 37.582%. Improvements can well be implemented by co generation and reducing losses [5].

## 4 Boiler furnace combustion

In furnaces and combustors, it is most essential to represent adequately the characteristics of heat transfer and energy balance in a mathematical model of the flow and reaction processes. This is done to enable determination of the actual heat flux distribution to the furnace walls and to predict the local gas temperature distribution. In real furnaces and combustors, two modes of heat transfer exist, namely radiation and convection. A fundamental calculation of heat transfer requires the simultaneous solution of fluid flow, chemical reaction, and energy transfer. The inter-dependence and interaction of these processes make the problem extremely complex; the general approach adopted is to develop simplified models that deal with the turbulence and the reaction characteristics. Radiative and convective heat transfer is discussed here in terms of available models, their assumptions, formation and validation in furnaces and combustion chambers.

### 4.1 Turbulent combustion

#### 4.1.1 Introduction

Turbulent combustion modeling is an essential tool for the design of furnaces and combustors. Such models yield closure to the fuel mass fraction conservation

Table 1: Steam turbine power plant performance table (300 MW, July 2007) [3].

| Day | Fuel | Fuel consumption (ton/day) | Fuel equ. power (MW hr) | Power generated (MW hr) | Average power (MW) | Gen. power factor (%) | Thermal efficiency (%) |
|---|---|---|---|---|---|---|---|
| July 1 | Oil | 1177.67 | 14518.0541 | 5341.98 | 226.3325 | 68.59 | 37.41534 |
| July 2 | Oil | 1240.8 | 15296.3067 | 5711.29 | 237.9704 | 72.11 | 37.33771 |
| July 3 | Oil | 1233 | 15200.15 | 5710.95 | 237.9563 | 72.11 | 37.57167 |
| July 4 | Oil | 1182 | 14571.4333 | 5457.28 | 227.3867 | 68.91 | 37.45191 |
| July 5 | Gas | 1099.8 | 13558.09 | 5103.89 | 212.6621 | 64.44 | 37.64461 |
| July 6 | Gas | 1033.47 | 12740.3885 | 4837.2 | 201.55 | 61.08 | 37.96745 |
| July 7 | Gas | 1271.3 | 15672.3039 | 5984.45 | 249.3521 | 75.56 | 38.18488 |
| July 8 | Gas | 1203.2 | 14832.7822 | 5648.5 | 235.3542 | 71.32 | 38.08119 |
| July 9 | Gas | 1063.5 | 13110.5917 | 5207.6 | 216.9833 | 65.75 | 39.72056 |
| July 10 | Gas | 1154.6 | 14233.6522 | 5272.15 | 219.6729 | 66.57 | 37.04004 |
| July 11 | Gas | 1291.4 | 15920.0922 | 6089.47 | 253.7279 | 76.89 | 38.25022 |
| July 12 | Gas | 1225.2 | 15103.9933 | 5799.96 | 241.665 | 73.23 | 38.40018 |
| July 13 | Gas | 1081.6 | 13333.7244 | 5126.27 | 213.5946 | 64.73 | 38.4459 |
| July 14 | Gas | 1091.7 | 13458.235 | 5079.07 | 211.6279 | 64.13 | 37.7395 |
| July 15 | Gas | 1147.2 | 14142.4267 | 5418.56 | 225.7733 | 68.42 | 38.31422 |
| July 16 | Gas | 1243.95 | 15335.1392 | 5846.9 | 243.6208 | 73.82 | 38.12747 |

(*Continued*)

Table 1: Continued

| Day | Fuel | Fuel consumption (ton/day) | Fuel equ. power (MW hr) | Power generated (MW hr) | Average power (MW) | Gen. power factor (%) | Thermal efficiency (%) |
|---|---|---|---|---|---|---|---|
| July 17 | Gas | 1343.3 | 16559.9039 | 6156.6 | 256.525 | 77.73 | 37.17775 |
| July 18 | Gas | 1264 | 15582.3111 | 5764.53 | 240.1888 | 72.78 | 36.99406 |
| July 19 | Oil | 1212.6 | 14948.6633 | 5586.84 | 232.785 | 70.54 | 37.37351 |
| July 20 | Oil | 1045.5 | 12888.6917 | 4851.38 | 202.1408 | 61.25 | 37.64059 |
| July 21 | Oil | 1156.6 | 14258.3078 | 5255.03 | 218.9596 | 66.35 | 36.85592 |
| July 22 | Oil | 1035.6 | 12766.6467 | 4563.67 | 190.1529 | 57.62 | 35.74682 |
| July 23 | Oil | 1117.9 | 13781.2228 | 4941.78 | 205.9075 | 62.40 | 35.85879 |
| July 24 | Oil | 1227.86 | 15136.7852 | 5602.9 | 233.4542 | 70.74 | 37.01513 |
| July 25 | Oil | 1292.14 | 15929.2148 | 5943.79 | 247.6579 | 75.05 | 37.31377 |
| July 26 | Oil | 1341.3 | 16535.2483 | 6105.64 | 254.4017 | 77.09 | 36.925 |
| July 27 | Oil | 1114.3 | 13736.8428 | 5323.69 | 221.8204 | 67.22 | 38.75483 |
| July 28 | Oil | 1294.12 | 15953.6238 | 5964.42 | 248.5175 | 75.31 | 37.38599 |
| July 29 | Oil | 1375.5 | 16956.8583 | 6321.11 | 263.3796 | 79.81 | 37.2776 |
| July 30 | Oil | 1346.3 | 16596.8872 | 6201.6 | 258.4 | 78.30 | 37.36604 |
| July 31 | Oil | 1444.95 | 17813.0225 | 6709.62 | 279.5675 | 84.72 | 37.66694 |
| Average values | | 1204.915 | 14853.922 | 5581.230 | 232.551 | 70.470 | 37.582 |

equations and those relating to the progress of reaction. The turbulence – combustion interactions and density fluctuation correlations play an important role in the adequacy of the predicted flow pattern and heat transfer [6].

### 4.1.2 Mathematical formulation

Three time averaged velocity components in *X*, *Y*, and *Z* coordinate directions were obtained by solving the governing equations using a "SIMPLE numerical algorithm" (semi-implicit method for pressure linked equation). The turbulence characteristics were represented by a modified and appropriately extended two-equation $k-\varepsilon$ model [7, 8] to account for normal and shear stresses and near-wall functions. Fluid properties such as densities, viscosity and thermal conductivity were obtained from references [9–13]. The present work uses the computer program 3DHVAC [7, 8]. The program solves the differential equations governing the transport of mass, three momentum components, energy, relative humidity, and the air age in 3D configurations. The different governing partial differential equations are typically expressed in a general form as:

$$\text{Div}\,(\rho V\Phi - \Gamma_{\Phi,\text{eff}} \cdot \text{grad}\,\Phi) = S_\Phi \tag{4}$$

where $\rho$ is the air density (kg/m$^3$), $\Phi$ is the dependent variable, $V$ is the velocity vector, $\Gamma_{\Phi,\text{eff}}$ is the effective diffusion coefficient, $S_\Phi$ is the source term of $\Phi$.

The effective diffusion coefficients and source terms for the various differential equations are listed in Table 2.

The solution of the governing equations can be realized through the specifications of appropriate boundary conditions. The values of velocity, temperature, kinetic energy, and its dissipation rate should be specified at all boundaries.

Table 3 highlights the boundary conditions.

Table 2: Coefficients and constants of the partial differential equations.

| | $\Phi$ | $\Gamma_{\Phi,\text{eff}}$ | $S_\Phi$ |
|---|---|---|---|
| Continuity | 1 | 0 | 0 |
| *X*-momentum | $U$ | $\mu_{\text{eff}}$ | $-\partial P/\partial x + S_U$ |
| *Y*-momentum | $V$ | $\mu_{\text{eff}}$ | $-\partial P/\partial y + S_V$ |
| *Z*-momentum | $W$ | $\mu_{\text{eff}}$ | $-\partial P/\partial z + S_W + \rho g\beta\Delta t$ |
| *H*-equation | $H$ | $\mu'_{\text{eff}}$ | $S_H$ |
| *k*-equation | $K$ | $\mu'_{\text{eff}}$ | $G - \rho\varepsilon$ |
| $\varepsilon$-equation | $\varepsilon$ | $\mu'_{\text{eff}}$ | $C_1\varepsilon G/k - C_2\rho\varepsilon^2/k$ |

$\mu'_{\text{eff}} = \mu_{\text{lam}}/\sigma_{\text{lam}} + \mu_t/\sigma_\Phi$ and for momentum $\mu_{\text{eff}} = \mu_{\text{lam}} + \mu_t$

$\mu_t = \rho C_\mu k^2/_\varepsilon$

$G = \mu[2\{(\partial U/\partial x)^2 + (\partial V/\partial y)^2 + (\partial W/\partial z)^2\} + (\partial U/\partial y + \partial V/\partial x)^2 + (\partial V/\partial z + \partial W/\partial y)^2 + (\partial U/\partial z + \partial W/\partial x)^2]$

$C_1 = 1.44$, $C_2 = 1.92$, $C_\mu = 0.09$.

$\sigma_{\text{lam}} = 0.7$, $\sigma_H = 0.9$, $\sigma_k = 1.0$, $\sigma_\varepsilon = 1.3$, for other equations $\sigma_\Phi = 0.7$.

Table 3: Boundary conditions.

| | |
|---|---|
| Walls | A non-slip condition at all solid walls is applied to the velocities. The logarithmic law of the wall (wall function) was used here, for the near wall boundary layer. |
| Supply outlets | At inlets, the air velocity was assumed to have a uniform distribution; inlet values of the temperature were assumed to be of a constant value and uniform distribution. The kinetic energy of turbulence and its dissipation rate are commonly estimated as follows: $k_{in} = 3(0.5(I_{in}U_{in})^2),\ \varepsilon_{ini} = C_\mu(k_{in})^{1.5}/l_e,$ where $I_{in}$ is the intensity of disturbance at air inlet and $l_e$ is the dissipation length at air inlet. |
| Initial values | All velocity components were set as zeros initially, and temperatures were assumed to be equal to the steady state value of the comfort condition. The kinetic energy and its dissipation are estimated as follows: $k_{initial} = 11 \times 10^{-5},\ \varepsilon_{initial} = C_\mu(k_{initial})^{1.5}/cd,$ where $c$ is a constant and $d$ is the distance to nearest sidewall. |

### 4.2 Combustion models

A total of 11 combustion models have been used together with equations for $U$, $V$, $W$, $k$, and $\varepsilon$. They are summarized on the following panels and their relative merits identified. All models require the solution of equations, of the general form (4).

Here $\Phi = H$ and $f$, where

$$H = M_{fu}H_{fu} + \sum M_i C_{p_a} T_a + \tfrac{1}{2}(U^2 + V^2 + W^2) \tag{5}$$

and

$$f = f_1/f_2$$

$$f_1 = [M_{fu} - (M_{ox}/i)] - [M_{fu} - (M_{ox}/i)]_{\text{air stream}}$$

$$f_2 = [M_{fu} - (M_{ox}/i)]_{\text{fuel stream}} - [M_{fu} - (M_{ox}/i)]_{\text{air stream}}$$

where $i$ is the stoichiometric ratio; $M_{fu}$ is the fuel mass fraction; $M_{ox}$ is the oxidant mass fraction; $U$, $V$, $W$ are the mean velocity components in the $X$, $Y$, and $Z$ coordinate directions; $C_{pa}$ is the specific heat at constant pressure of the species; $T_a$ is the mean temperature of species.

#### 4.2.1 Fast chemical reactions models

##### 4.2.1.1 Combustion model 1

1. Fuel and oxidant do not coexist at same place at any time, reaction rate is infinitely fast, equilibrium is attained [1, 2].

2. An equation for mixture fraction ($f$) that has no source term is solved.
3. There are no fluctuations for ($f$).

$$C_pT = H - M_{fu}H_{fu} + f(H_{fu} + C_pT_{\text{fuel stream}}) + (1-f)(C_pT_{\text{air stream}})$$

with $C_0 = \Sigma G_a C_{0a} / \Sigma G_a$ and $C_{0a} = a_0 + a_1T + a_2T^2 + a_3T^3$

#### 4.2.1.2 Combustion model 2

1. The effect of concentration fluctuations is considered here in terms of ($g$) which is the concentration fluctuations expressed as $g$ = square of ($f'$) fluctuations.
2. The modeled form of the scalar transport equation has a source term expressed in terms of generation term of concentration fluctuations and its dissipation rate.
3. Two delta functions at $f = 0$ and $f = 1$ were assumed.

$$C_pT = H - M_{fu}H_{fu} + f(H_{fu} + C_pT_{\text{fuel stream}}) + (1-f)(C_pT_{\text{air stream}})$$

unless $f_+ > 0$, where $f = \alpha f_+ + (1-\alpha)f_-$, $T = \alpha T_+ + (1-\alpha)T_-$ ($\alpha$ is a fraction less than unity), and

$$T'^2 = \alpha\,(T_+ - T)^2 + (1-\alpha)\,(T - T_-)^2 \tag{6}$$

Computing time = 1.25 × that for model 1.

#### 4.2.1.3 Combustion model 3

1. The modeled form of the scalar transport equation has a source term expressed in terms of generation term of concentration fluctuations and its dissipation rate.
2. A clipped Gaussian probability distribution of mixture fraction ($f$) is incorporated instead of the two delta functions at $f = 0$ and $f = 1$.
3. The model requires the solution of transport equations for $f$ and $g$ as well as those of the mass, momentum and energy.

Equation (7) is expressed as:

$$T = AT_{\text{air stream}} + BT_{\text{fuel stream}} + \frac{1}{\sigma\sqrt{2R}}\int_0^L \left[\frac{H - M_{fu}H_{fu}}{C_p\mu}\exp\left[-\frac{1}{2}\left(\frac{f-\mu}{\sigma}\right)^2\right]\right]df$$

$$M_{fu} = \int_{01}^{1} M_{fu}(f)P(f)\,df \tag{7}$$

Equation (7) above gives the fuel mass fraction and similar equations for other scalar properties

Computing time = 1.4 × that for model 1.

### 4.2.2 Finite chemical reaction rate models

#### 4.2.2.1 Combustion model 4

1. In this situation, fuel and oxidant are mixed prior to combustion in a single step; finite rate is assumed.
2. The modeled form of the scalar transport equation has a source term expressed in an Arrhenius form for the reaction rate.

3. This model requires the solution of transport equations for fuel mass fraction ($M_{fu}$). The effect of turbulence on reaction rates may be introduced as: $m_{ox}^2$, $m_{fu}^2$, $m_{ox}m_{fu}$, etc. [14–16].

$$R_{fu} = A\rho^2 M_{ox} M_{fu} \exp(-E/RT) \quad \text{Arrhenius} \tag{8}$$

or

$$R_{fu} = C_R g_{fu}^{1/2} \rho\varepsilon / k \quad \text{Eddy break up} \tag{9}$$

where $R_{fu}$ is the rate of fuel consumption, $A$ is the pre-exponential constant, $E$ is the activation energy, $R$ is the universal gas constant, $g$ is the square of concentration fluctuations, $C_R$ is a constant.

The source term in the $g$ equation is taken be $(\partial M_{fu}/\partial x_i)^2$ and not $(\partial f/\partial x_i)^2$.
Computing time = 1.7 × that for model 1.

#### 4.2.2.2 Combustion model 5

1. In this situation, fuel and oxidant are mixed prior to combustion in a multi-steps; Finite rate is assumed.
2. The modeled form of the scalar transport equation has a source term expressed in an Arrhenius form for the reaction rate for each intermediate reaction step.
3. This model requires the solution of transport equations for fuel mass fraction ($M_{fu}$). The effect of turbulence on reaction rates may be introduced as: $m_a^2$, $m_b^2$, $m_c^2$, $m_a m_b$, etc.

$$\begin{gathered}
CH_4 + OH \rightarrow CH_3 + H_2O \\
CH_4 + H \rightarrow CH_3 + H_2 \\
CH_2 + O_2 \rightarrow CH_2O + OH \\
CH_2O + O \rightarrow CHO + OH \\
CHO + OH \rightarrow CO + H_2O \\
CO + OH \rightarrow CO_2 + H \\
H + O_2 \rightarrow O + OH \\
O + H_2O \rightarrow OH + OH
\end{gathered}$$

Computing time = 3 × that for model 1.

#### 4.2.2.3 Combustion model 6

1. Model 6 incorporates the effect of turbulence on the finite chemical reaction rates.
2. This is carried out by the solution of an extra transport equation for the correlation $m_{ox}m_{fu}$.
3. The Damkohler number defined as $N_D = t_s/t_k$.
4. $t_s$ is defined as the stretching time scale of flame eddies.
5. $t_k$ is defined as the chemical kinetics time scale.

$$\begin{aligned}
R_{fu} &= A\rho^2 M_{ox} M_{fu} \exp(-E/RT) + R'_{fu} \\
R'_{fu} &= A\rho^2 m_{ox} m_{fu} \exp(-E/RT)
\end{aligned} \tag{10}$$

This obviates the need for an Eddy break up term.
Computing time = 1.9 × that for model 1.

#### 4.2.2.4 Combustion model 7

1. Model 7 incorporates the effects of turbulence and temperature on the finite chemical reaction rates.
2. This is carried out by the solution of an extra transport equation for the correlation $m_{ox}T'm_{fu}T'$.
3. Correlations relating temperature and concentration fluctuations are solved in the general form of the transport equations.
4. The effect of density fluctuation correlations is also considered [14–16].

$$R'_{fu} = A\rho^2 m_{ox} m_{fu} \exp(-E/RT)[(m_{ox}m_{fu})/(M_{ox}M_{fu})$$
$$+ a_1(T'^2)/(T^2) + a_2(......) + .....] \tag{11}$$

and $a_2 = E/RT$, $a_1 = 0.5(E/RT)^2 - E/RT$.

Computing time = 2.5 × that for model 1.

Model 7 incorporates the effects of turbulence and temperature on the finite chemical reaction rates. This is carried out by the solution of an extra transport equation for the correlation $m_{ox}T'm_{fu}T'$; correlations relating temperature and concentration fluctuations are solved in the general form of the transport equations. The effect of density fluctuations correlations should also be considered.

**4.2.2.5 Probability density approach** The present model presented a new feature of characterizing the effect of turbulence on reaction rate through solutions of the transport equations of fuel and oxygen mass fraction, square of the fluctuations of mass fractions, their correlation and the probability density $P(\Phi)\varphi$ given as:

$$\partial\rho UP(\Phi)/\partial x + \partial\rho VP(\Phi)/\partial y + \partial\rho WP(\Phi)/\partial z$$
$$= \partial/\partial x(\Gamma\partial P(\Phi)/\partial x) + \partial/\partial y(\Gamma\partial P(\Phi)/\partial y) + \partial/\partial z(\Gamma\partial P(\Phi)/\partial z) \tag{12}$$
$$+ \partial/\partial\Phi(\rho P(\Phi)C2\varepsilon/k\Phi 2) - (P(\Phi)S(\Phi)$$

where $C2$ is a constant = 0.925 (after Khalil), $\Phi$ is a scalar entity that can be $M_{fu}$, $M_{ox}$, and $S(\Phi)$ is rate of formation or disappearance of the entity $\Phi$.

The present approach to model reacting flames solves the transport equation for the probability density function, eqn. (12) [14, 15] .The time averaged reaction rate is, consequently, obtained from the integrated product of the instantaneous rate and the local probability, as previously proposed by Khalil [1] and Borghi [14], neglecting diffusion terms reported simpler forms of eqn. (12). Further work was reported among others by Hutchinson *et al.* [16] and Patankar [17].

**4.2.2.6 Grid distribution and timing** A staggered grid system is employed for the velocities to avoid the decoupling effects between the velocity and the pressure that are frequently observed with the non-staggered grid (Fig. 11).

A hyperbolic grid distribution is employed, with the grid points, for instance in the $Z$ direction given by the functions as suggested by Henkes [18]. All the computations at the production level were run on a PC and required an average

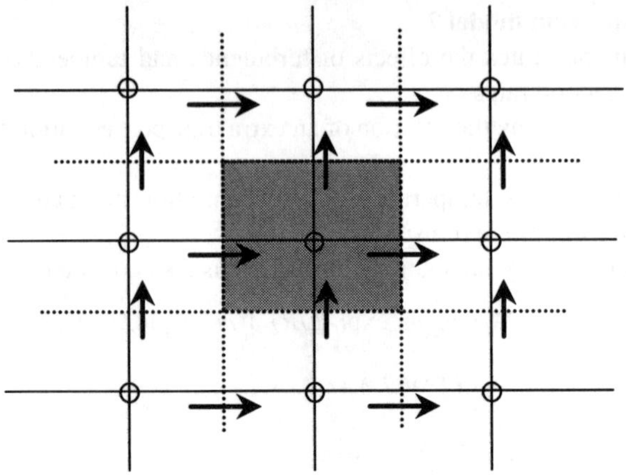

Figure 11: Grid arrangement.

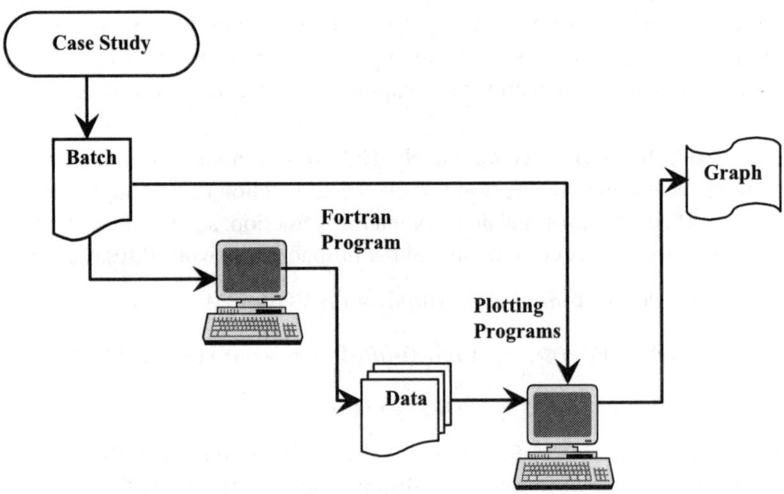


Figure 12: Program flow chart.

of $2.48 \times 10^{-5}$ s CPU time/one iteration/grid node/equation. The total CPU time for the results presented here depends on the grid size and number of iterations.

**4.2.2.7 Program procedure** The program produced massive amount of the data during each of the present investigated cases (Fig. 12) . Obtained data were about the $10^5$ data point for each case, which were utilized to plot the appropriate vector plots or contours.

**4.2.2.8 Program convergence** To determine whether convergence to steady state has been attained, a relative error, defined as the difference between the values at two successive iteration levels ($n$) and ($n + 1$) over the value at level ($n$) for each dependant variable, is monitored with iterations. This continues until the error attains a value $\leq 10^{-3}$ while, in addition, the relative error in the overall energy balance becomes $\leq 0.005$. Numerical computations were obtained for convergence criteria of residuals less than a value of $10^{-3}$ of the variable in question [19, 20].

### 4.3 Boiler furnace computations

The computational technique is used to calculate the flow pattern and temperatures in real boiler furnaces as reported by Kameel and Khalil [19, 20]. Figures 13–17

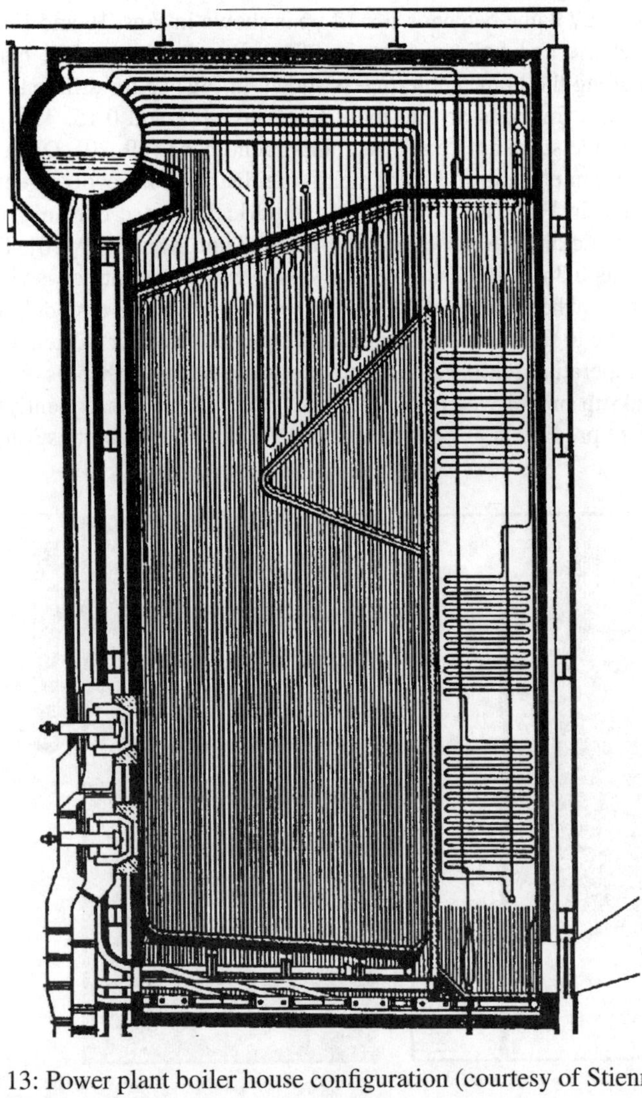

Figure 13: Power plant boiler house configuration (courtesy of Stienmuller).

demonstrate the capabilities of the numerical methods to predict various furnace performance parameters.

### 4.3.1 Examples of engineering simulations

The furnace of the 80 ton/hour steam at 64 bars is shown in Fig. 13 and the corresponding calculations are shown in Figs. 14 and 15. These include temperatures and flow velocities at start up and during running.

The predicted velocity vectors and thermal contours clearly identified the flames locations and characteristics. Mixing and interaction between the different burners and flames are clearly shown at start up. In Fig. 15 the corresponding predictions outline the emergence of the various flame envelopes and the extent of the flames.

The unsteady flame behavior at start up is shown in Figs. 16 and 17 in terms of the heat release and fuel consumption. The time-dependent fuel mass fraction depletion along the flame centerline is shown in Fig. 16 for different volumes of rich fuel mass fractions at fuel mass fractions >0.1 and 0.15. The predictions identified flame squashing and stretching phenomena [19, 20]. Detailed predictions of the flame shape and the temperature distributions evolvement with time are indicated in Fig. 18 for a vertical cylindrical furnace burning Natural gas. The Furnace details are those listed by Kameel and Khalil [19, 20]. The furnace diameter was 0.3 m and of a length of 0.9 m. In the present non-swirl combusting flow, the Arrhenius model yields very good representative model. Such model is so sensitive to the boundary conditions especially near or at the flame regions. In high temperature regions, the switching between the Arrhenius model and the eddy-break-up model was frequently observed, which consequently influences the transient prediction and the steady sate results. When combustion model 4 is

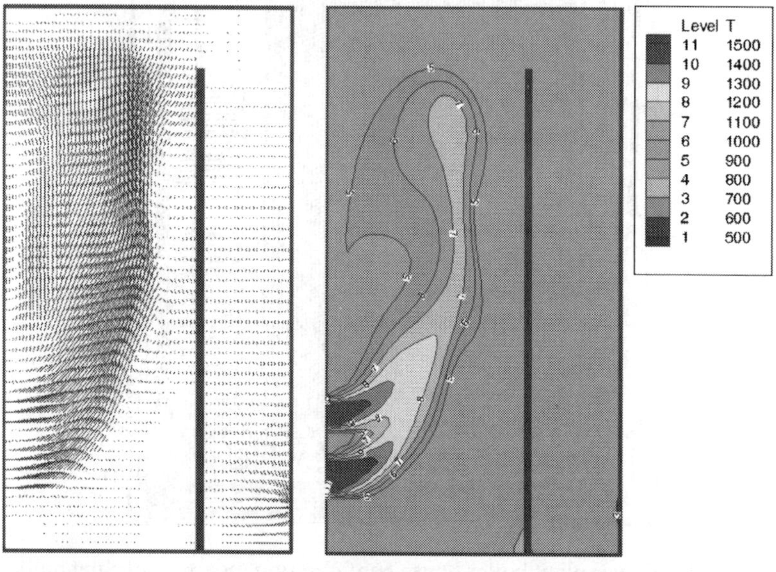


Figure 14: Velocity vectors and temperature contours at boiler start-up.

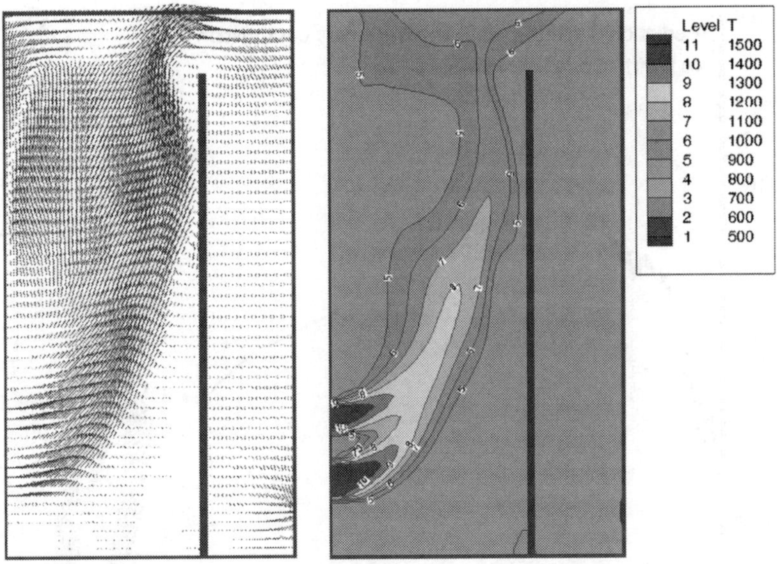


Figure 15: Velocity vectors and temperature contours after boiler running up.

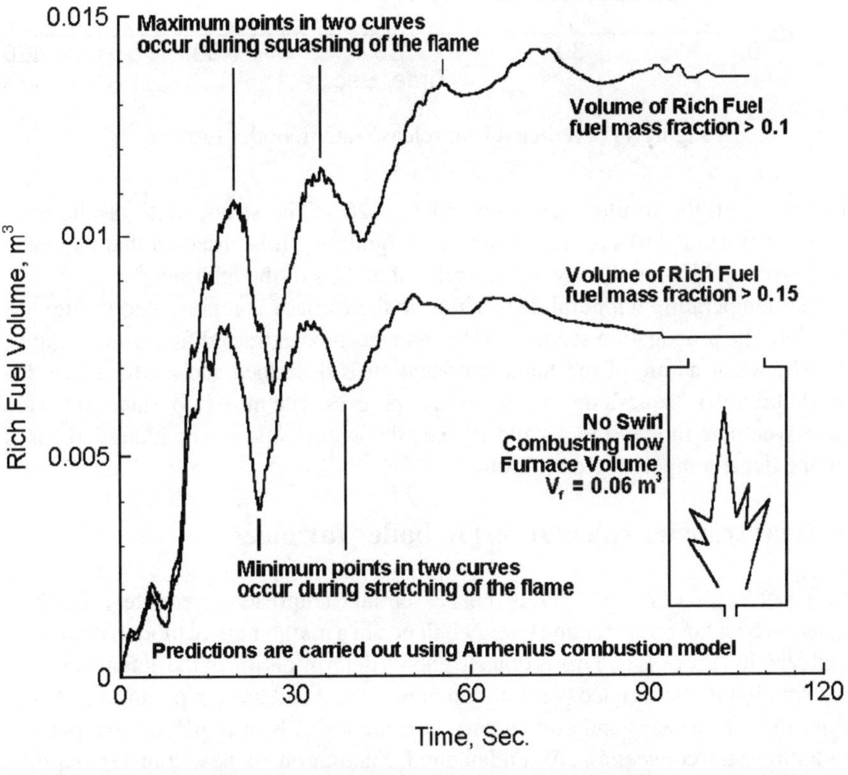


Figure 16: Predicted fuel depletion along furnace centerline at no swirl.

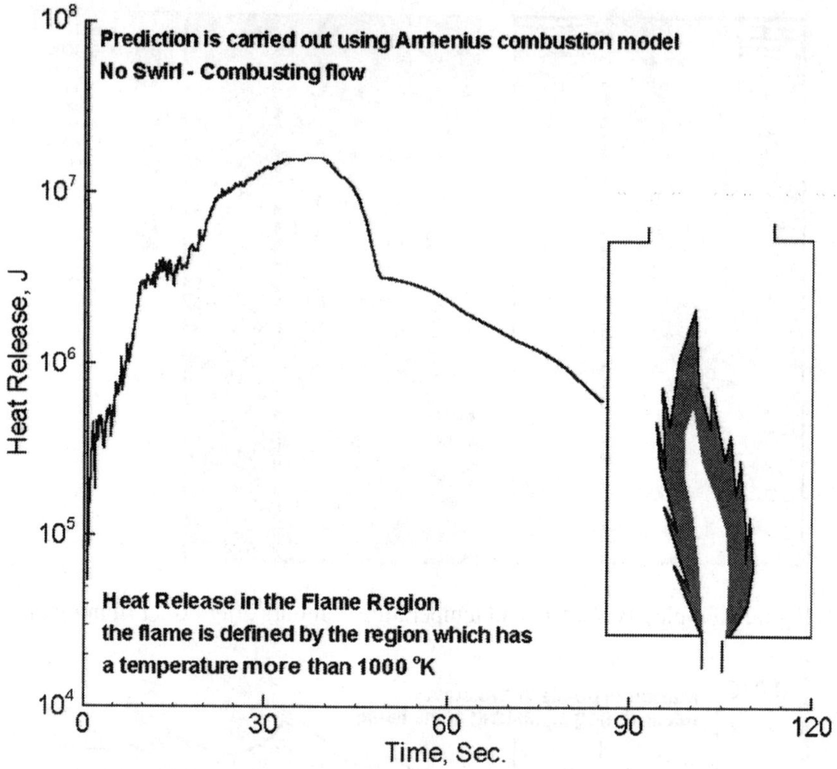


Figure 17: Predicted heat release rate in boiler furnace.

incorporated, the results were recorded for 120 s. The steady state results were attained at about 110 s approximately after ignitions. It is observed that the flame had expanded longitudinally and laterally after 11 s of the ignition,

The temperature temporal distribution in the furnace is represented in Figs. 18 and 19. The propagation scenario of the flame can be extracted from these figures. The transient nature of the flame propagation is described and explained in this work through a "squashing and stretching" process. The most important events are displayed here in the Figs. 18 and 19. The flame envelope can be identified based on the flame temperature definition.

## 5 Heat transfer calculations in boiler furnaces

In furnaces and combustors, it is most essential to represent adequately the characteristics of heat transfer and energy balance in a mathematical model of the flow and reaction processes. This is done to enable determination of the actual heat flux distribution to the furnace walls and to predict the local gas temperature distribution. In real furnaces and combustors, two modes of heat transfer exist, namely radiation and convection. A fundamental calculation of heat transfer requires the simultaneous solution of fluid flow, chemical reaction, and energy transfer.

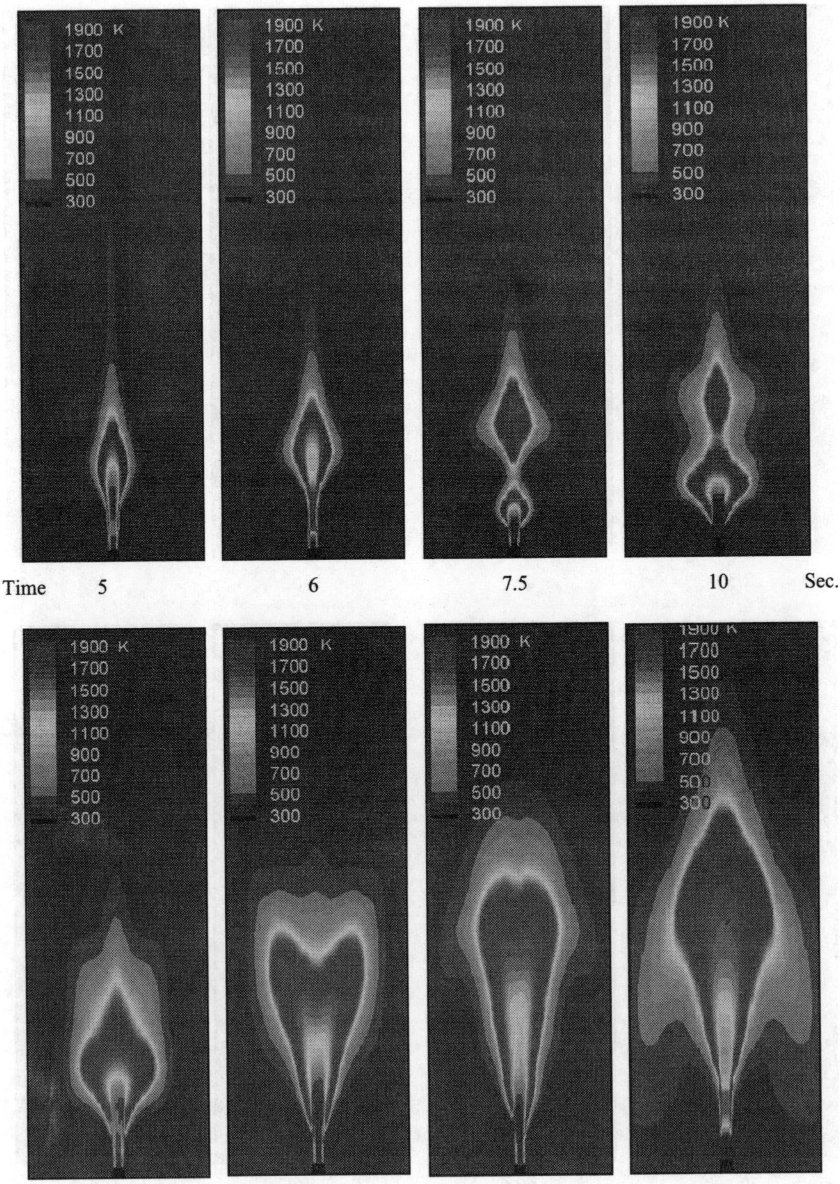


Figure 18: Temperature distribution in a confined flame.

The inter-dependence and interaction of these processes make the problem extremely complex; the general approach adopted is to develop simplified models that deal with the turbulence and the reaction characteristics. Radiative and convective heat transfer is discussed here in terms of available models, their assumptions, formation and validation in furnaces and combustion chambers.

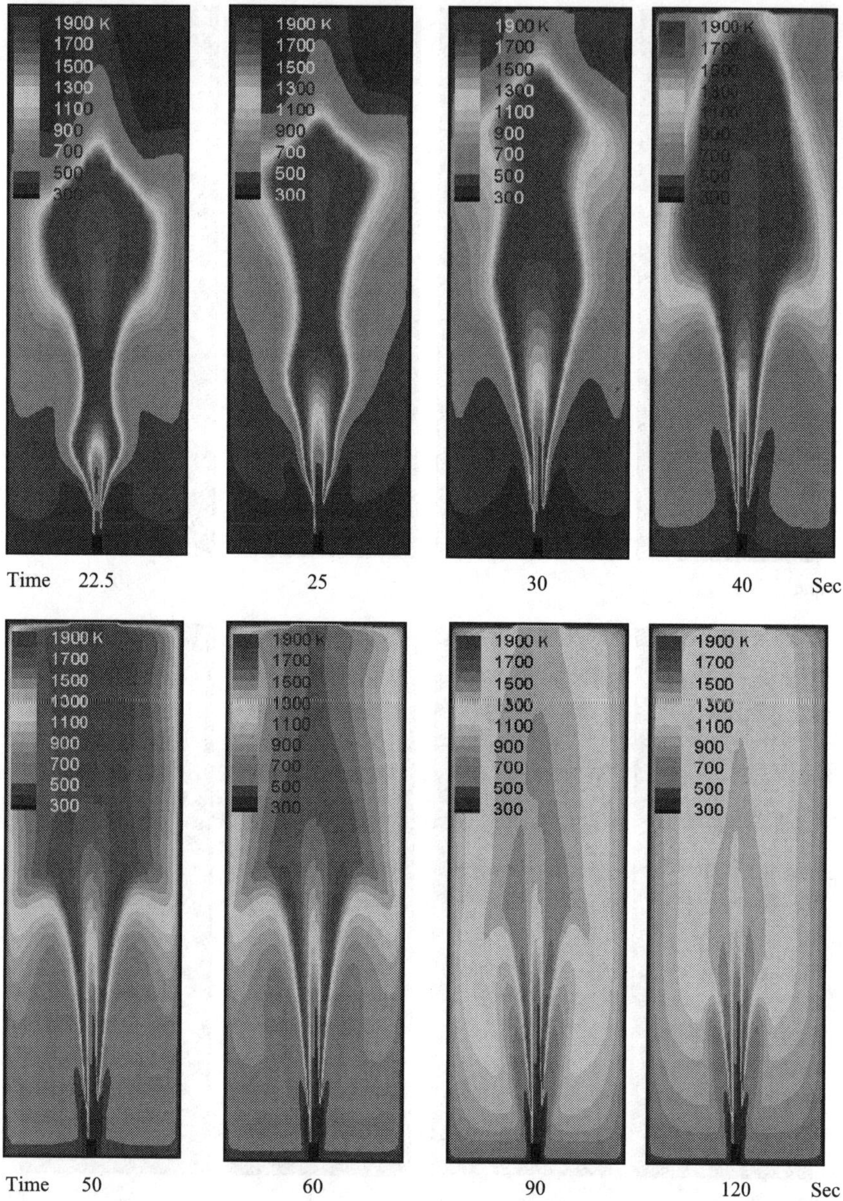


Figure 19: Temperature distribution in a confined flame.

## 5.1 Equation of radiant energy transfer

The basis on which all the methods of solving the radiation problems stand is the equation of radiant energy transfer. It is driven by writing a balance equation

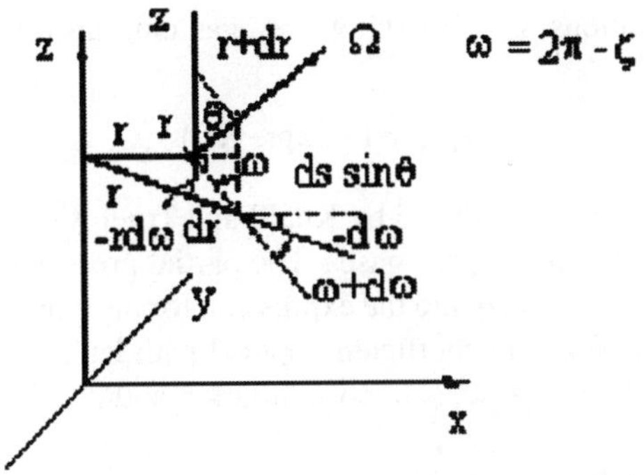


Figure 20: Radiation intensity distribution.

for monochromatic radiant energy, passing in specified direction, through a small volume, in an emitting-absorbing-scattering medium. For steady-state conditions, and for coherent isotropic scattering, this equation is expressed as following,

$$(\Omega \cdot \nabla) I_\lambda(r, \Omega) = \mu_0 \frac{\partial I}{\partial r} - \frac{\eta_0 \partial I}{r \partial \xi} + \xi_0 \frac{\partial I}{\partial x} + \frac{\eta_0}{r} \frac{\partial I}{\partial \theta} \tag{13}$$

and $I = I(r, \theta, x, \zeta)$, $\zeta$ is an angular variable that equals 2π-ω. $\mu_0$, $\eta_0$, $\zeta_0$ are the directional cosines, and are defined as shown in Fig. 20; $\mu_0 = \sin\theta\cos\zeta$; $\eta_0 = \sin\theta\sin\zeta$; $\zeta_0 = \cos\theta$.

### 5.2 Representation of real furnace gas

In natural gas and oil fired boiler furnaces, only three species contribute significantly, in the infrared region, to radiation [1, 21]. These species are

- carbon dioxide,
- water vapor, and
- hot soot particles within the flame.

The gases radiate in several bands, while the soot emits continuously over a wide range of the wavelength.

$$\varepsilon(T, L) = \frac{1}{\sigma T^4} \int_0^\infty E_\lambda(t)[1 - \exp(-K_{a,\lambda} L)] d\lambda \tag{14}$$

where $L$ is the radiation path length, $E\lambda(T)$ is the Blanck spectral distribution of emissive power, $\sigma$ is the Stefan–Boltzmann constant.

For real gas situations, the following grey gas emittance-path length relation does not hold:

$$\varepsilon_{grey} = 1 - \exp(-KL) \tag{15}$$

A modified expression was deduced by Khalil and Truelove [21], by representing the real gases by a mixture of grey gases. The partial pressures of carbon dioxide and water vapor are introduced into the expression for the emittance of gas mixture $\varepsilon_g$ as a function of absorption coefficient, optical path length, partial pressures of water vapor and carbon dioxide, soot concentration and density.

## 5.3 Radiation models

Various types of mathematical models that were proposed to adequately represent the radiation heat source in the energy equations are found in the literature. The more conveniently applied models are those listed here:

- zone method,
- spherical harmonic model,
- discrete ordinate model, and
- flux models.

The basic concept is to solve the radiation intensity equation with adequate physical and boundary conditions with full introduction of water vapor, carbon dioxide and soot radiating proportions. On one hand, the Spherical harmonics approach solves the radiation equation with retaining of adequate terms of harmonics terms. On the other hand, the basic concept of the Discrete Ordinate model Proposed by Khalil and Truelove [21] is very simple, the angular integral in the radiant energy transfer equation are approximated using a numerical quadrature scheme, as shown in Fig. 21.

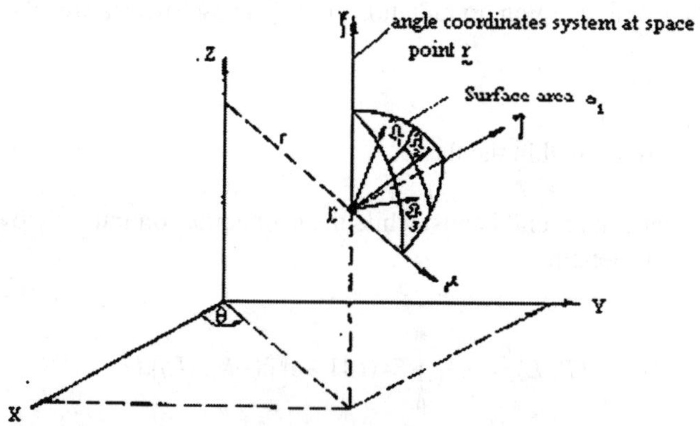


Figure 21: Angular quadrature weights $a_i$ as elements of area on the surface of the unit sphere in angle space. One octant is shown here for clarity.

The radiation source term $S_H$ in the energy equation is time-averaged as

$$S_H = K\sum_{i=1}^{N} a_i I_i - 4K\sigma\overline{T}^4 \tag{16}$$

where $i = 1, \ldots, N$ and $I_i = I(r, \zeta, x, \Omega_i)$.

A sample of the predicted wall heat flux in the furnace of Pai *et al.* [22], at the International Flame Research Foundation, is shown here in Fig. 22.

The discrete ordinate model is computationally simpler and more economical (although, of course, somewhat less accurate), requiring very little computer storage and run time. The model can be easily coupled to the equations for fluid flow and chemical reaction in a complete mathematical model used for furnace flow prediction. In spite of these facts, the present model has not been widely applied to furnace and combustor flows.

Flux models are based on the use of some simplifying assumptions of the angular variation of the radiant intensity in space. These assumptions allow the exact

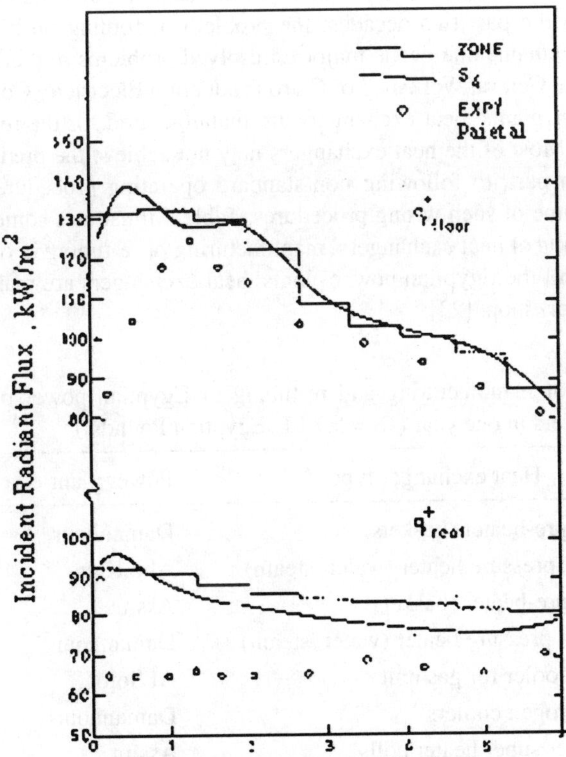


Figure 22: Comparisons between zone model and the discrete ordinate model ($S_4$) and the experiments of Pai *et al.* [22].

integro-differential radiation transfer equations to be reduced to a system of approximate partial differential equations as shown by Khalil [1] and others. The assumptions of this model category allow the energy transfer in four principal coordinate directions. The modeled equations are obtained by formulating the radiant energy balances on a small volume of the medium. There are two objections to this model approximation:

- The first is that there is no inter-linkage (coupling) between the axial and the radial fluxes, a feature which is physically unrealistic.
- The second objection lies in the form of the intensity distribution, from which the flux equations are obtained.

# 6 Power plant water problems

## 6.1 Introduction

Despite the great technical achievements in the design and manufacture of heat exchangers in the past two decades, the problem of fouling on heat exchanger surfaces still remains one of the major unresolved problems in thermal sciences. In the Egyptian Central Workshop of Cairo Production Electricity Company, many types of power plants' heat exchangers are manufactured for the replacement of failing items. Most of the heat exchangers may not achieve the predicted validity period due, in part, to following non-standard operating procedures: fouling is one bad outcome of such wrong procedures. Table 4 illustrates some of the basic cost components of heat exchangers' manufacturing or re-tubing in only one year; however, not all the Egyptian power plants' heat exchangers are manufactured in the Central Workshop. [23]:

Table 4: Cost of manufacturing and re-tubing of Egyptian power plant heat exchangers in one year (1$ = 5.7 LE[Egyptian Pounds]).

| No. | Heat exchanger type | Power plant | Cost (LE) |
|---|---|---|---|
| 1 | Air pre-heater (baskets) | Damanhour | 5292169 |
| 2 | Low pressure heater (water–steam) | Abu Kair | 2200000 |
| 3 | Air pre-heater (baskets) | Assiut | 1833862 |
| 4 | High pressure heater (water–steam) | Damanhour | 708400 |
| 5 | Air cooler for gas unit | Al Soiof | 253000 |
| 6 | Hydrogen coolers | Damanhour | 163482 |
| 7 | Boiler super heater coils | Assiut | 7860 |
| Total cost | | | 10458773 |

## 6.2 What is fouling?

Fouling is generally defined as *the accumulation of undesired deposits of materials on the surfaces of processing equipment.* It has been recognized as a nearly universal problem in heat exchangers design and operation. It affects the operation of equipment in two ways:

- The fouling layer has a low thermal conductivity. This increases the resistance to heat transfer and reduces the effectiveness of heat exchangers – increasing temperature.
- As deposition occurs, the cross-sectional area is reduced, which causes an increase in pressure drop across the apparatus.

Measurement of the fouling resistance is typically performed by measuring the total thermal resistance ($1/UA$) for the clean and fouled conditions. The fouling resistance ($R_f$) is obtained by subtraction of the $1/UA$ values for the fouled and clean conditions, respectively, giving:

$$R_f/A = 1/(UA)_f - 1/(UA)_c \qquad (17)$$

where $A$ is the surface area on which the $R_f$ is based. For tube side fouling, the $UA$ values for the clean and fouled conditions are defined by eqns. (18) and (19), respectively:

$$1/(UA)_c = 1/h_iA_i + t_w/k_wA_w + 1/h_oA_o \qquad (18)$$

$$1/(UA)_f = (1/h_i + R_f)(1/A_i) + t_w/k_wA_w + 1/h_oA_o \qquad (19)$$

The inside ($h_i$) and outside ($h_o$) heat transfer coefficients must be equal in the dirty and clean tube conditions. Otherwise, the measured fouling resistance $R_f$ will be erroneous.

## 6.3 Types of fouling

Fouling research has resulted in the definition of six different types of fouling that may occur with liquid or gases:

- *Precipitation fouling (scaling)* is the most common form of fouling and is associated with inverse solubility salts. Examples of such salts are $CaCO_3$, $CaSO_4$, $Ca_3(PO_4)_2$, $CaSiO_3$, $Ca(OH)_2$, $Mg(OH)_2$, $MgSiO_3$, $Na_2SO_4$, $LiSO_4$, and $Li_2CO_3$. The characteristic which is termed inverse solubility is that, unlike most inorganic materials, the solubility decreases with temperature. The most important of these compounds is calcium carbonate, $CaCO_3$. Calcium carbonate exists in several forms, but one of the more important is limestone. Running primarily through openings in limestone rock, it becomes saturated with calcium carbonate. Water pumped from the ground and passed through a water heater becomes supersaturated as it is heated, and so $CaCO_3$ begins to crystallize on the internal passages. Similar results occur when ground water is used in any industrial cooling process. The material frequently crystallizes in a form closely resembling marble, another form of calcium carbonate. Such materials are extremely difficult to remove mechanically and may require acid cleaning.

- *Corrosion fouling* is classified as a chemical reaction, which involves the heat exchanger tubes. Many metals, copper and aluminum being specific examples, form adherent oxide coatings which serve to passivate the surface and prevent further corrosion. Metal oxides are a type of ceramic and typically exhibit quite low thermal conductivities. Even relative thin coatings of oxides may significantly affect heat exchanger performance and should be included in evaluating overall heat transfer resistance.
- *Chemical reaction fouling* involves chemical reactions in the process stream which results in deposition of material on the heat exchanger tubes. When food products are involved, this may be termed scorching but a wide range of organic materials are subject to similar problems. This is commonly encountered when chemically sensitive process fluids are heated to temperatures near that for chemical decomposition. Because of the non-flow conditions at the wall surface and the temperature gradient, which exists across this laminar sub–layer, these regions will operate at somewhat higher temperatures than the bulk and are ideally suited to promote favorable conditions for such reactions.
- *Freezing fouling* is said to occur when a portion of the hot stream is cooled to near the freezing point for one of its components. This is most notable in refineries where paraffin frequently solidifies from petroleum products at various stages in the refining process, obstructing both flow and heat transfer.
- *Biological fouling* is common where untreated water is used as a coolant stream. Problems range from alga or other microbes to barnacles. During the season when such microbes are said to bloom, colonies several millimeters deep may grow across a tube surface virtually overnight, impeding circulation near the tube wall and retarding heat transport. Viewed under a microscope, many of these organisms appear as loosely intertwined fibers – much like the form of fiberglass insulation. Traditionally, these organisms have been treated which chlorine, but the present day concerns on possible contamination to open water bodies have severely restricted the use of oxidizers in open discharge systems.
- *Particulate fouling (sedimentation)* results from the presence of Brownian-sized particles in solution. Under certain conditions, such materials display a phenomenon known as thermophoresis in which motion is induced as a result of a temperature gradient. Thermodynamically this is referred to as a cross-coupled phenomenon and may be viewed as being analogous to the sea beck effect. When such particles accumulate on a heat exchanger surface they sometimes fuse, resulting in a build-up having the texture of sandstone. Like scale, these deposits are difficult to remove mechanically. Most of the actual data on fouling factors are tightly held by a few specialty consulting companies. The data which are commonly available are sparse. An example is demonstrated in Table 5.

Table 5: Representative fouling factors [24].

| Fluid | $R''_f$ (m$^2$ K/W) |
|---|---|
| Seawater and treated boiler feed water (below 50°C) | 0.0001 |
| Seawater and treated boiler feed water (above 50°C) | 0.0002 |
| River water (below 50°C) | 0.0002–0.001 |
| Fuel oil | 0.0009 |
| Refrigerating liquids | 0.0002 |
| Steam (non-oil bearing) | 0.0001 |

### 6.4 Fouling fundamentals

Fouling is a rate-dependent phenomenon. The net fouling rate is the difference between the solids deposition rate and their removal rate:

$$dm_f/d_t = m_d - m_r \tag{20}$$

Depending on the magnitudes of the deposition and removal terms, several fouling rate characteristics are possible. Figure 23 illustrates the increase of the fouling factor ($R_f$) with time for different possible fouling situations. Crystallization fouling has an initial delay period ($t_d$), during which nucleation sites are established; then, the fouling deposit begins to accumulate. A linear growth of the fouling deposit occurs if the removal rate ($m_r$) is negligible or if $m_d$ and $m_r$ are constant with $m_d > m^{\cdot}_{\alpha}$. Crystallization, chemical reaction, corrosion, and freezing fouling will show behavior in the linear or falling rate categories. The fouling resistance will attain an asymptotic value if $m_d$ is constant and $m_r$ eventually attains a constant value. Particulate fouling typically shows asymptotic behavior. Which type of fouling characteristics occurs depends on the fouling mechanism.

References [25–28] provide a detailed discussion of the different fouling mechanisms, these also describe the models to predict the deposition and removal rates for the different fouling mechanisms.The deposition model is a function of the fouling mechanism. The removal rate model depends on the re-entrainment rate, which is proportional to the shear stress at the surface. The heat exchanger operating parameters that may influence the deposition or removal rates are as follows:

1. The bulk fluid temperature will typically increase chemical reaction rates and crystallization deposition rates.
2. Increasing surface temperature will increase chemical reaction rates, or crystallization from inverse solubility salts. Reducing surface temperature will increase solidification fouling.
3. The combination of surface material and fluid will influence corrosion fouling. Copper surfaces act as a biocide to biological fouling. Rough surfaces may promote nucleation based phenomena.

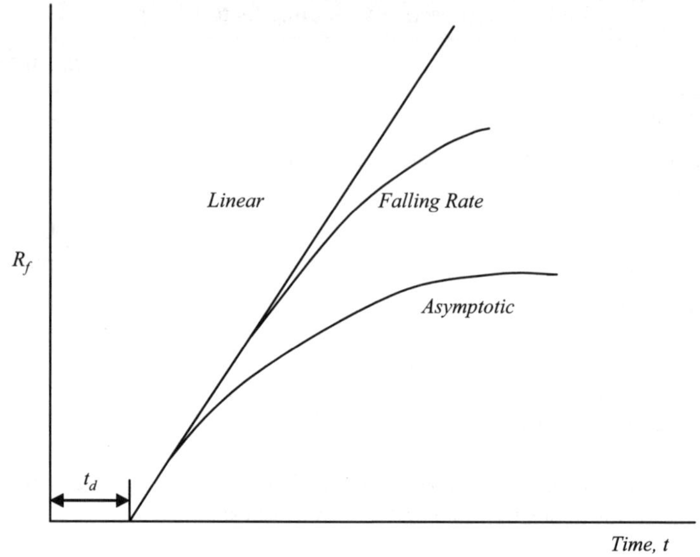


Figure 23: Characteristic fouling curves.

4. The liquid or gas velocity may influence both the deposition and removal rates. Removal of soft deposits is affected by surface shear stress, which increases with velocity. Increased velocity will increase the mass transfer coefficient, and hence the deposition rate for diffusion controlled particulate fouling.

### 6.4.1 Particulate fouling

We will take an example of particulate fouling, which is important for both gases and liquids. The particulate fouling deposition rate term is given by:

$$m_{\mathrm{d}} = SK_{\mathrm{m}}(C_{\mathrm{b}} - C_{\mathrm{w}}) \tag{21}$$

The deposition process is controlled by the mass transfer coefficient ($K_{\mathrm{m}}$) and the concentration difference between that in the bulk fluid ($C_{\mathrm{b}}$) and that in the liquid at the foulant surface ($C_{\mathrm{w}}$). Typically, $C_{\mathrm{w}} = 0$. The term $S$ is called the "sticking probability." This is the probability that a particle transported to the wall will stick to the wall.

The foulant removal rate model was proposed by Taborek and Hewitt [29] and is assumed to be proportional to surface shear stress and foulant deposit thickness ($x_{\mathrm{f}}$) and inversely proportional to the deposit bond strength factor ($\gamma$). The removal rate is given by:

$$m_{\mathrm{r}} = m_{\mathrm{f}}\tau_{\mathrm{w}}/\gamma = \rho_{\mathrm{f}}x_{\mathrm{f}}\tau_{\mathrm{w}}/\gamma \tag{22}$$

Mathematical manipulation of the above equation and using $R_{\mathrm{f}} \equiv x_{\mathrm{f}}/k_{\mathrm{f}}$ and $R_{\mathrm{f}}^* = dR_{\mathrm{f}}/dt$ give the differential equation to be solved:

$$dR_{\mathrm{f}}/dt = SK_{\mathrm{m}}C_{\mathrm{b}} - BR_{\mathrm{f}} \tag{23}$$

where

$$B = \rho_f x_f \tau_w / \gamma \tag{24}$$

The resulting fouling resistance is given as:

$$R_f = R_f^*(1 - e^{-Bt}) \tag{25}$$

where $R_f^*$ is the asymptotic fouling resistance defined by:

$$R_f^* = SK_m C_b / B \tag{26}$$

Equations (25) and (26) show that the $R_f$ and $R_f^*$ can be predicted if $K_m$, $\tau_w$, $S$, and $\gamma$ are known. One must also know the foulant density ($\rho_f$) and thermal conductivity ($k_f$). The values of $S$ and $\gamma$ were unity. Three different regimes for particle deposition these are diffusion, inertia, and impaction regimes. Which regime controls the deposition process is determined by the dimensionless particle relaxation time defined by:

$$t^+ = \rho_p d^2 (u^*)^2 / 18\mu v \tag{27}$$

The values of $t^*$ associated with the diffusion, inertia, and impaction regimes are $t^+ < 0.10$ (diffusion). $0.10 \leq t^+ \leq 10$ (inertia), and $t^+ > 10$ (impaction). Fine-particle transport will probably be diffusion controlled. For example, the deposition process will be diffusion controlled for $d_p \leq 10$ μm for Re = 30,000 in a 15-mm-diameter tube. If the particle transport is diffusion-controlled, the mass transfer coefficient may be predicted from heat transfer data on the enhanced surface, using the heat-mass transfer analogy, as reported [28, 29] on rough surfaces. If the heat transfer coefficient for the enhanced surface ($h$) is known, the mass transfer coefficient can be calculated. Small particles result in high Schmidt numbers; a high heat transfer coefficient will result in a high mass transfer coefficient. Hence, enhanced heat transfer surfaces should result in higher foulant deposition rates than those occurring with plain surfaces at the same operating velocity.

## 6.5 Fouling mitigation, control and removal techniques

For many years, fouling has been considered an unavoidable and unresolved problem besides being human responsibility. Owing to the enormous cost associated with fouling, a considerable number of fouling mitigation strategies have been developed. Fouling is a function of many variables. For example, fouling in crude oil heat exchangers is affected by oil composition, inorganic contaminates, process conditions (temperature, pressure flow rate, etc), exchanger and piping configuration and surface temperature, etc. Therefore, as an effective control of these variables certain conditions may minimize fouling. Generally, effective fouling control methods should involve:

- Preventive fouling formation.
- Preventive foulant from adherence to heat transfer surfaces.
- Removal of deposits from these surfaces.

### 6.5.1 Configuration of heat exchanger

An excess heat transfer surface area that is considered in design to cope with fouling can be as much 10–500%. However, this does not prevent deposition of fouling. At best, it increases the operation time between cleaning cycles. The deposition of dirt on heat transfer surfaces can be greatly reduced by proper design and selection of heat exchangers.

### 6.5.2 Reduction of fouling concentration

Fouling generally increases with increasing foulant concentration in flowing fluids. Filtration, flocculation and sedimentation can remove particulate matters. Reducing the attractive forces of deposition by dispersants and surfactants is also possible. Scaling species can be removed by ion exchange or by chemical treatments. For example, crude desalting to remove most of the solids and salts can reduce fouling in crude oil exchangers. Biofouling in cooling water systems can be reduced by micro-mesh filters to filter out the eggs and larvae of organisms.

### 6.5.3 Use of chemical additives

Antifoulants are widely used to control fouling. Most antifoulants have several functions, such as oxygen scavengers, metal deactivators and dispersants. For auto-oxidation including fouling, antioxidants can be added to absorb oxygen or react with oxidation products in a way to prevent the chain reaction of the auto-oxidation process. Metal deactivators are added to chelate metal ions, thereby preventing their catalytic effect on the auto-oxidation process. Once insoluble is formed by either auto-oxidation or thermal decomposition, dispersants can be added to minimize agglomeration of small insoluble polymeric or coke-like particles or deposit, or sticking of particles to the tube wall. Antifouling chemicals are formulated from several materials. Some prevent foulant forming, while others prevent fouling from depositing on heat transfer equipment. Materials that prevent deposit formation include antioxidants, scale inhibitors, biocides, and corrosion inhibitors. Compounds that prevent deposition are surfactants that act as detergents or dispersants. Different combinations of these properties are blended together to maximize results for different applications. These poly-functional antifoulants are generally more versatile and effective since they can be designed to combat various types of fouling presented in any given system. Antifoulants are designed to prevent equipment surfaces from fouling, but they are not designed to clean up existing foulant. Therefore, antifoulants should be used immediately after cleaning of equipment. Many kinds of biocides, such as biguanides, phenol, formaldehyde and chlorine, etc. have been used to control bio fouling in industrial water systems. For different kinds of bacteria, it is essential to apply the correct biocides and dosages at the correct frequency.

### 6.5.4 High flow rate and low surface temperature

A well-established method of reducing fouling is to increase the wall shear stress by raising the flow velocity or by increasing the turbulence level. Temporarily, increasing the flow velocity, reversing the flow direction and air rumbling are effective and inexpensive mitigation techniques, as long as the deposits are not too hard and adherent. The increase of fouling as the heat transfer temperature goes up is caused by increasing supersaturating, reaction rate, stickability or biological growth. Reducing the temperature usually leads to lower heat fluxes and, therefore, larger heat exchangers. Increasing the velocity does not lead to high wall shear stress, but it leads to high film coefficients and hence a desired lower surface temperature.

### 6.5.5 Chemical or mechanical cleaning of fouled process equipment

Chemical cleaning often more effective than mechanical cleaning. In some cases, it can be done while the equipment is still in service. But the main disadvantage of chemical cleaning is the inability of the chemical solution to penetrate plugged tubes.

Mechanical cleaning, on the other hand, is a commonly used method to remove deposits from the shell side of heat exchanger tube bundles. In general, the bundles are first pulled out, and then immersed in various chemical liquids to loosen or soften deposits. They are then subjected to a combination of high pressure hydro-blasting, rodding, sawing, scrapping, scratching, and in some cases an occasional light sandblasting [28].

### 6.5.6 Surfaces coatings and treatments

The relationship between surface energy and foulant adhesion has been intensively studied. It has been shown that the poorest foulant adhesion occurs on materials with low surface energies. Up to present, several low-energy coatings, for examples, fluoropolymers and silicone polymers, have been developed. Although the coated surfaces do accumulate fouling, the coatings themselves provide a significant additional resistance to heat transfer due to their low thermal conductivity, their use in heat exchangers limited. An effective method of bio fouling control is the application of antifouling coatings containing biocides with toxic properties to fouling organisms. The mechanism of surface protection by these paints upon control leads to poisonous ingredients.

Rough surfaces provide ideal nucleation sites for the growth and promote all types of fouling. The use of electro-polished stainless steel evaporator tubes in pulp and paper, chemical and food processing industries allows longer operation periods between cleaning cycles. If the heat transfer resistance of the tube material is not the major problem, polypropylene tubing, or glass-lined tubing can be used.

**6.5.6.1 Scheduling cleaning** The functions of cleaning exchangers are to permit thorough inspection and to remove foulant and scale that have reduced their performance. Plant designers try to match deterioration and fouling rates of system components so that no single one limits output. However, corrosion, erosion, and fouling rates vary with the fluids flowing, and their concentrations, flow rates, and temperatures. Furthermore, since temperatures and concentrations in a given fluid nearly always change as the fluid passes through the system, deterioration

Figure 24: Example of fouled water tube furnace wall [23].

rates and fouling build-up often vary with the location of a piece of equipment. In a tubular heat exchanger, the rates of deterioration and fouling may vary across the whole unit. An example is a multipass unit with hot fluid entering the tubes that suffer more corrosion in the inlet pass than in the outlet one. An example of a fouled boiler tube surface is shown in Fig. 24.

Cleaning the tube and shell sides of heat exchangers requires taking into account the fouling. The following procedure is suggested as a basis for scheduling cleaning in which fouling is deposited from cooling-tower water as indicated in Fig. 25:

1. Achieve steady-state operation.
2. Measure the four terminal temperatures and pressures and the flow rate on each side.
3. Calculate the clean overall coefficient and actual shell- and tube-side pressures drop.
4. Repeat the measurements and calculations at regular intervals. Measurements and calculations may be made manually, or automatically using a fouling monitor setup.
5. Calculate the fouling rate.
6. Plot the fouling resistance against time.
7. If there is a time delay before the onset of measurable fouling, relocate the origin of the plot to set time zero at the onset point.

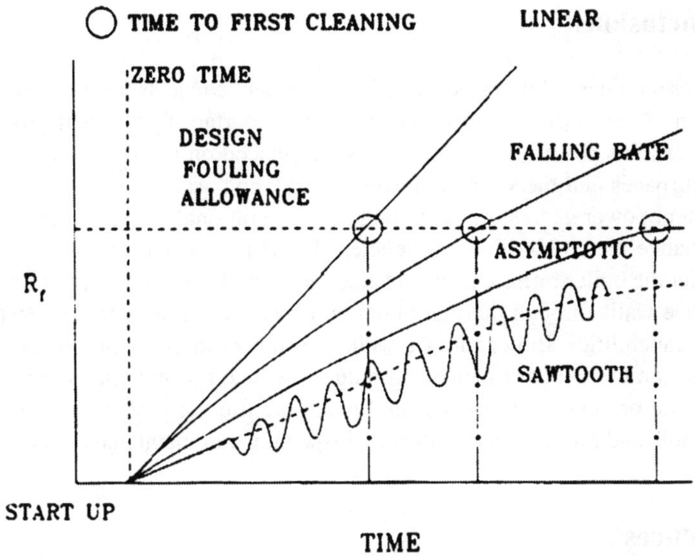


Figure 25: Variation of apparent differential fouling with time for industrial cooling waters.

8. Extrapolate the plot as a linear falling rate or asymptotic rate to the point of intersection with the design fouling allowance.
9. Plan to clean at the time shown by the intersection point.
10. Continue to monitor to refine the extrapolation.
11. After cleaning, repeat the process to establish a cycle.
12. Alternatively, plot the rate of pressure-drop build-up.

Figure 25 is a plot of fouling resistance with time for three kinds of industrial cooling water fouling. In linear fouling, resistance to heat flow increases linearly with time. This indicates that the shear force of the water on the foulant never becomes great enough to peel off accreted layers. In falling-rate fouling, shear forces due to flow make it increasingly difficult for layers to build up. Asymptotic fouling is the desired type. Here, the rate of shear approaches the rate of accumulation at the design fouling allowance. The fourth curve is a generalization for any of the types when acid cleaning fluid is injected into cooling-tower water to remove deposits.

To set a schedule, the scheduler needs some idea of the work to be done, who will do it, and how long it will take. Furthermore, a logical sequence of doing the work must be prepared. The first step is to make an on-line external inspection, including non-intrusive thickness measurements by ultrasonic scanning; the second is to examine the heat transfer characteristics of the equipments and compare with design values.

## 7 Conclusions

The present chapter describes briefly the basic concepts of powder generation from Steam Power Plants and the more important factors that govern their performance as power cycles. Emphasis was placed on fossil fuel combustion in boiler furnaces and the various prevailing modeling assumptions. The main goal of efficient power generation is, among others to rationalize the use of fossil fuels and enhance the combustion efficiencies. The ability of numerical computations to predict the boiler furnace thermal behavior is an ultimate goal. The heat transfer to furnace walls through thermal radiation is reviewed briefly to demonstrate the present capabilities. Boiler furnace walls are subject to major problem of fouling that deteriorate the performance and reduce the heat transfer characteristics drastically. The present work briefly highlighted the fouling problem in Power plant water walls and proposes a monitoring, inspection and maintenance schedule.

## References

[1] Khalil, E.E., *Power Plant Design*, Gordon and Breach: New York, USA, 1990.

[2] Khalil, E.E., *Modelling of Furnaces and Combustors*, Abacus Press: Tunbridge Wells, UK, 1983.

[3] Khalil, E.E., *Flow, Combustion & Heat Transfer in Axisymmetric Furnaces*, PhD Thesis, London University, 1977.

[4] World Consumption of Primary Energy by Energy Type and Selected Country Groups, 1980–2006 (XLS), Energy Information Administration, U.S. Department of Energy (July 31, 2006). Retrieved on 2007-01-20.

[5] *Ministry of Electricity Annual Report*, Egyptian Government, August 2007.

[6] Khalil, E.E., Spalding, D.B. & Whitelaw, J.H., The calculation of local flow properties in two-dimensional furnaces. *International Journal of Heat and Mass Transfer*, **18**, pp. 775–792, 1975.

[7] Khalil, E.E., Numerical calculations of turbulent reaction rates in combustors, ASME 86-WA/HT-37, 1986.

[8] Khalil, E.E., Effect of swirl on flame characteristics in furnaces, AIAA paper AIAA-2004-5539, 2004.

[9] Launder, B.E. & Spalding, D.B., The numerical computation of turbulent flows. *Computer Methods in Applied Mechanics and Engineering*, **3(2)**, pp. 269–289, 1974.

[10] Gosman, A.D., Khalil, E.E. & Whitelaw, J.H., *The Calculations of Two Dimensional Turbulent Recirculating Flows, Turbulent Shear Flows*, Springer Verlag, pp. 13–35, 1979.

[11] Rodi, W.G., *Turbulence Models and Their Applications in Hydraulics – A State of the Art Review*, IAHR: Delft, Netherlands, 1984.

[12] Spalding, D.B. & Patankar, S.V., A calculation procedure for heat, mass and momentum transfer in three dimensional parabolic flows. *International Journal of Heat and Mass Transfer*, **15**, pp. 1787–1794, 1974.

[13] Gupta, A.K. & Lilley, D.G., *Flow Field Modelling and Diagnostics*, 1st edn, Abacus Press: Tunbridge Wells, UK, 1985.

[14] Borghi, R., Computational studies of turbulent flow with chemical reaction, Project SQUID on Turbulent Mixing, AGARD CP125, 1974.

[15] Khalil, E.E., On the modelling of reaction rates in turbulent premixed confined flames, AIAA–80-0015, 1980.

[16] Hutchinson, P., Khalil, E.E. & Whitelaw, J.H., The calculations of flow and heat transfer characteristics of gas fired furnaces. *Proc. 18th Symp. (Int.) on Combustion*, pp. 1927–1931, 1980.

[17] Patankar, S.V., *Numerical Heat Transfer and Fluid Flow*, Hemisphere Publishing Corporation, Washington, DC, 1980.

[18] Henkes, R.W.M., *Natural Convection Boundary Layers*, PhD Thesis, Delft University, Netherlands,1990.

[19] Kameel, R. & Khalil, E.E., Heat transfer characteristics in furnaces: effect of combustion and heat transfer modelling, AIAA Paper, AIAA-2004-0803, January 2004.

[20] Kameel, R.A. & Khalil, E.E., Predictions of energy losses in furnaces under transient conditions. *Proc. of the 3rd IECEC*, San Francisco, AIAA-2005-5524, August 2005.

[21] Khalil, E.E. & Truelove, J.S., Calculations of heat transfer in a large gas fired furnacc. *Letters in Heat and Mass Transfer*, **4**, pp. 353–365, 1977.

[22] Bartelds, H., Lewis, T.M., Michelfelder, S. & Pai, B.R., Prediction of radiant heat flux distribution in furnaces and its experimental testing, IFRF Doc. G02/a/23, 1973.

[23] Central Electrical Laboratory, Ministry of Electricity and Power, Egyptian Government, July 2005.

[24] ASME, *Mechanical Engineering Handbook*, ASME, 2005.

[25] Panchal, C.B., Heat transfer monitor for measurements of fouling of industrial heat exchangers, *Symp. on Condenser Biofouling Control*, Lake Buena Vista, Fla., Argonne National Lab., IL, June 1985.

[26] Somerscales, E.F.C., *Fouling of Heat Transfer Equipment*, ed. J.G. Knudsen, Hemisphere, December 1981.

[27] Förster, M. & Rumpf, B., Low fouling heat transfer surfaces for industrial applications. *Proceedings of the Heat Exchanger Fouling and Cleaning Fundamentals and Applications Conference*, Santa Fe, New Mexico, USA, May 2003.

[28] Pugh, S., Hewitt, G.F. & Müller-Steinhagen, H., Fouling during the use of seawater as coolant – the development of a 'User Guide'. Keynote Lecture, *Proceedings of the Heat Exchanger Fouling and Cleaning Fundamentals and Applications Conference*, Santa Fe, New Mexico, USA, May 2003.

[29] Taborek, J. & Hewitt, G.F., Advances in heat exchangers. *Heat Exchangers: Theory and Practice*, eds J. Taborek, G. Hewitt, & N.H. Afgan, Hemisphere Publishing Corp. and McGraw-Hill, pp. 979–990, 1983.

# CHAPTER 5

# Enhancement of nuclear power plant safety by condensation-driven passive heat removal systems

K. Vierow
*Department of Nuclear Engineering, Texas A&M University, USA.*

## Abstract

In response to increasingly higher demands on nuclear power plant safety systems, the nuclear community is responding by developing "passive" safety systems that are inherently safe. Passive systems are those that do not rely on power supplies, moving parts or human intervention. Condensation heat transfer is a primary driving force for many of these systems. These passive condensation heat transfer systems are reliable and inherently safe because they rely on basic physical forces such as gravity and small pressure differences to transport energy and maintain the plant within design specifications. In recent years, due to advancements in our understanding of the phenomena and in analysis capabilities, designs for essential heat removal systems such as the containment long-term cooling systems have come to rely extensively on passive condensation heat exchangers. The development of innovative systems for future reactors promises to further increase the safety, reliability and economic competitiveness of nuclear power. While significant advancements have been made, challenges remain particularly with regard to accurate, detailed analytical evaluations of system performance. This chapter discusses the numerous ways in which passive condensation heat transfer enhances nuclear power plant safety in current and future nuclear power plants. The discussion focuses on U.S.-design light water reactors and U.S. reactor safety codes although there are many commonalities to reactors of other designs. The physical phenomena are described and the state-of-the art in analysis methods are presented. Challenges for improved analysis are summarized.

# 1 Passive systems with condensation heat transfer

## 1.1 Definition

Adoption of passive systems for heat removal during design basis and hypothetical accident scenarios is one of the strategies for achieving simplification and improving safety and reliability of future nuclear reactors. Passive systems are those that do not require any external input such as AC power sources or operator action to function. Compared with active systems, the passive designs are much simpler because they do not depend on the availability of large power supplies and they do not rely on safety-grade containment cooling systems, both of which add cost and complexity. Yadigaroglu [1] and Juhn *et al.* [2] provide reviews of the various passive designs.

The driving forces for these systems are relatively small forces such as natural circulation for cooling and gravity for condensate return. In particular, the heat transfer processes are driven by small pressure and temperature differences. Thus, to achieve the needed cooling rates, heat transfer with phase change is necessary. In one of the first passive concepts, General Electric designed the Simplified Boiling Water Reactor (SBWR) Passive Containment Cooling Systems (PCCS) with vertical heat exchangers that condense containment steam and transfer the heat to a pool outside the containment [3, 4]. The design has evolved [5] and become the basis for passive systems of several current plant designs in the U.S., Europe and Japan.

Another leading design takes advantage of the large heat transfer area on containment walls to condense steam on the inner surface and remove heat through the outer surface by natural circulation of air [6].

There is a strong move towards passive safety systems because the equipment is driven by failsafe forces or mechanisms such as gravity and natural circulation. Adoption of passive systems with efficient heat removal mechanisms will promote future installation of additional nuclear power plants that will increase both reliability and security to our energy supply.

## 1.2 Goals and requirements

Key goals of these condensation-driven passive heat removal systems include:

1. enhanced safety systems,
2. improved reliability, and
3. greater economic competitiveness via simplification.

Development of passive condensation heat removal mechanisms is particularly timely because the licensing of future nuclear reactors will require a stronger safety case than for previously licensed reactors. The requirements for these systems in U.S. reactors are specified in the U.S. Nuclear Regulatory Commission's (NRC) Code of Federal Regulations (CFR). Appendix A of 10 CFR Part 50 defines several General Design Criteria such as GDC 35, "Emergency Core Cooling," which stipulates that the emergency core cooling system (ECCS) be capable of heat removal to ensure the core maintains a coolable geometry [7]. For example, the Isolation Condenser System (ICS), a passive condenser in the ECCS of some BWR designs, must satisfy the criteria relevant to ECCS.

In addition, the Electric Power Research Institute (EPRI) led an industrial effort in the 1990s to define technical standards for advanced light water reactor designs that fulfill both the NRC criteria and utility requirements for operational and safety improvements [8, 9]. The Utility Requirements Document addressed both evolutionary and passive plant designs.

The designers may add additional requirements. For example, General Electric designed the Economic Simplified Boiling Water Reactor (ESBWR) ECCS to function successfully without any operator action for 72 hours following accident initiation [10].

### 1.3 Challenges

Skeptics of passive systems suggest that accident termination may be slower than for active systems. It will be shown in later sections that passive systems satisfy safety requirements, in part because of the extremely efficient phase change heat transfer that is possible with condensation. Additional margin is derived from the power densities of passive plants, which are generally lower than of plants with active systems.

System reliability has also been questioned in the past with regard to condenser heat removal initiation and performance over a wide range of conditions. A concern has been for system activation by small passive forces. Significant effort has been invested to demonstrate that noncondensable gases initially residing along condensation surfaces are quickly purged, enabling timely and efficient system startup. Regarding the range of operability, the performance of condenser in safety systems is confirmed experimentally and analytically prior to awarding the U.S. Design Certification.

Safety analyses of the AP600 and AP1000 indicate improved performance over active systems with respect to peak clad temperatures, peak pressures and Departure from Nucleate Boiling (DNB) margin [6]. ESBWR loss of coolant accident (LOCA) analyses of the ECCS performance using the ICS similarly show compliance with all of the applicable criteria [10].

A challenge remains to further clarify the details of the heat transfer phenomena. A more detailed understanding of the physics, including the condensate film behavior and noncondensable gas redistribution, will enable improved predictive capabilities and reduce the margins currently imposed to cover for analytical uncertainties.

## 2 Roles of passive condenser systems in nuclear power plants

The passive condensation heat transfer systems are divided into three classifications herein.

1. In-vessel decay heat removal during normal shutdown or refueling.
2. In-vessel decay removal under hypothetical accident conditions.
3. Containment heat removal under hypothetical accident conditions.

"In-vessel" refers to occurrence inside the reactor pressure vessel. "Containment" refers to occurrence in the containment atmosphere surrounding the reactor pressure vessel. Decay heat is the energy released by decay of radioactive fission products following reactor shutdown.

Current safety systems in U.S. Pressurized Water Reactors (PWR) and Boiling Water Reactors (BWR) and their possible passive condenser heat exchanger replacements are described below.

## 2.1 In-vessel decay heat removal during normal shutdown or refueling

The Residual Heat Removal (RHR) system, an active system, maintains the core in a safe condition during normal shutdown and refueling by pumping cooling water through the reactor core and removing core decay heat. Under hypothetical conditions, the RHR may be lost and alternative active cooling sources may not be available. The loss of RHR situation could lead to core boil-off, fuel rod heatup, and eventually core damage and has therefore been widely studied [11].

Steam condensation in the steam generator U-tubes can become one of the major heat removal mechanisms. As a passive mechanism, steam generated by coolant boil-off would enter the steam generator tube primary side via the hot leg, along with any noncondensable gases ingested during preparation for maintenance and/or refueling. Reflux condensation is the mode of phase change heat transfer occurring when a condensing gas and the liquid condensate flow in opposite directions. In the situation under investigation, steam flows vertically upward in steam generator U-tubes and water forms a condensate film that returns downward to the tube inlet plenum.

BWR safety systems in currently operating reactors rely on a Reactor Core Isolation Cooling system (RCIC) to maintain an adequate water supply in the reactor pressure vessel for core cooling and maintain the reactor in a standby condition. The RCIC also allows for complete shutdown under a loss of normal feedwater system [12]. The RCIC relies on active systems to carry out its functions [13].

In advanced BWRs such as the ESBWR, the ICS is the most comparable system to the RCIC [13]. Designed to passively remove reactor decay heat following reactor shutdown and isolation, the ICS can also prevent the initiation of other safety systems that could perform the RCIC functions. One of the design goals of ICS is to "achieve and maintain safe stable conditions" for 72 hours without operator action following a non-LOCA event. The ICS limits reactor pressure and maintains core cooling by condensation of steam in vertical tube bundles and return of condensate to the reactor pressure vessel. This passive heat removal proceeds independent of any active ICS initiation under reactor isolation events, loss of AC power, transients or hypothetical accidents.

## 2.2 In-vessel decay removal under postulated accident conditions

Active systems are present in current PWR and BWR that provide sufficient core cooling to prevent core damage and release of radioactive materials. An ECCS injects coolant flow into the primary piping to provide in-vessel decay heat removal during normal shutdown or refueling.

Condensation heat transfer is typically not a major decay heat removal mechanism in PWRs under postulated accident conditions. Reflux condensation in the steam generator tubes would be ineffective in meeting larger heat removal requirements because flooding would inhibit condensation at higher steam mass flow rates.

The active ECCS of the latest BWRs in commercial operation (BWR/6) consists of a high-pressure core spray, an automatic depressurization system, a low-pressure core spray and low-pressure core injection systems. The RCIC system and the RHR system may also be employed [14].

The ICS was designed for the ESBWR to passively remove sensible heat and decay heat following reactor shutdown and isolation by condensation heat transfer. It is part of the ECCS and acts in combination with other new passive features to replace the active components of current ECCS. It activates following sudden reactor isolation, station blackout (loss of all AC power), LOCAs and Anticipated Transient Without Scram (ATWS) [13]. By removing heat and returning condensate to the reactor pressure vessel, the ICS aims to avoid unnecessary reactor depressurization and actuation of other safety systems. The ICS is also considered the first line of defense against a severe accident [10].

## 2.3 Containment heat removal under postulated accident conditions

Containment systems are designed to contain the steam and radioactive materials released from the reactor pressure vessel following a postulated accident, assuming that the ECCS performs as designed.

PWR active systems include fan coolers and containment spray systems by which water is sprayed from the upper part of the containment. The falling drops condense steam to lower the pressure and cool the containment atmosphere. In the AP600 and AP1000 passive systems [6], steam condenses on a steel containment vessel and the heat is transferred outside containment. Air, circulating by natural convection along the outer surface of the steel vessel, removes the condensation heat. The steel containment vessel is a large heat transfer surface to provide sufficient heat removal even in the presence of noncondensable gases on the condensation side.

For containment heat removal under hypothetical accident conditions, BWRs suppress the pressure by forcing any released steam to flow through a large pool of water. The depressurization system that initiates flow is an active system. Some PWRs of Western design employ the vapor suppression concept by use of “ice condensers”. Conventional refrigeration equipment maintains ice in several large chests placed around the periphery of the containment. The chests are arranged such that any steam released from a reactor system rupture must pass through the ice chests before flowing into the free volume of the containment. Both BWRs and PWRs employ active containment spray systems where water is pumped from a water source to spray nozzles at the top of the containment vessel.

The BWR pressure suppression system is completely passive in future reactor designs and does not require replenishment of water on the secondary side for a minimum of 72 hours. For the initial blowdown from full pressure, vertical vents direct steam from the drywell to a large pool in the wetwell where most of the vapor is condensed. Once the blowdown has progressed and the drywell-to-wetwell pressure difference is insufficient to clear the vents, the PCCS removes heat via condensation in vertical tube bundles. The PCCS loops are extensions of the containment pressure boundary. Each loop includes a condenser that sits in a pool of water and transfers the heat of condensing steam from the primary side to the coolant pool on the secondary side. The pools are vented to the atmosphere.

## 3 Description of scenarios and phenomena

### 3.1 General description of condensation heat transfer

The classical Nusselt analysis of condensation on an isothermal vertical plate in a stagnant steam environment serves as the starting point for describing other scenarios. Figure 1 shows the film condensation situation of the isothermal plate cooling the surrounding vapor, which condenses and creates a thin condensate film on the wall. The film is very thin for steam/water situations. A small natural convection current develops within the vapor region near the condensate film due to the downward flow of the film.

The phenomena occurring in the nuclear power plant safety systems at hand may be described by referring to the Nusselt condensation situation and superimposing complications relevant to the particular scenario. The condensation heat transfer scenarios that arise in passive condenser systems may be classified as:

1. "reflux condensation" in vertical tubes with steam/noncondensable gas inflow from the tube bottom end;
2. condensation in vertical tubes with steam/noncondensable gas inflow from the tube top end;
3. condensation in horizontal tubes with steam/noncondensable gas inflow at one end and condensate draining at the other end;
4. condensation in large water pools;
5. condensation on large vertical walls.

### 3.2 Reflux condensation in vertical tubes with steam/noncondensable gas inflow from the tube bottom end

During routine shutdown of a PWR for maintenance and refueling, the coolant system may be in a partially drained state known as "midloop operation". In midloop

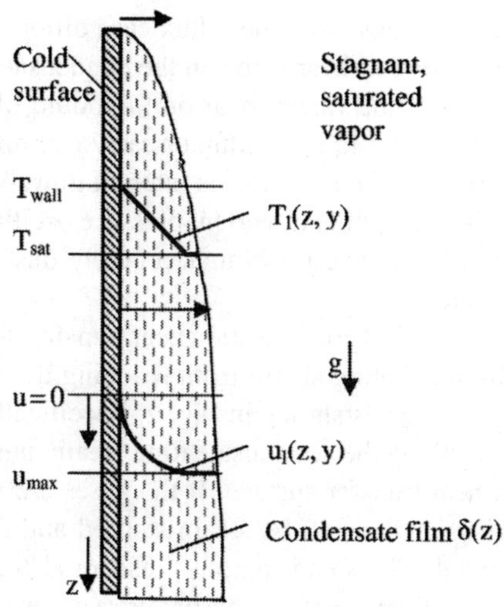


Figure 1: Film condensation along a vertical flat plate.

operation, the water level is above the core, but below the hot leg elevation. Various primary system closures such as the reactor vessel head may not be secured and air or nitrogen can be drawn into the system. Under hypothetical conditions, the RHR may be lost and alternative active cooling sources are not available.

Nitrogen injection from the accumulator into the reactor Coolant System (RCS) following a LOCA also corresponds to midloop operation with noncondensable gases present. If the accumulator valve were to fail following water injection, nitrogen would enter the RCS and degrade the heat transfer performance of the steam generator. Although the possibility of such a scenario occurring is highly unlikely, the consequences must be known.

Assuming loss of AC power, the possible alternative cooling sources are:

1. gravity drain from the Reactor Water Storage Tank,
2. core boil-off and steam venting, and
3. reflux condensation heat transfer in the steam generators.

Reflux steam condensation in the steam generator U-tubes with noncondensable gases present can become one of the major heat removal mechanisms. Some or all of the steam produced by coolant boil-off would condense inside the hot leg risers of the steam generator U-tubes and the condensate would drain to the hot leg and back to the core region. For reflux condensation to be an effective heat removal mechanism, the RCS must be closed and inventory must be maintained on the steam generator secondary side.

The reference situation is the classical Nusselt condensation of stagnant vapor on a vertical flat plate. Nusselt analysis provides a solution for the film thickness as a function of distance from the top of the plate and for the local heat transfer coefficient.

Compared to Nusselt condensation, the reflux case differs in that the steam has an upward velocity that exerts a shear force on the condensate film in the opposite direction of film draining. If the vapor shear on the liquid film is sufficient, condensate draining is prohibited. Liquid holdup or carryover upward by the vapor is known as “flooding”. Further, the tube flow is internal flow, which provides a constrained geometry that is susceptible to violent pressure oscillations upon flooding. The pressure oscillations would result in unacceptably unsteady performance of the U-tube steam generators.

Noncondensable gases inhibit condensation heat transfer by accumulating along the liquid film, thereby preventing steam from reaching the heat transfer surface. The situation in PWR U-tubes is shown in Fig. 2. Specifically, with noncondensable gases accumulating along the condensate film, steam must diffuse through the gas layer to reach the heat transfer surface. If the gases are not removed, the gas layer adjacent to the liquid film will be steam-starved and the condensation rate will be severely degraded. The steam partial pressure at the condensate surface will be significantly lower than the steam partial pressure in the bulk of the steam/gas mixture and most of the steam will pass through the condenser tube riser without condensing.

With regard to flooding, at a given distance from the condenser tube inlet, a vapor/gas mixture will have more upward momentum than a pure vapor flow. Thus, the condensate film will experience a greater shear upward. Conversely, since less vapor condenses when noncondensable gases are present, there will be less condensate and the effects of flooding on stable operation may therefore be less. Additional investigations are necessary to clarify the effects of noncondensable gases on flooding.

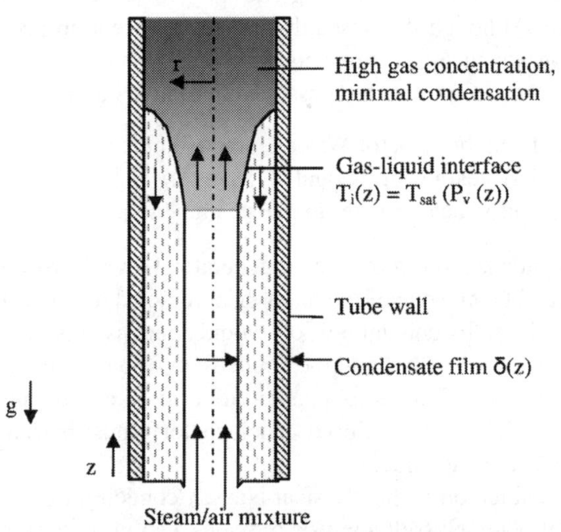


Figure 2: Reflux condensation.

Reflux condensation does not occur on the downflow side of PWR U-tubes because the steam/gas mixture and the condensate liquid are flowing in the same direction.

## 3.3 Condensation in vertical tubes with steam/noncondensable gas inflow from the tube top end

### 3.3.1 Condensation in IC tubes under normal shutdown and isolation conditions

The ICS consists of multiple loops, each with an isolation condenser unit that condenses steam on the primary side and transfers heat through the tube wall to the secondary side water pool. Condensed steam drains from the tubes back to the reactor pressure vessel to maintain core cooling. Each ICS loop has noncondensable gas vent lines from the top and bottom headers that can purge noncondensable gas from the unit and assure good heat transfer performance. The ICS is expected to have an inflow of essential pure steam under most operating conditions. During a severe accident, hydrogen from a core coolant interaction may enter the tubes. Noncondensable gases from the containment can enter the ICS units after the reactor pressure vessel depressurization.

The inlet to the ICS condenser tubes is open during normal reactor operation while the exit is valved shut. Thus, the condenser tubes are filled with condensate water and the noncondensable gas vent lines are shut. The condensate return valves open upon LOCA initiation, providing a source of coolant inventory to the core. During refueling, the ICS loops are isolated from the reactor, with all steam supply, condensate return and vent line valves closed [13].

The physical situation inside a vertical tube, such as an IC tube, is shown in Fig. 3. A steam–noncondensable gas mixture enters the condenser and vapor begins to condense at the inlet. As steam is drawn to the cooling surface, the gas experiences a force similar to suction through a permeable wall. The noncondensable gas concentration at the film interface becomes higher than that in the central core and a gas–vapor boundary layer develops adjacent to the liquid boundary layer. Between the annular gas boundary layer and tube centerline, the steam concentration is constant. However, the cross-section average of noncondensable gas concentration increases with distance along the tube axis as the boundary layer thickens. At some axial location, the boundary layer bridges the tube so that there is no longer a central core of inlet composition.

With steam continuing to condense, but at diminishing rates, a fully developed condition may not be achieved in the gas phase. Downstream from the point of nearly complete condensation, the gaseous mixture contains steam in equilibrium with the condensate and noncondensable gas, with a partial pressure that maintains tube total pressure at nearly the system pressure level.

A distinguishing feature of forced convection condensation inside tubes from the Nusselt flat plate case and reflux condensation is the effect of interfacial shear. The gas phase has a higher velocity than the condensate film, producing

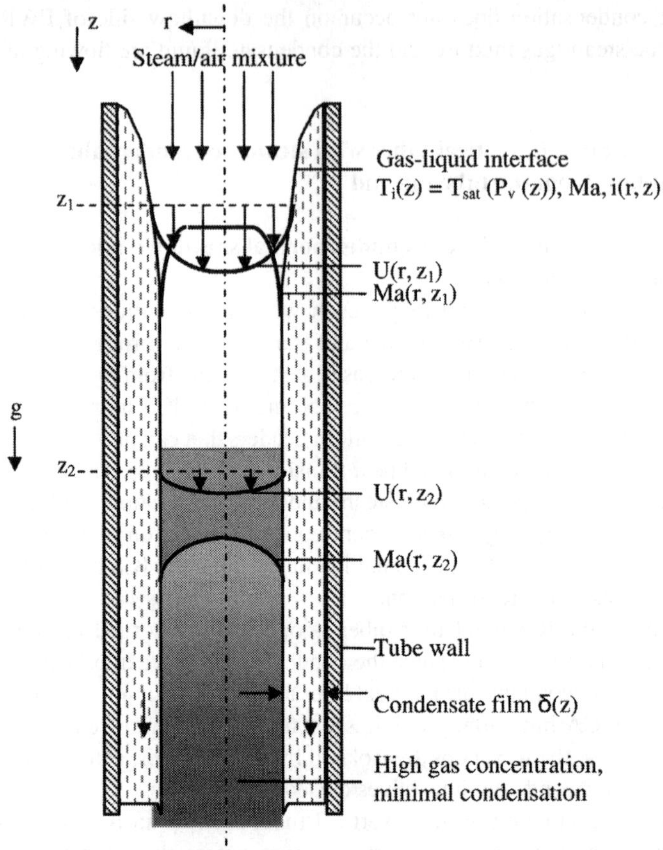


Figure 3: Film condensation in a vertical tube with downward flow.

interfacial shear and increasing condensate film turbulence. With the gas and liquid film both flowing downward, turbulence thins the film and reduces film resistance to heat transfer. Additionally, the increased interfacial shear promotes turbulence and mixing in the gas core. The resistance of the noncondensable gas boundary layer to condensation heat transfer is reduced. A second distinction concerns the axial variation of the cross-section average temperature and species concentration for the internal flow case. In contrast to the stagnant atmosphere case, the temperature decreases with distance along the condenser tube and the noncondensable gas concentration increases. Axial variations of the conditions driving heat and mass transfer arise, but they do not occur in the stagnant gas, vertical flat plate situation.

Sparrow and Lin's work [15] show similarity in the boundary layers and a uniform interface temperature. For a constant, free stream gas mass fraction, Denny, *et al.* [16] note nonsimilarity in the three-phase problem. With a variable centerline gas mass fraction, the problem becomes even more nonsimilar in character. An exact analytical solution would require solving radially and axially dependent boundary layer conservation equations that could not be reduced to ordinary differential equations.

### 3.3.2 Condensation in IC and PCC tubes under hypothetical accident conditions

During a postulated accident in which there is a large energy release from the primary system, steam and flashing water can cause a significant containment pressure rise. For the passive plants, new systems for removal the energy sources to outside of the containment without releasing radioactive materials are needed. A condensation heat transfer mechanism of long-term decay heat removal, the PCCS has been incorporated into the design of the ESBWR[5].

The PCCS is a post-LOCA, low pressure, decay heat removal system designed for the SBWR [3, 4] and upgraded for the larger power rating ESBWR [10]. The function of the PCCS is to remove heat from the reactor containment system (drywell and suppression chamber) and maintain pressure below the design value by passive means. The primary components of the PCCS are several condensation heat exchanger units. Located in a large pool of water outside of the containment, the PCCS units consist of an upper steam plenum, condenser tubes, and a lower condensate plenum (Fig. 4).

On the primary side of the PCCS units, noncondensable gases and decay heat steam enter from the drywell. Steam condenses inside the tubes and condensate drains to the GDCS pools. These pools are elevated relative to the reactor and serve to maintain reactor water inventory following an emergency depressurization of the primary system. These pools drain to the reactor pressure vessel when the sum of drywell pressure plus water head in the drain lines is greater than the sum of the reactor vessel pressure plus vessel water head above the drain line inlet nozzles. As the condensation proceeds, gases and a small amount of steam exiting the condenser tubes are vented to the suppression pool. The steam, drain, and vent flows are driven by natural circulation and the pressure difference between the drywell and suppression chamber.

On the secondary side of the PCCS units, heat is carried away from the condenser tubes by water in the surrounding natural circulation pools. Pool water inventory is maintained sufficient to prevent tube uncovery for at least 72 hours. The pool initially is subcooled at atmospheric pressure, and rises to saturation temperature at just above one atmosphere after a few hours of decay heat removal.

The distribution of noncondensable gases throughout the containment system is a key factor in determining the post-LOCA conditions and behavior of the PCCS units. During normal reactor operation, the SBWR containment is inerted with nitrogen, and operates at just above atmospheric pressure. The containment gas mixture is composed primarily of nitrogen and steam. The low oxygen content prevents buildup of a potentially combustible gas mixture. However, the noncondensable gas inventory has the potential to degrade the performance of the heat exchangers.

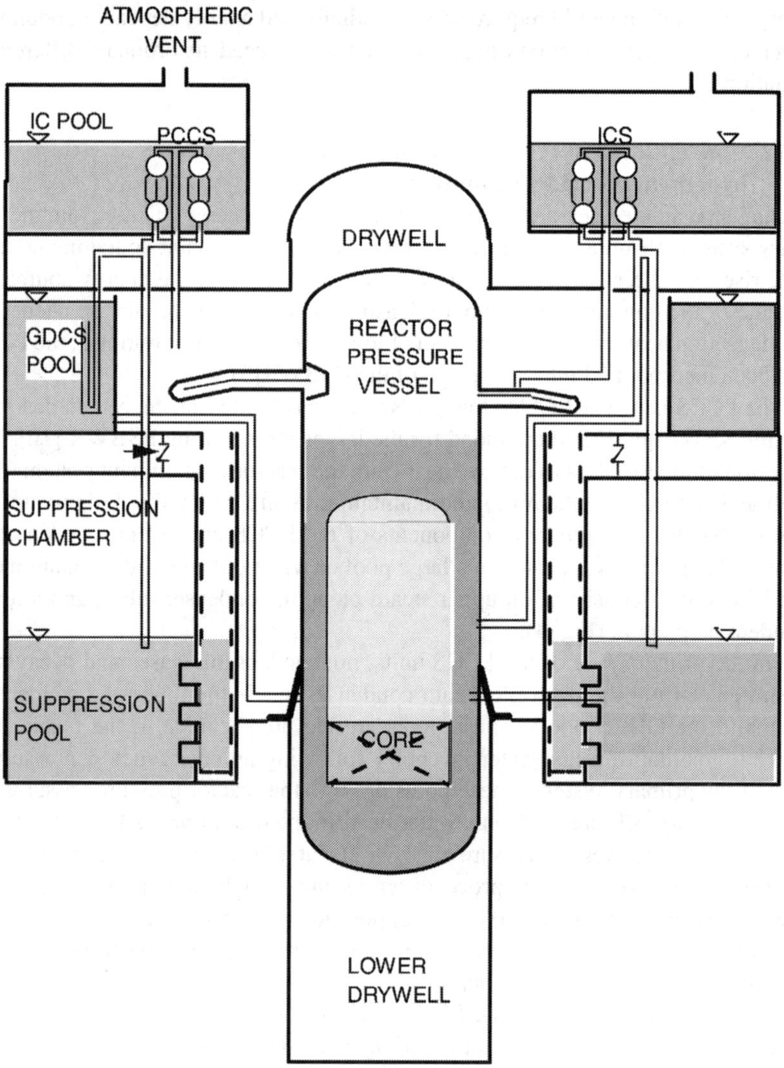


Figure 4: Economic Simplified Boiling Water Reactor Isolation Condenser System and Passive Containment Cooling System.

Early in the blowdown transient, most of the drywell gas is transferred to the suppression chamber gas space, along with the blowdown steam, via the main drywell-to-suppression chamber vent system. The PCCS was designed for long-term heat removal and does not have the capacity to remove all of the blowdown heat. In the longer term, steam produced by decay heat is released by the reactor pressure vessel into the drywell for intake by the PCCS. Residual noncondensable gas in the drywell is vented to the suppression pool via the PCCS units and, in the process, degrades heat removal performance by inhibiting steam condensation within the tubes. The ICS ingests steam produced by decay heat and any noncondensable gases that have entered the reactor pressure vessel. The phenomena inside the tubes are the same as described for the ICS in the previous section.

### 3.3.3 Condensation in horizontal tubes with steam/noncondensable gas inflow at one end and condensate draining at the other end

As one of the most widely used types of heat exchangers, horizontal condensation heat exchangers have traditionally found many industrial applications, including in the process industry, the air conditioning and refrigeration industry, and for condensation of mixed vapors for distillation of hydrocarbons [17].

In the nuclear industry, horizontal heat exchangers are also widely used. Recently, a horizontal heat exchanger design has been proposed for a PCCS of future light water reactors [18]. Tujikura *et al.* [19] and Ueno *et al.* [20] also consider the participation of horizontal steam generators as condensers in PWRs in the event of postulated accidents. Current PCCS designs typically employ a vertical condenser [5]. The horizontal design is proposed because horizontal heat exchangers have a potentially higher heat removal capability than vertical heat exchangers [21] and higher seismic resistance.

The details of condensation heat transfer in the presence of a noncondensable gas in horizontal condenser tubes are not well understood. Compared to the well-studied case of in-tube condensation for vertical downflow [22, 23], the complication of this phenomenon mainly comes from several factors. The first is the non-symmetrical two-phase phase distribution in the tube cross section (Fig. 5). Also, within the gas phase, when the velocity is low, the asymmetrical noncondensable gas concentration caused by the density difference between the noncondensable gas and vapor could affect the heat transfer [24]. For flow regimes with low gas-phase superficial velocity, the heat transfer characteristics at the top and bottom of the tube are quite different, and thus the heat transfer coefficients are also different. Second, the interfacial phenomena including film surface instabilities and droplet entrainment/deposition may play a larger and different role than in vertical tubes. Under low gas velocity conditions, condensate droplets may detach from the condensate film due to gravity, interacting with the gas core and exposing a cooler condensate film surface to the vapor. The third factor is the mode of condensate drainage. In a horizontal tube, condensate must be carried out by its own momentum or by shear from the vapor/gas mixture whereas gravity draining occurs in a vertical tube.

Gravitational and interfacial shear forces determine the local flow regime. The flow regimes are commonly classified into three categories, as reviewed by Wu [25]:

- annular and annular-mist flow, which are driven primarily by shear forces.
- stratified flow, wavy flow and slug/plug flow, which are gravity driven.
- wavy-annular flow, which have strong contributions from both the gravity and shear forces.

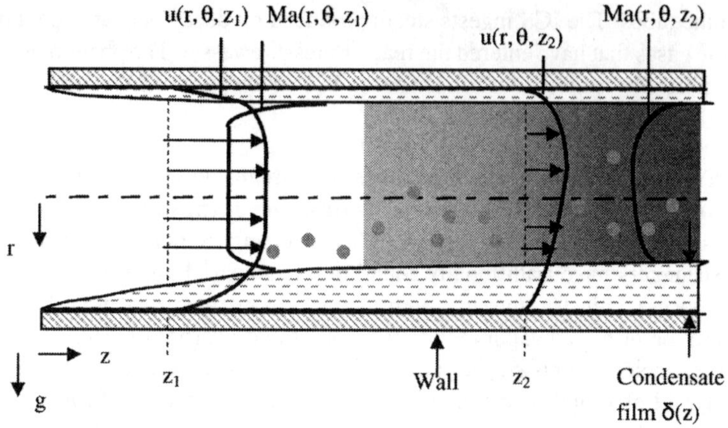


Figure 5: Film condensation in a horizontal tube.

For shear-driven flows, the heat transfer mode across the condensate film is forced convection heat transfer. For gravity-driven flows, the condensate film experiences largely conduction at the top of the tube and a combination of forced convection and conduction at the bottom of the tube.

### 3.4 Condensation in large water pools

Condensation in large water pools differs from the other condensation modes discussed herein because the condensation surface is not a liquid film. Rather, vapor undergoes direct contact condensation on the surrounding subcooled water in a complicated variety of interfacial configurations and thermal hydraulic interactions. The vapor phase may be discontinuous (bubbly flow) or discrete (jet flow) depending on whether the vapor transport is due to blowdown of the drywell into the suppression chamber or venting from the PCCS heat exchangers.

During the initial blowdown of a BWR, the drywell steam/gas mixture is discharged into a large, annular water volume termed the suppression pool. The hydrodynamic phenomena associated with the steam/gas jets are rapid and violent due to the large driving pressure difference between volumes. Pressure waves against the outer wall of the suppression pool are expected along with a high degree of turbulence due to condensation chugging and mixing of the entire suppression pool.

In the longer-term containment cooling by the PCCS, the PCCS condensers purge a vapor/noncondensable gas mixture into the suppression pool via a vent line with relatively shallow submergence at the exit. Internal circulation of the noncondensable gas within the vapor bubbles affects the condensation rate. The vent line is submerged shallow enough to vent, yet deep enough for vapor bubble condensation below the water surface.

A detailed description of the phenomena, with emphasis on the noncondensable gas effect, is provided by Xu [26].

### 3.5 Condensation on large vertical walls

The PCCS provides one method for passively removing energy from the containment to the atmosphere, thereby avoiding a potential threat to the structural integrity of the containment. A second method for containment heat removal relies on thermodynamics principles. During a hypothetical LOCA, the flashing coolant released from the primary side break will condense on the containment walls if the surface temperature is below the dew point. Because the wall heat transfer area is large, condensation on structures can be an important heat removal mechanism.

Heat removal by condensation on large vertical containment walls is an AP-1000 and AP-600 passive safety feature. A secondary side air circulation system has been incorporated into the design to take advantage of the wall heat transfer. Other systems without the secondary side may take smaller credit for this heat removal mechanism. Dropwise condensation will occur on the upper regions of the condensation surfaces. Most of the surface is expected to be covered with a condensate film.

Departures from the Nusselt analysis scenario include the presence of noncondensable gases and the effect of the liquid–gas interface on the heat transfer rates. A wavy interface on the film increases transport of heat, mass and momentum by increasing the turbulence intensity, decreasing the gas viscous sublayer thickness and by reducing the laminar film thickness.

## 4 State-of-the-art analysis methods

### 4.1 Basic approaches

Rigorous safety analysis is needed to design and license a nuclear power plant. The basic approach for analysis of new systems is to apply the techniques used for standard plants to the passive plant systems, modifying the analyses as experimental data becomes available.

Many empirical formulations apply correction factors to the local heat transfer coefficient or Nusselt number obtained from the Nusselt analysis. For pure vapor condensation on a vertical plate, the vapor boundary layer and the film are typically thin enough for the boundary layer assumptions to be valid. Other assumptions in the Nusselt analysis are listed below. These are reflected in the velocity and temperature profiles of Fig. 1.

1. Stagnant, pure vapor at $T_{sat}$ (no noncondensable gases).
2. Negligible shear stress at the liquid–vapor interface.
3. Laminar flow and constant properties in the condensate film.
4. Condensate film heat transfer by conduction only, no momentum or energy transfer by advection in the condensate.

Nusselt solved the momentum equation for the condensate film to obtain the film thickness as a function of distance from the top of the plate. The film thickness was converted to a thermal resistance, from which local and plate-averaged heat transfer coefficients were formulated.

$$h(z) = \frac{k_l}{\delta(z)} \tag{1}$$

where the local film thickness is:

$$\delta(z) = \left[ \frac{4k_l \mu_l \left(T_{\text{sat}} - T_{\text{surf}}\right) z}{g \rho_l \left(\rho_l - \rho_v\right) h_{\text{fg}}} \right]^{1/4}$$

The original work of Nusselt has been extended by Rohsenow [27] to account for the thermal advection contribution by adding a term to the latent heat of vaporization. The local film heat transfer coefficient is expressed as:

$$h(z) = \left[ \frac{g \rho_l \left(\rho_l - \rho_v\right) k_l^3 h'_{\text{fg}}}{4\mu_l \left(T_{\text{sat}} - T_{\text{surf}}\right) z} \right]^{1/4} \tag{2}$$

Theoretical models are preferable because they better reflect the physical phenomena and are not restricted to ranges of conditions of particular experimental facilities. Current modeling techniques for relevant condensation heat transfer scenarios are discussed below.

## 4.2 Reflux condensation in vertical tubes with steam/noncondensable gas inflow from the tube bottom end

Reflux condensation in nuclear engineering applications is generally evaluated as part of a reactor system analysis using a reactor safety code. The heat transfer models that the codes call upon for reflux condensation analysis were developed for condensation of vapor/condensate co-current downward flow on flat plates, in tubes or for other situations and were shown to have deficiencies in predicting reflux condensation heat transfer. The capabilities of reactor safety codes such as RELAP5 [28, 29] and TRACE [30] have been recently evaluated to predict reflux condensation. Moon and Park noted these deficiencies in RELAP5, while Queral presented a system evaluation without details on the reflux condensation model or results.

Some of these evaluations have used the correlations of the codes inappropriately. For example, Park *et al.* [29] used correlations in RELAP5 developed for downward steam flow in a reflux condensation analysis. The gas mixture Reynolds number is the highest where the condensate film is thickest, which renders correlations developed for other in-tube situations inapplicable for reflux condensation analysis.

Several recent experimental programs have provided new data on reflux condensation heat transfer rates for development of empirical correlations [28, 31–35]. These correlations are generally a function of Reynolds number of the gas mixture and/or condensate, noncondensable gas concentration, and dimensionless property quantities such as the Prandtl number and Jacob number. They are simple and formulated to be compatible with reactor safety code implementation. The correlations have been implemented in reactor safety codes such as RELAP5/SCDAPSIM/MOD3.2 [33] and verified against independent data sets.

Moon *et al.* [28] correlated a degradation factor to account for degradation by noncondensable gases and enhancement by vapor shear. The local Nusselt heat transfer coefficient, when multiplied by this factor, provides a local heat transfer coefficient for reflux condensation.

$$F(z) = \frac{h_{\mathrm{exp}}(z)}{h_{\mathrm{f}}(z)} = 2.58 \times 10^{-4}\, Re_{\mathrm{v}}^{0.200}\, Re_{\mathrm{l}}^{0.502}\, W_{\mathrm{air}}^{-0.244}\, Ja^{-0.642} \tag{3}$$

where $W_{\mathrm{air}}$ is the local air mass fraction. The stated range of validity is $6{,}110 < Re_{\mathrm{v}} < 66{,}586$, $1.2 < Re_{\mathrm{l}} < 166.6$ and $0.140 < W_{\mathrm{air}} < 0.972$.

Nagae *et al.* [33] derived a local Nusselt number for the interfacial heat transfer coefficient, $h_{\mathrm{i}}$, using the data of Vierow *et al.* [32].

$$Nu_{\mathrm{i}} = \frac{h_{\mathrm{i}} d_{\mathrm{w,i}}}{k_{\mathrm{v}}} = 120\left(\frac{P_{v}}{P_{\mathrm{air}}}\right)^{0.75} \max\left(1.0,\, aRe_{\mathrm{v}}^{b}\right) \tag{4}$$

where $Re_{\mathrm{v}} = j_{\mathrm{v}} d_{\mathrm{w,i}} / \upsilon_{\mathrm{v}} \le 5{,}000$, $a = 0.0012$ and $b = 1.0$. This correlation is valid for steam–air mixtures under laminar flow conditions.

Most recently, Lee *et al.* [35] correlated a local heat transfer coefficient correction factor as:

$$F(z) = \frac{h_{\mathrm{exp}}(z)}{h_{\mathrm{f}}(z)} = 4.88 \times 10^{-4}\, Re_{\mathrm{l}}^{0.59}\, W_{\mathrm{air}}^{-0.29}\, Ja^{-0.89} \tag{5}$$

where $1.5 < Re_{\mathrm{l}} < 246$, $0.02 < W_{\mathrm{air}} < 0.96$ and $0.014 < Ja < 0.123$.

The lack of mechanistic models for reflux condensation may be traced to insufficient understanding of factors not present in the classic Nusselt situation. First, the vapor velocity and vapor-condensate film interactions must be considered. Interfacial waves induce film turbulence and Nusselt's assumption of a linear film temperature distribution becomes invalid. The local heat transfer coefficient may be a strong function of the vapor velocity [36] Second, noncondensable gases are generally present in nuclear power plant applications. Third and unique to reflux condensation, flooding limits the range of conditions under which reflux condensation is stable.

Gross and Philipp [36] provide a summary of the development of mechanistic models for reflux condensation heat transfer starting from the Nusselt analysis. They noted that the theoretical Nusselt model had been extended to include vapor flow effects such as drag coefficients.

The Gross and Philipp study of the shear stress effect on the local Nusselt number provides insight for new mechanistic model development. Under pure vapor conditions, the shear stress was shown to decrease the Nusselt number for low film Reynolds numbers and to enhance heat transfer at high film Reynolds number. A semi-empirical model was presented for local Reynolds numbers of very thin films, eqns (6) and (7). For thicker films, the effect of shear stress on Nusselt number was successfully correlated, eqn. (8). Increased understanding of the film behavior at the film-gas interface and the noncondensable gas flow is needed for further mechanistic model development.

$$\delta^{+} = 0.59\tau^{*} + \left(3^{1/3} - 0.0086\tau^{*}\right)\left(Re_{1} + 0.28\tau^{*3}\right)^{1/3} \tag{6}$$

where $\delta^{+}$ is a dimensionless film thickness and $\tau^{*}$ is a dimensionless shear stress at the vapor–liquid interface.

$$Nu = \left[0.59\tau^{*} + \left(3^{1/3} - 0.0086\tau^{*}\right)\left(Re_{1} + 0.28\tau^{*3}\right)^{1/3}\right]^{-1} \tag{7}$$

$$Nu_{\mathrm{corr}} = Nu_{\tau\to 0}\left[1 + \left(0.18\tau^{*} + \tau^{*2.2}\right)\left(-0.25 + 0.016Pr_{1}^{1.08}\right)\right] \tag{8}$$

### 4.3 Condensation in vertical tubes with steam/noncondensable gas inflow from the tube top end

For PCCS analysis, empirical correlations have been developed from the early 1990s [22, 37]. Multiplication factors were obtained for the local Nusselt heat transfer coefficient to account for shear enhancement and noncondensable gas degradation. These enabled efficient analysis with reactor safety codes.

Several researchers have extended the Nusselt analysis for pure vapor condensation on a vertical plate in a stagnant gas atmosphere, providing valuable insights into the phenomena occurring in vertical tube geometries such as the Isolation Condenser and PCCS heat exchangers.

The first series of analyses were for condensation with noncondensables on a flat plat with a laminar condensate film. Sparrow and Eckert [38] performed a boundary layer analysis which showed the large degrading effect of a noncondensable gas on heat transfer rates for a laminar condensate film. Comparing the results with experimental data, they noted that free convection to/from the condensation surface plays an important role when noncondensable gases are present although this effect was not included in their analysis. Also using boundary layer analysis, Koh *et al.* [39] considered the shear forces at the liquid–vapor interface due to induced motions of the vapor. The shear effect on heat transfer rates was small for Prandtl numbers of one or greater but large for Prandtl numbers in the range of liquid metals. Sparrow and Lin [15] derived rigorous sets of conservation equations for the liquid and vapor boundary layers that considered the liquid shear forces on the vapor and free convection. Minkowycz and Sparrow [40] investigated the interfacial resistance phenomena in greater deal and found that formulation of the case with interfacial shear does not yield a similarity solution. This analysis considered the roles of temperature in the diffusion equation and variable properties in the boundary layers. Rose [41] provided an algebraic relation between the heat and mass transfer properties and obtained good agreement with the boundary layer analyses and experimental data.

Several of the free convection boundary layer analyses underpredicted experimentally measured heat transfer coefficients. The validity of the boundary layer analyses was demonstrated by Al-Diwany and Rose [42] who performed experiments with particular care to avoid unintentional forced convection flows and obtained good agreement between analysis and experiment.

Sparrow *et al.* [43] solved the boundary layer equations for the vapor/noncondensable gas condensation with forced convection and a laminar condensate film. The vapor energy equation was not considered due to relatively small energy flow in the vapor and a numerical procedure was employed to solve the equations. The heat transfer rate in a forced convective flow was shown to be less sensitive to the noncondensable gas concentration than in a free convection situation. Denny *et al.* [16] observed nonsimilarity of the situation and solved the equations with a finite difference technique.

While the boundary layer approach has been solved for laminar flow conditions and great insight into the phenomena was obtained, incorporation of the boundary layer equations into reactor safety analysis is not practical. A method based on the heat and mass transfer analogy allows for simpler formulations while retaining good accuracy.

Colburn and Hougen [44] first applied the heat and mass transfer analogy to vapor condensation using the stagnant film model for vapor–air mixtures. The overall heat transfer conductance between the condenser tube wall and vapor–air mixture is the sum of the conductance of the condensate film and the vapor/noncondensable gas boundary layer. The heat transfer through the gas boundary layer includes sensible and latent heat transfer. The latent heat transfer is evaluated by using the stagnant film model combined with the heat and mass transfer analogy.

Peterson's diffusion theory [45] significantly improved modeling capabilities by providing a theoretical analysis method compatible with reactor safety codes. Peterson introduced the concept of a "condensation thermal conductivity" in developing the diffusion layer model, wherein the overall gas-side conductivity is formulated as a combination of the condensation thermal conductivity for latent heat transfer and the standard thermal conductivity for sensible heat transfer. Treating the gas film as a diffusion layer, Peterson [45, 24] derived a condensation thermal conductivity in the diffusion layer model on a molar basis:

$$k_{\mathrm{c}}^{\mathrm{p}} = \frac{1}{\phi} \frac{h'_{\mathrm{fg}} h_{\mathrm{fg}} P D M_{\mathrm{v}}^2}{R^2 \overline{T}^3} \tag{9}$$

where

$$\phi = \frac{\ln[(1 - x_{\mathrm{gb}})/(1 - x_{\mathrm{gi}})]}{\ln(x_{\mathrm{gi}} / x_{\mathrm{gb}})} \tag{10}$$

$k_c^P$ is the condensation thermal conductivity, $P$ is pressure, $D$ is the diffusion coefficient, $M$ is the molecular weight and $x$ is molar fraction. The total heat transfer coefficient is conveniently given by combining the parallel gas-side latent heat transfer and sensible heat transfer coefficients in series with the liquid-side heat transfer coefficient:

$$\frac{1}{h_t} = \frac{1}{h_c + h_s} + \frac{1}{h_l} = \frac{L}{Sh \cdot k_c^P + Nu \cdot k} + \frac{\delta_l}{k_l} \quad (11)$$

Peterson's diffusion theory has been incorporated into several reactor safety codes such as TRACE and CONTAIN. The model is successful in predicting condensation heat transfer rates in vertical tube experiments prototypic of the PCCS and other vertical tube heat exchangers with a trend to slightly underpredict the data. Since the underprediction is not extreme and in the conservative direction, the results are generally acceptable. There have been, however, several attempts to modify the theory formulation.

The diffusion layer model for condensation heat transfer of vapor with noncondensable gases was originally derived on a molar basis. Developed from an approximate formulation of mass diffusion, the effect of variable vapor–gas mixture molecular weights across the diffusion layer on mass diffusion was neglected. This is valid for a gas with a molecular weight close to that of the vapor or for low vapor mass transfer rates, however a large error may be incurred if a steep gradient in the gas concentration exists across the diffusion layer. Liao and Vierow [46] show from the kinetic theory of gases that Fick's law of diffusion should be expressed on a mass basis rather than on a molar basis. A generalized diffusion layer model was derived on a mass basis with an exact formulation of mass diffusion. The generalized model considers the effect of variable mixture molecular weights across the diffusion layer on mass diffusion and fog formation effects on sensible heat. Under certain limiting conditions, the generalized model reduces to that developed by Peterson.

The condensation thermal conductivity derived on the mass basis was recast to a form comparable to that derived on a molar basis by using the Clausius–Clapeyron equation and the ideal gas law.

$$k_c = \frac{\phi_2}{\phi_1} \frac{h_{fg}' h_{fg} PDM_v M_g}{R^2 \overline{T}^3} \quad (12)$$

where

$$\phi_1 = \frac{\ln[(1-m_{gb})/(1-m_{gi})]}{\ln(m_{gi}/m_{gb})} \quad (13)$$

and

$$\phi_2 = \frac{\overline{M}_m^2}{M_{mb} M_{mi}} \quad (14)$$

$\phi_1$in eqn. (13) is analogous to $\phi$ in eqn. (10) and takes into account the effects of noncondensable gas and suction on vapor condensation. $\phi_2$ in eqn. (14) is a new factor, which takes into account the effects of variable mixture molecular weights on vapor diffusion resulting from large concentration differences through the diffusion layer. Comparisons with a variety of experimental data show that the generalized diffusion layer model were shown to better predict the data than molar-based diffusion layer models. The generalized diffusion layer model is currently being implemented and tested in reactor safety codes.

### 4.4 Condensation in horizontal tubes with steam/noncondensable gas inflow at one end and condensate draining at the other end

Similar to condensation in a vertical tube, condensation in a horizontal tube is of the film condensation mode except very close to the steam inlet where the droplet mode may be observed. The tube orientation, however, introduces complications that render analysis of horizontal tube condensation far more than a mere extension of vertical tube analysis. First, while annular flow occurs in vertical tubes, the flow regime in horizontal tubes depends on the vapor–film shear and the peripheral and axial location. Therefore a methodology for flow regime determination must be added. Second, because the condensate film thickness varies with peripheral location, the fluid and heat transfer phenomena are highly three-dimensional. The conservation equations for the boundary layer analysis should be expanded to include the peripheral coordinate. Third, the interfacial phenomena including film surface instabilities and droplet entrainment may play a larger and different role than in vertical tubes. Models should be added for these phenomena.

The method for flow regime determination is discussed first because selection of the heat transfer models depends on the local flow regime. Wu and Vierow [47] and Thome *et al.* [48] provide recent reviews of methods for flow regime determination in horizontal tubes with condensation. In addition to investigations for nuclear safety systems, Thome *et al.* are conducting experimental and theoretical work to better understand refrigeration and other industrial systems. Thome *et al.* [48] presented a flow regime mapping based on estimations of the "logarithmic mean void fraction", which lead to liquid and vapor mass velocities and a flow regime determination. A validated heat transfer model for condensation inside horizontal plain tubes was developed using this flow regime mapping [49].

Gravitational and interfacial shear forces determine the local flow regime. The flow regimes are commonly classified into three categories, as reviewed by Wu [25]:

- annular and annular-mist flow, which are driven primarily by shear forces;
- stratified flow, wavy flow and slug/plug flow, which are gravity driven;
- wavy-annular flow, which have strong contributions from both the gravity and shear forces.

Wu [25] adopted flow regime criteria from Jaster and Kosky [50] and Soliman [51–53] for theoretical modeling of steam in larger diameter tubes representing the PCCS tubes. Both criteria were developed for condensation inside tubes and apply to larger tube sizes than those used by Thome. Jaster and Kosky expressed flow regime criteria for annular flow-to-transition flow and transition flow-to-stratified flow as a ratio of the axial shear force to the gravitational force. Soliman predicted the mist-to-annular flow transition based on a modified Weber number that represents a balance between forces inhibiting liquid entrainment and vapor inertia enhancing mist formation. For the annular-to-wavy flow transition, Soliman [51] adopted a modified Froude number, which is a balance between inertial and gravitational forces on the liquid film.

Following flow regime identification, the local heat transfer calculation may be formulated. The pure vapor formulations and vapor/noncondensable gas formulations are presented below.

Chato [54] proposed a Nusselt number formulation for condensation of pure vapor in low-velocity stratified flow. The Nusselt number has the same functional dependencies as Nusselt's formula for condensation on flat plates but with a lower coefficient. For higher velocities in stratified flow, Rosson and Meyers [55] showed that the Nusselt number at the top of the tube has the same functional dependencies with a $Re^n$ multiplier to account for vapor shear effects. The Nusselt number at the bottom of the tube depends on the *Re*, *Pr* and the Lockhart–Martinelli parameter *X*.

For shear-driven, pure-vapor, annular flow situations, Cavallini and Zecchin [56], Shah [57] and Dobson and Chato [58] obtained Nusselt number correlations in the form of the Dittus–Boelter correlation times a two-phase multiplier. These are the most widely used correlations for reactor safety analysis, although approaches based on evaluation of shear on the laminar sublayer and interfacial surface have also been developed. Levich [59] adopted a boundary layer approach based on a laminar sublayer at the wall and a second viscous sublayer near the liquid–gas interface in which surface tension arising from turbulent deformation damps the turbulent interfacial fragmentation.

Thome *et al.* [49] present a heat transfer calculation method for pure vapor condensation that covers all of the flow regimes expected in reactor applications and is in a format compatible with reactor safety codes.

To account for the noncondensable gas effect on reducing heat transfer efficiency, several reactor safety codes employ Peterson's [45] diffusion theory. The boundary layer work presented in earlier sections was formulated for flat surfaces and the available empirical correlations are restricted to the ranges tested.

In summary, numerous models and correlations exist for various aspects of the problem; however the call remains for a robust and verified model applicable to nuclear safety system scenarios.

## 4.5 Condensation in large water pools

The phenomena associated with vapor bubble condensation in large water pools have been studied most recently by Xu [26]. Xu presents an experimental investigation of the direct condensation of bubbles in a subcooled water pool, with and without noncondensable gas. Although these phenomena are important for evaluation of containment pressures, the discussion of analysis techniques herein is limited to condensation in the film mode.

### 4.6 Condensation on large vertical walls

The heat and mass transfer analogy is widely employed by reactor safety codes for evaluation of condensation heat transfer rates on containment walls. To evaluate the heat transfer rates, correlations are implemented for natural convection, forced convection and mixed convection. Invoking the heat and mass transfer analogy, the Sherwood number is assumed equal to the Nusselt number for the mass transfer rate calculation. The degrading effect of noncondensable gases on the condensation heat transfer rate is accounted for by incorporating diffusion theory to provide the appropriate log mean pressure in the mass transfer equations. Peterson [60] demonstrated the improvement of this diffusion theory representation over earlier methods that employed correlations based on the Uchida data.

To complete the models, CONTAIN [61], MELCOR [62] and COMMIX [57] have adopted the same film tracking model to calculate the condensate film thickness on containment walls. Summarized by Sha [57], the film tracking model assumes filmwise condensation and computes the film thickness, film temperature and velocity. Because the film is generally $<1$ mm in thickness, laminar flow predominates and the classic Nusselt analysis is the basis for the formulation of the boundary layer equations.

## 5 Analysis challenges

Remaining challenges for analysis of passive systems driven by condensation heat transfer arise from incomplete understanding of phenomena, uncertainty in boundary conditions and numerical issues. These issues impact analysis at the fundamental heat transfer level and at the reactor system level.

Major challenges include:

1. Better understanding of condensate film behavior in reflux and horizontal condensers, including both film sublayers and the film surface interactions. Boundary layer analysis may be a useful tool for this purpose.
2. Peripheral characteristics of condensate film behavior in horizontal and inclined tubes. Additional experimental study with film visualizations would be beneficial.
3. Mixing in large volumes to predict noncondensable gas concentrations at condenser tube inlets.
4. Accurate calculations over a long duration.
5. Accounting for the tighter coupling of the primary loop pressure with the containment pressure in passive plants than in active plants.

## 6 Summary

Passive systems driven by condensation heat transfer represent a return to simplicity while achieving significant strides in reactor safety. These passive condensation heat transfer systems are reliable and inherently safe because they rely on basic physical forces such as gravity and small pressure differences to transport energy and maintain the plant within design specifications. While significant advancements have been made, challenges remain particularly with regard to accurate, detailed evaluations of heat transfer rates.

The numerous roles for passive condensation heat transfer systems have been summarized in this chapter with a focus on U.S.-designed light water reactors and U.S. reactor safety codes.

The state-of-the-art in analysis methods was presented along with analysis challenges at the phenomenological level and system level. Suggestions with respect to advancement of the condensation modeling were provided. In particular, boundary layer analysis was shown to have provided great insight into the details of the phenomena, particularly for laminar condensate films. Extension of the boundary layer analysis for turbulent flows under additional boundary conditions and geometries is desirable. Secondly, condensate film behavior in inclined and horizontal tubes is not well understood under nuclear safety system conditions. Additional experimental studies could clarify the phenomena and provide a theoretical basis for modeling.

## Acknowledgments

Several of the research efforts reported in this paper were performed by Dr. Tiejun Wu and Dr. Yehong Liao. The author is indebted to their hard work which made these projects successful.

## Nomenclature

### Symbols

| | |
|---|---|
| $D$ | diffusion coefficient |
| $d$ | diameter |
| $F$ | multiplication factor |
| $g$ | gravity vector |
| $h$ | local heat transfer coefficient |
| $h_c$ | condensation heat transfer coefficient |
| $h_{exp}$ | experimentally measured heat transfer coefficient |
| $h_f$ | Nusselt heat transfer coefficient, film heat transfer coefficient |
| $h_{fg}$ | latent heat of vaporization |
| $h_s$ | sensible heat transfer coefficient |
| $j$ | superficial velocity |

| | |
|---|---|
| *Ja* | Jacob number |
| *k* | thermal conductivity |
| *L* | characteristic length |
| *M* | molecular weight |
| *m* | mass fraction |
| *Ma* | noncondensable gas mass fraction |
| *Nu* | Nusselt number |
| *P* | pressure |
| *Pr* | Prandtl number |
| *R* | universal gas constant |
| *r* | radial coordinate |
| *Re* | Reynolds number |
| *Sh* | Sherwood number |
| *T* | temperature |
| *u* | velocity |
| *W* | noncondensable gas mass fraction |
| *x* | mole fraction |
| *y* | coordinate normal to the condensation surface |
| *z* | coordinate parallel to the condensation surface |

## Greek symbols

| | |
|---|---|
| $\delta$ | condensate film thickness |
| $\theta$ | circumferential coordinate |
| $\mu$ | dynamic viscosity |
| $\nu$ | kinematic viscosity |
| $\rho$ | mass density |
| $\tau$ | shear stress |

## Subscripts

| | |
|---|---|
| air | air |
| corr | correlation |
| g | gas |
| gb | gas in the bulk |
| gi | gas at the interface |
| i | interface |
| l | liquid |
| max | maximum |
| sat | saturation |
| surf | surface |
| v | vapor |
| wall | wall |

## References


[1] Yadigaroglu, G., Passive core and containment cooling systems: characteristics and state of the art. *Proc. of the Ninth International Topical Meeting on Nuclear Reactor Thermal Hydraulics (NURETH-9)*, San Francisco, CA, October, 1999.

[2] Juhn, P.E., Kupitz, J., Cleveland, J., Cho, B. & Lyon, R.B., IAEA activities on passive safety systems and overview of international development. *Nuclear Engineering and Design*, **201**, pp. 41–59, 2000.

[3] Vierow, K., Townsend, H.E., Fitch, J.R., Andersen, J.G.M., Alamgir, Md. & Schrock, V.E., BWR passive containment cooling system by condensation-driven natural circulation. *Proc. of the First ASME/JSME Joint International Conference on Nuclear Engineering – 1 (ICONE-1)*, Tokyo, Japan, pp. 289–294, 1991.

[4] Shiralkar, B.S., Alamgir, M. & Andersen, J.G.M., Thermal hydraulic aspects of the SBWR design. *Nuclear Engineering and Design*, **144**, pp. 213–222, 1993.

[5] Challberg, R.C., Cheung, Y.K., Khorana, S.S. & Upton, H.A., ESBWR evolution of passive features. *Proc. of the Sixth International Conference on Nuclear Engineering (ICONE-6)*, San Diego, USA, pp. 1–10, 1998.

[6] Schulz, T., Westinghouse AP1000 advanced passive plant. *Nuclear Engineering and Design*, **236**, pp. 1547–1557, 2006.

[7] U.S. Nuclear Regulatory Commission, Emergency core cooling system (Chapter 6.3). *Standard Review Plan*, NUREG-0800, Office of Nuclear Reactor Regulation, Draft Rev. 3, 1996. Available at http://www.nrc.gov/reading-rm/doc-collections/nuregs/staff/sr0800/ch6/.

[8] Leaver, D.E.W. & Additon, S.L., ALWR utility requirements document containment performance requirements. *Nuclear Engineering and Design*, **145**, pp. 307–319, 1993.

[9] Advanced Light Water Reactor Utility Requirements Document, ALWR Passive Plant, Vol. III, Rev. 8, March 1999 (as referenced by T. Schulz, 2006).

[10] GE Nuclear Energy, *ESBWR Design Control Document, Tier 2, Chapter 6, Engineered Safety Features*, 26A6642AT, Rev. 01, January 2006, available at the U.S. NRC Electronic Reading room, accession number ML060520258.

[11] Naff, S.A. *et al.*, *Thermal-hydraulic Processes during Reduced Inventory Operation with Loss of Residual Heat Removal*, NUREG/CR-5855, EGG-2671, 1992.

[12] General Electric, Nuclear Energy Group, *BWR/6 General Description of a Boiling Water Reactor*, San Jose, CA, Rev. September 1980.

[13] GE Nuclear Energy, *ESBWR Design Control Document, Tier 2, Chapter 5, Reactor Coolant System and Connected Systems*, 26A6642AR, Rev. 01, January 2006, available at the U.S. NRC Electronic Reading room, accession number ML060520257.

[14] Lahey, R.T., Jr. & Moody, F.J., *The Thermal Hydraulics of a Boiling Water Nuclear Reactor*, American Nuclear Society: La Grange Park, IL, 1993.

[15] Sparrow, E.M. & Lin, S.H., Condensation heat transfer in the presence of a noncondensable gas. *Journal of Heat Transfer*, **86**, pp. 430–436, August 1964.

[16] Denny, V.E., Mills, A.F. & Jusionis, V.J., Laminar film condensation from a steam-air mixture undergoing forced flow down a vertical surface. *Journal of Heat Transfer*, **93**, pp. 297–304, 1971.

[17] Kakac, S. & Liu, H., Condensers and evaporators. *Heat Exchangers: Selection, Rating and Thermal Design*, CRC Press: New York, pp. 355–400, 1998.

[18] Nakamura, H., Single U-tube testing and RELAP5 code analysis of PCCS with horizontal heat exchanger. *Proc. of the NTHAS2: Second Japan-Korea Symposium on Nuclear Thermal Hydraulics and Safety*, Fukuoka, Japan, pp. 336–343, 2000.

[19] Tujikura, Y., Oshiba, T., Kijima, K. & Tabuchi, K., Development of passive safety systems for next generation PWR in Japan. *Nuclear Engineering and Design*, **201**, pp. 61–70, 2000.

[20] Ueno, T. *et al.*, Study on core cooling of hybrid safety system for next-generation PWR during LOCA. *Proc. of the Third ASME/JSME Joint International Conference on Nuclear Engineering (ICONE-3)*, Kyoto, Japan, Vol. 2, pp. 963–969, 1995.

[21] Rohsenow, W.M., Film condensation (Section 12, Part A). *Handbook of Heat Transfer*, McGraw-Hill: New York, 1973.

[22] Vierow, K., *Behavior of Steam-Air Systems Condensing in Concurrent Vertical Downflow*, MS Thesis, University of California, Berkeley, California, 1990.

[23] Blangetti, F., Krebs, R. & Schlunder, E.U., Condensation in vertical tubes – experimental results and modelling. *Chemical Engineering Fundamentals*, **1(2)**, pp. 20–42, 1982.

[24] Peterson, P.F., Diffusion layer modeling for condensation with multicomponent noncondensable gases. *Journal of Heat Transfer*, **122**, pp. 716–720, 2000.

[25] Wu, T., *Horizontal In-tube Condensation in the Presence of a Noncondensable Gas*, PhD Thesis, Purdue University, West Lafayette, IN, 2005.

[26] Xu, Y., *Direct Contact Condensation with and without Noncondensable Gas in a Water Pool*, PhD Thesis, Purdue University, West Lafayette, IN, 2004.

[27] Rohsenow, W.M., Heat transfer and temperature distribution in a laminar film. *Trans. ASME*, **78**, pp. 1645–1648, 1956.

[28] Moon, Y.M., No, H.C., Park, H.S. & Bang, Y.S., *Assessment of RELAP5/MOD3.2 for Reflux Condensation Experiment*, NUREG/IA-0181, April. 2000.

[29] Park, H.S., No, H.C. & Bang, Y.S., Analysis of experiments for in-tube steam condensation in the presence of noncondensable gases at a low pressure using the RELAP5/MOD3.2 code modified with a non-iterative condensation model. *Nuclear Engineering and Design*, **225**, pp. 173–190, 2003.

[30] Queral, C., Gonzalez, I. & Exposito, A., Analysis of the cooling capability of steam generators during loss of residual heat removal system at midloop operation with closed primary system. *Annals of Nuclear Energy*, **33**, pp. 1102–1115, 2006.

[31] Chun, M.-H. *et al.*, An experimental investigation of reflux condensation phenomena in multiple U-tubes with and without noncondensible gas. *Proc. of the IMECE' 01, 2001 International Mechanical Engineering Congress and Exposition*, New York, NY, 2001.

[32] Vierow, K., Nagae, T. & Wu, T., Experimental investigation of reflux condensation heat transfer in PWR steam generator tubes in the presence of noncondensible gases. *Proc. of the 10th International Topical Meeting on Nuclear Reactor Thermal Hydraulics (NURETH-10)*, Seoul, Korea, October 5–9, 2003.

[33] Nagae, T., Murase, M., Wu, T. & Vierow, K., Evaluation of reflux condensation heat transfer of steam-air mixtures under gas-liquid countercurrent flow in a vertical tube. *Journal of Nuclear Science and Technology*, **42(1)**, pp. 50–57, 2005.

[34] Tanrikut, A. & Yesin, O., Experimental research on in-tube condensation under steady-state and transient conditions. *Nuclear Technology*, **149**, pp. 88–100, 2005.

[35] Lee, K.W., No, H.C., Chu, I.C., Moon, Y.M. & Chun, M.H., Local heat transfer during reflux condensation mode in a U-tube with and without noncondensible gases. *International Journal of Heat and Mass Transfer*, **49**, pp. 1813–1819, 2006.

[36] Gross, U. & Philipp, Ch., 'Conjugated shear stress and Prandtl number effects on reflux condensation heat transfer inside a vertical tube. *International Journal of Heat Mass Transfer*, **49**, pp. 144–153, 2006.

[37] Vierow, K. & Schrock, V.E., Condensation in a natural circulation loop with noncondensable gases. Part I – Heat transfer. *Proc. of the International Conference on Multiphase Flow' 91 (Japanese Society of Multiphase Flow, ANS)*, Tsukuba, Japan, pp. 183–186, September 1991.

[38] Sparrow, E.M. & Eckert, E.R.G., Effects of superheated vapor and noncondensable gases on laminar film condensation. *AIChE Journal*, **7(3)**, pp. 473–477, 1961.

[39] Koh, J.C.Y., Sparrow, E.M. & Hartnett, J.P., The two phase boundary layer in laminar film condensation. *International Journal of Heat and Mass Transfer*, **2**, pp. 69–82, 1961.

[40] Minkowycz, W.J. & Sparrow, E.M., Condensation heat transfer in the presence of noncondensables, interfacial resistance, superheating, variable properties, and diffusion. *International Journal of Heat and Mass Transfer*, **9**, pp. 1125–1144, 1966.

[41] Rose, J.W., Condensation of a vapour in the presence of a non-condensing gas. *International Journal of Heat and Mass Transfer*, **12**, pp. 233–237, 1969.

[42] Al-Diwany, H.K. & Rose, J.W., Free convection film condensation of steam in the presence of non-condensing gases. *International Journal of Heat and Mass Transfer*, **16**, pp. 1359–1369, 1973.

[43] Sparrow, E.M., Minkowycz, W.J. & Saddy, M., Forced convection condensation in the presence of noncondensables and interfacial resistance. *International Journal of Heat and Mass Transfer*, **10**, pp. 1829–1845, 1967.

[44] Colburn, A.P. & Hougen, O.A., Design of cooler condensers from mixtures of vapors with noncondensing gases. *Ind. Eng. Chem.*, **26**, pp. 1178–1182, 1934.

[45] Peterson, P.F., Schrock, V.E. & Kageyama, T., Diffusion layer theory for turbulent vapor condensation with noncondensable gases. *Journal of Heat Transfer*, **115**, pp. 998–1003, 1993.

[46] Liao, Y. & Vierow, K., A generalized diffusion layer model for condensation of vapor with non-condensable gases. *Transactions of the ASME, Journal of Heat Transfer*, **129(8)**, pp. 988–944, 2007.

[47] Wu, T. & Vierow, K., Local heat transfer measurements of steam/air mixtures in horizontal condenser tubes. *International Journal of Heat and Mass Transfer*, **49**, pp. 2491–2501, 2006.

[48] Thome, J.R., El Hajal, J. & Cavallini, A., Condensation in horizontal tubes. Part 1: Two-phase flow pattern map. *International Journal of Heat and Mass Transfer*, **46**, pp. 3349–3363, 2003.

[49] Thome, J.R., El Hajal, J. & Cavallini, A., Condensation in horizontal tubes, Part 2: New heat transfer model based on flow regimes. *International Journal of Heat and Mass Transfer*, **46**, pp. 3365–3387, 2003.

[50] Jaster, H. & Kosky, P.G., Condensation heat transfer in a mixed flow regime. *International Journal of Heat and Mass Transfer*, **19**, pp. 95–99, 1976.

[51] Soliman, H.M., A general heat transfer correlation for annular flow condensation. *Journal of Heat Transfer*, pp. 267–276, 1968.

[52] Soliman, H.M., On the annular-to-wavy flow pattern transition during condensation inside horizontal tubes. *Canadian Journal of Chemical Engineering*, **60**, pp. 475–481, August 1982.

[53] Soliman, H.M., The mist-annular transition during condensation and its influence on the heat transfer mechanism. *International Journal of Multiphase Flow*, **12(2)**, pp. 277–288, 1986.

[54] Chato, J.C., Laminar condensation inside horizontal and inclined tubes. *ASHRAE Journal*, **4**, p. 52, 1962.

[55] Rosson, H.F. & Meyers, J.A., Point values of condensing film coefficients inside a horizontal tube. *Chemical Engineering Progress Symposium Series*, **61(59)**, pp. 190–199, 1965.

[56] Cavallini, A. & Zecchin, R., A dimensionless correlation for heat transfer in forced convection condensation. *Proc. of the Fifth International Heat Transfer Conference*, September 3–7, pp. 309–313, 1974.

[57] Shah, M.M., A general correlation for heat transfer during film condensation inside pipes. *International. Journal of Heat and Mass Transfer*, **22**, pp. 547–556, 1979.

[58] Dobson, M.K. & Chato, J.C., Condensation in smooth horizontal tubes. *Journal of Heat Transfer*, **120**, pp. 193–213, 1998.

[59] Levich, V.G., Chapter XII. *Physicochemical Hydrodynamics*, Prentice-Hall Inc.: Englewood Cliffs, NJ, 1962.

[60] Peterson, P.F., Theoretical basis for the Uchida correlation for condensation in reactor containments. *Nuclear Engineering and Design*, **162**, pp. 301–306, 1996.

[61] Murata, K.K. *et al.*, *User's Manual for CONTAIN 1.1: A Computer Code for Severe Nuclear Reactor Accident Containment Analysis*, Sandia National Laboratories Report SAND87-2309, April, 1993.

[62] Gauntt, R.O. *et al.*, *MELCOR Computer Code Manuals, Vol. 2: Reference Manuals, Version 1.8.5*, Prepared by Sandia National Laboratories for the U.S. Nuclear Regulatory Commission, Office of Nuclear Regulatory Research, NUREG/CR-6119, Vol. 2, Rev. 2, May, 2000.

# CHAPTER 6

# Modern CFD application on aerothermal engineering aspects of natural draft cooling towers

D. Bohn[1] & K. Kusterer[2]
[1]*Institute of Steam and Gas Turbines, RWTH Aachen University, Aachen, Germany.*
[2]*B&B-AGEMA GmbH, Aachen, Germany.*

## Abstract

Natural draft cooling towers are a very efficient way to provide the cold end of thermal power plants with a steam turbine cycle. The cooling tower shape, details of construction and arrangement and type of cooling fill have a significant influence on the cooling tower efficiency, in particular, on the cold water temperature that can be reached. Furthermore, influences of cross wind situations on the performance of the cooling tower are of high interest. Modern computational tools for three-dimensional flow simulations (CFD) can be coupled by local heat and mass transfer calculation schemes with the cooling water film calculation and thus allow a comprehensive numerical investigation of the aerodynamic and thermodynamic behaviour of natural draft cooling towers under specific operating conditions. After validating the numerical approach, the numerical method has been used for determination of the influences of fill types, of the inner rim structure at the tower outlet and of an additional flue gas discharge on the cooling tower performance. Furthermore, numerical calculations of several cross wind operating conditions have been performed. The numerical simulations show clearly that the equipment of the cooling tower with a modern fill-type achieve higher efficiencies. Cross wind situations lead to a significant reduction in the cooling tower efficiency and thus to higher cold water temperatures. A major result of the simulations is that under unfavourable circumstances, a periodic cold air ingestion at the tower outlet might appear that leads to unstable operating conditions with reduced efficiencies. The inner rim structure at the tower outlet has a stabilizing effect in cross wind situations.

# 1 Introduction

Cooling towers provide the cold end for steam turbine cycles mainly used in thermal power plants for producing electricity. The objective of the cooling tower and condenser operation is to condense steam exiting from the low-pressure turbines at a pressure, which should be as low as possible to maximize the power output from the plant. The condenser pressure is thermodynamically linked to the cooling water temperature offered by the cooling tower. The closer it is to ambient temperature, the better it is for the cycle performance.

High efficiencies are typically achieved with wet cooling towers. In this design, cooling water heated in the condenser is cooled by convection and evaporation in contact with ambient air. The contact time and area between air and water are increased by spraying the water over a fill (a grid of bars or plates) and by passing air through the fill. The water to be cooled then trickles from the top through the cooling fills. The water is cooled in two ways when the cooling air is passing by. A small amount of water evaporates so that evaporative heat is extracted from the cooling water flow. Furthermore, heat is transferred from the water to the cooling air by convection.

Most efficient designs, although comparatively expensive, are based on buoyancy driven air flow. Natural draft cooling towers exist in two flow arrangements, a counterflow and a crossflow arrangement (see Fig. 1). In mechanical draft cooling towers, a fan provides the necessary air amount to the fill. Due to higher air velocities the heat transfer is enhanced and, thus, mechanical draft cooling towers are much smaller in size than natural draft cooling towers, but the necessary fan power reduces the overall power output of the plant. Modern natural draft cooling towers can be as high as 200 m or even more with base diameters of approximately 150 m. Figure 2 shows a typical natural draft cooling tower with a modern design including an additional flue gas discharge in the centre of the tower.

Natural draft cooling towers are very sensitive to special operation conditions, e.g. additional flue gas discharge and environmental conditions, e.g. cross wind effects. The incoming air in case of a cross wind situation is unevenly distributed to the fill and the buoyancy driven air flow in the tower might be affected by cold

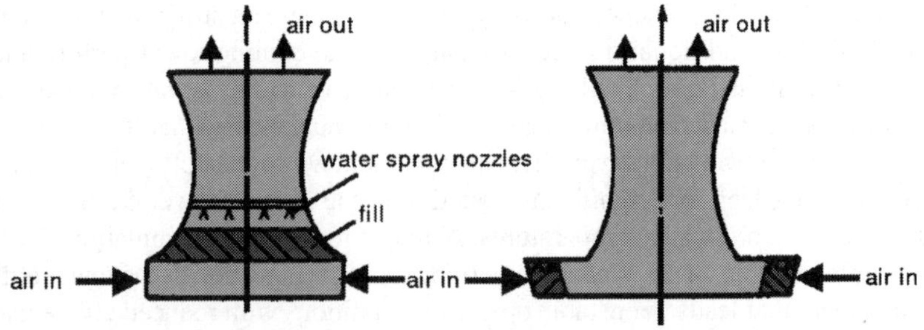


Figure 1: Natural draft cooling towers with counterflow (left) and crossflow (right) arrangement.

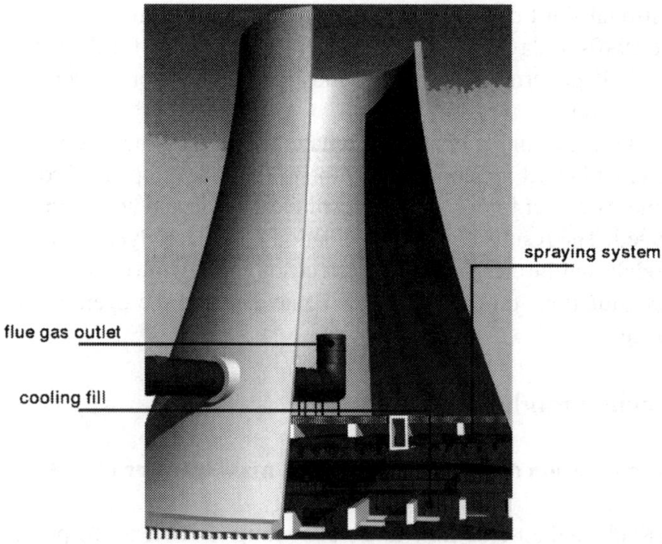


Figure 2: Modern natural draft cooling tower (counterflow arrangement).

air entering the tower from the top. The consequence of both influences is a distorted heat transfer distribution across the cooling tower cross section, leading to an overall reduction in the cooling tower efficiency. As a result, the cold water temperature increases and thus the efficiency and performance of the complete steam cycle are reduced.

Further improvements of natural draft cooling towers that lead to reduced cold water temperatures need not only a precise knowledge of the local heat transfer and flow conditions but also a detailed modelling of the interactions of heat- and mass transfer and the cooling tower aerodynamics with a comprehensive numerical approach. The numerical approach should combine the abilities of modern computational fluid dynamics (CFD) tools for calculation of three-dimensional steady and unsteady flow conditions in a cooling tower and the empirical heat and mass transfer correlations for special fill arrangements in the cooling tower as provided by the manufacturers.

Early collections of correlations and investigations on typical influences of parameters on the heat and mass transfer in cooling towers can be found in [1, 2]. With respect to the numerical calculation, an early approach that treats the air flow to be two-dimensional, while the water flow is considered to be one-dimensional, has been presented in the 1980s by Majumdar *et al.* [3, 4]. They have applied the numerical scheme to several examples for mechanical and natural draft cooling towers of both counterflow and crossflow designs. Major limitations in that time can be related to the lack of computational performance that did not allow using three-dimensional flow solvers and high-resolution numerical grids. Due to the significant progress in commercial software development (computational fluid dynamics) and the great increase in computational abilities in the 1990s, today tools and computers are available for the comprehensive solution of all governing equations and their interactions, even for complex geometric models.

In this chapter, example results for a natural draft cooling tower application of commercial CFD tool [5] combined with specially developed subroutines on the heat and mass transfer for local coupling of the air-side and water-side calculations are presented. The results contain a validation of the aerodynamic and thermodynamic models and further present the effects of cooling tower design features and cross wind situations on the local flow behaviour and the overall cooling tower performance.

## 2 Numerical modelling

### 2.1 Implementation of a heat transfer and mass transfer model

The water-side and air-side flow in a wet cooling tower is a two-phase flow with complex heat transfer and mass transfer conditions between both fluids. Thus, the complete numerical simulation of all relevant conservation equations for the three-dimensional flow in cooling towers is still beyond the economical application for industrial use with respect to time and cost efforts. Therefore, a simplification is introduced by handling the wet air as a one-phase mixture of fluids consisting of the two components (1) dry air and (2) water steam. This is necessary to take into account the change in density due to the local wetness. In the following section, this one-phase model fluid is mentioned as "air" or "air-side."

The energy balance and mass balance between the cooling water film and the air can be established. Transferred heat and mass flow values are implemented as additional source terms to the governing equations for the calculations on the cooling film side and wet air flow. As the flow of the cooling water cannot be taken into account additionally to the air flow directly in the finite volume cells of the computational grid, information on inlet and outlet temperatures and the local mass flow for the cooling water flow have to be stored separately for every cell and have to be exchanged with the neighbouring cells in the flow direction of the water.

Figure 3 shows the important physical parameters of the mass and energy balance for the cooling water element and the air-side finite volume cell. For the energy balance of the cooling water element, the evaporative heat flux due to the evaporation process and the convective heat flux have to be taken into account. The local gradient of the temperature between the cooling water film and the air-side is determining the local convective heat flux, whereas the local gradient between the water content for the saturated moist air and the water content for the moist air is determining the local evaporation rate and, thus, the evaporative heat flux. For the convective heat flux calculation, the averaged temperature of the inlet and outlet temperature of the cooling water element is used. Furthermore, it is assumed that the local air condition directly in contact with the cooling water film is always

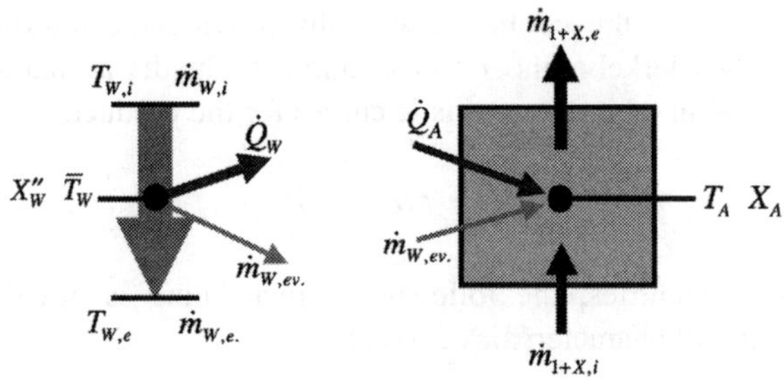


Figure 3: Physical parameters for heat and mass balances at a cooling water element (left) and a finite volume cell of the air-side (right).

saturated. Altogether, the local heat flux transferred from the cooling water element can be calculated as follows:

$$\dot{Q}_{\mathrm{W}} = \alpha \cdot A \cdot \left(\overline{T}_{\mathrm{W}} - T_{\mathrm{A}}\right) + \beta_x \cdot A \cdot \left(X''_{\mathrm{W}} - X_{\mathrm{A}}\right) \cdot \Delta h_{\mathrm{ev}} \qquad (1)$$

In case the air in the finite volume cell is saturated with water as well as the air in contact to the cooling water film, but a temperature gradient between the cooling water film and the air-side is established, the diffusion process continues further despite the saturated air condition due to the gradient of the water content. This leads to a re-condensation of water on the air-side. The re-condensed water and small droplets taken away by mechanical forces are the reason for the visible plume at the cooling tower exit. The re-condensation of steam on the air-side is not part of the used numerical models up to now.

The outlet temperature of the cooling water for the cooling water element can be calculated based on the first law of thermodynamic as following:

$$T_{\mathrm{W,e}} = \frac{\left(\dot{m}_{\mathrm{W,i}} \cdot c_{\mathrm{W}} \cdot T_{\mathrm{W,i}} - \dot{Q}_{\mathrm{W}} - \dot{m}_{\mathrm{W,ev}} \cdot c_{\mathrm{W}} \cdot \overline{T}_{\mathrm{W}}\right)}{\dot{m}_{\mathrm{W,e}} \cdot c_{\mathrm{W}}} \qquad (2)$$

The heat flux into the air-side finite volume cell is mainly determined by the convectively transferred heat:

$$\dot{Q}_{\mathrm{A}} = \alpha \cdot A \cdot \left(\overline{T}_{\mathrm{W}} - T_{\mathrm{A}}\right) + \dot{m}_{\mathrm{W,ev}} \cdot h''_{\mathrm{S}} \qquad (3)$$

## 2.2 Evaporation number and Lewis analogy

The theoretical approach on the calculation of local mass transfer and heat transfer is mainly based on an evaporation number $k_{\mathrm{V}}$ (often also mentioned as Merkel number) for the determination of a mass transfer coefficient and the Lewis analogy for determination of a convective heat transfer coefficient.

The Merkel number is defined:

$$k_{\mathrm{V}} = Me = \int_A \frac{\beta_x \, dA}{\dot{m}_{\mathrm{W}}} \qquad (4)$$

The manufacturers of the cooling tower fills provide the information on the dependency of the Merkel number on the ratio λ of the dry air mass flow and the cooling water flow in their characteristic curves for the product:

$$Me = Me_{(\lambda=1)} \cdot \lambda^m \tag{5}$$

In Anglo-Saxon countries, the following nomenclature is often used for the description of the fill characteristics curves:

$$\frac{KaV}{L} = f(L/G) = f(1/\lambda) \tag{6}$$

Therefore, it is the responsibility of the manufacturer to provide the designer with precise fill characteristics for the thermal performance of the fills based on high-quality measurements within special testing facilities by application of relevant test conditions for real operation.

The convective heat transfer coefficient is calculated based on the Lewis-analogy taking into account a correction for one-directional diffusion at the phase interface [6]:

$$\frac{a}{\beta_x} = c_{pm} \cdot Le^{(1-n)} \cdot \frac{\left(p''_{\mathrm{S,W}} - p_{\mathrm{S,A}}\right)}{\left(p - p''_{\mathrm{S,W}}\right) \cdot \ln\left(p - p_{\mathrm{S,A}} / p - p''_{\mathrm{S,W}}\right)} \tag{7}$$

Here, $c_{\mathrm{pm}}$ is the specific heat capacity of wet air related to the dry air mass:

$$c_{pm} = c_{p,\mathrm{A}} + c_{p,\mathrm{S}} \cdot X_{\mathrm{A}} \tag{8}$$

The Lewis number and the exponent $n$ depend on the flow type and are based on experimental investigations. The determination of the values for the investigated cooling tower is reported in Section 3.2.1.

For cooling tower performance calculations with conventional schemes, it is of great importance to calculate precisely the demand curve of the cooling tower, which depends on the specific design data of the cooling tower. The point of intersection of the demand curve and the cooling fill characteristic curve gives as a result the design value for the ratio λ as shown in Fig. 4. As the precise calculation of the cooling tower demand curve is a very difficult task because of the great number of influence parameters and the complex interaction of the cooling tower aerodynamics and thermodynamics, there is a great uncertainty within this approach.

The great advantage of the presented comprehensive numerical approach is that in a converged solution, the calculated operating point will automatically fulfil the necessary requirements. The inlet air mass flow is the result of the numerical simulation and it depends on the iterated heat transfer (based on the correlations), the pressure drop (based on further correlations), the cooling tower geometry, ambient conditions and inlet water flow conditions. The numerical results provide not only the integral values for transferred heat and the cold water temperature, but also the local data for the flow field, local heat and mass transfer values and local cooling down of the cooling water film. When the numerical model has been validated for a specific cooling

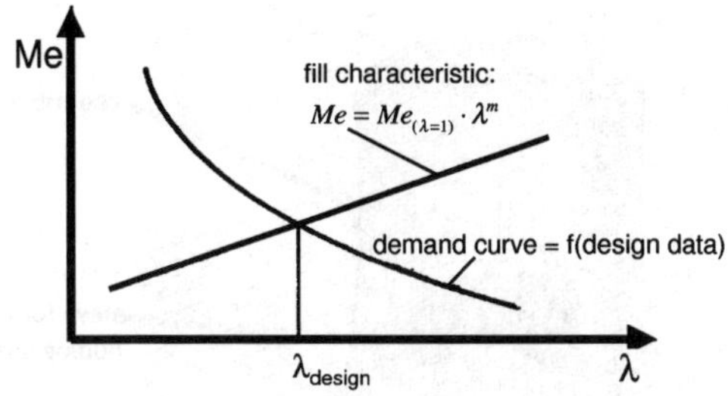


Figure 4: Conventional design approach based on characteristic curve and separated calculation of the cooling tower demand curve.

tower, parametric studies on the variation of the ambient conditions (e.g. ambient temperature, pressure and humidity) from the design values allow a precise determination of the effects for the cooling tower performance, e.g. the calculated cold water temperature. This includes cross wind situations or installation of different fill types or change of the fill arrangements. Another topic of interest might be an inhomogeneous distribution of the sprayed water instead of a homogeneous distribution. Thus, designers and operators of cooling towers get an improved decision basis for new cooling tower arrangements or retrofit measures for existing cooling towers.

# 3 Validation

## 3.1 Aerodynamic code validation

Based on experimental results [7, 8] for the aerodynamics of a cooling tower model, the validation of the basic CFD code has been performed. The original cooling tower has a total height of $H_{tot} = 120$ m, a waist height (smallest diameter) of $H_w = 0.79H_{tot}$, and a base diameter of $D_b = 100$ m. In the experiments, the geometry model of the cooling tower has a scaling factor of 1:275. For aerodynamic investigations of a group arrangement, further experiments include an additional cooling tower model that has been placed at a distance of $d = 1.87D_b$ behind the first model. The contour of the cooling towers has been modelled by a circular arc that can be derived from the given geometric parameters.

The three-dimensional grid consists not only of the cooling tower; it is also necessary to model the environment of the cooling tower with boundaries far away from the cooling tower such that undisturbed pressure distributions can be prescribed, which, under these conditions, have only a negligible influence on the calculated pressure distributions of the cooling tower. Here, all boundaries of the three-dimensional grid in all three coordinate directions have been placed at least at a distance of three times the base diameter away from the cooling towers. Figure 5 shows a close-up of the numerical grid of the cooling tower. The grid for the external flow part consists of approximately 550,000 finite volume cells. The internal flow region of the cooling tower that consists also of the air supply channel and a diffuser has been modelled with approximately 600,000 additional cells. Furthermore, the pressure losses of two sieves, one placed at the diffuser outlet and

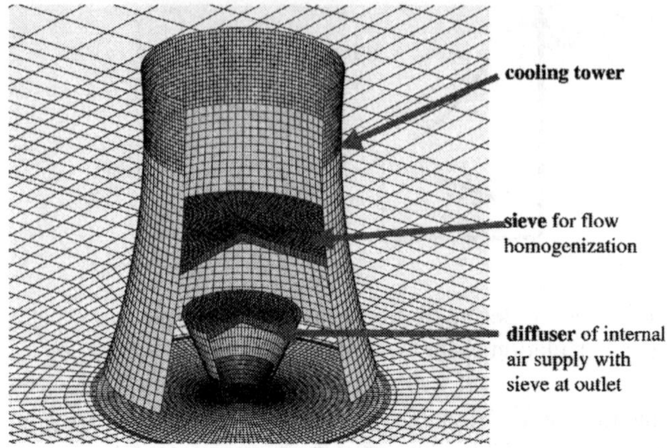


Figure 5: Numerical model and grid of single cooling tower.

the other one placed in front of the tower waist have been taken into account by a pressure loss model based on pressure loss coefficients.

A fully turbulent flow has been reached in the experiments by additional turbulators on the surface of the cooling tower. Thus, the standard-$k$, $\varepsilon$-model with an additional wall roughness parameter has been used in the numerical simulations. Furthermore, calculations with the original geometry size have been performed. In that case, the flow has a very high Re number and a turbulent boundary layer is established. Therefore, the calculations can be performed with a hydraulic even surface, means without an additional roughness parameter.

For the cross wind simulation, a velocity profile has to be prescribed at the inlet boundary, which should be an approximation of the atmospheric boundary layer. This is reached by a velocity distribution in dependency of the atmospheric height $z$ in the following exponential law [8]:

$$\frac{c(z)}{c_\infty} = \left(\frac{z}{z_\infty}\right)^{0.25} \tag{9}$$

In the investigated cases of a single cooling tower and a group arrangement of two cooling towers, the undisturbed velocity $c_\infty$ at the height of the cooling tower exit has been $c_\infty = 2.2$ m/s.

### 3.1.1 Results for single cooling tower

Based on the measured pressure distribution (timely averaged values) on the cooling tower surface at the waist height of $0.79H_{\text{tot}}$, a comparison of the pressure coefficients

$$c_p = \frac{p - p_\infty}{(1/2)\cdot \rho \cdot c_\infty^2} \tag{10}$$

with the calculated values can be given in Fig. 6.

The values are displayed in polar coordinates. The 0°-position is the stagnation point of the cooling tower. Measured values are available for the hydraulic even and hydraulic rough surfaces. It can be shown that the calculated pressure distribution is in good agreement with the measured values in both cases.

### 3.1.2 Results for cooling tower arrangement

The numerical grid is shown in Fig. 7. The arrangement of the cooling towers has been exposed to a cross wind with an offset angle of 15° to the centre connection line of the cooling towers. Thus, an asymmetric velocity and pressure distribution can be expected. This can be seen in the calculated pressure fields in Fig. 8a and b. Whereas Fig. 8a shows the experimental case with a rough

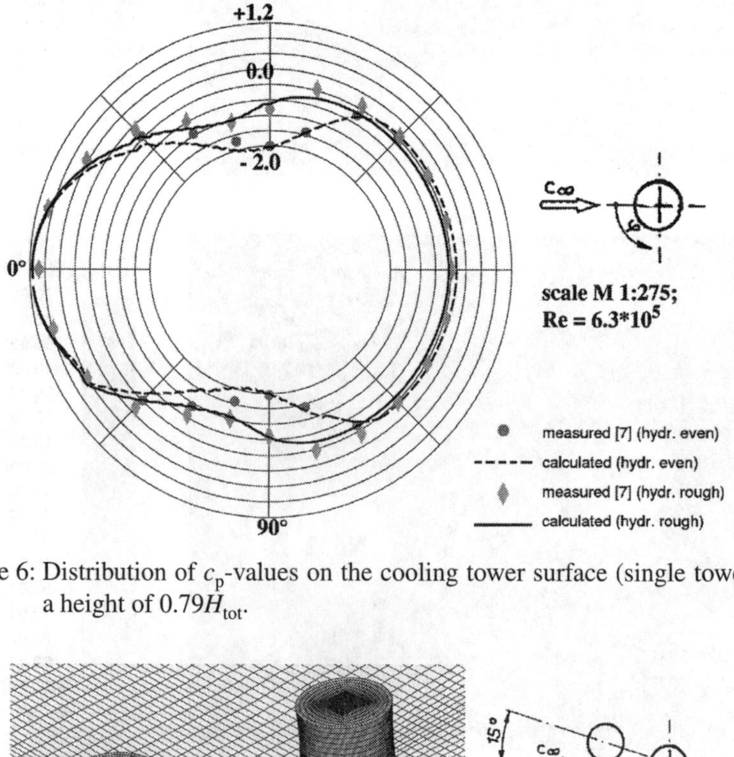


Figure 6: Distribution of $c_p$-values on the cooling tower surface (single tower) at a height of $0.79H_{tot}$.

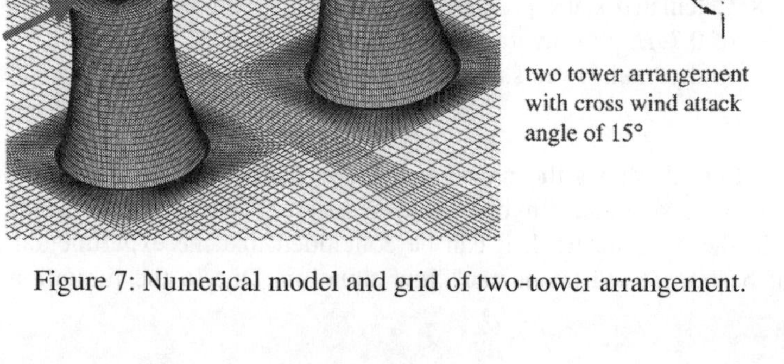


Figure 7: Numerical model and grid of two-tower arrangement.

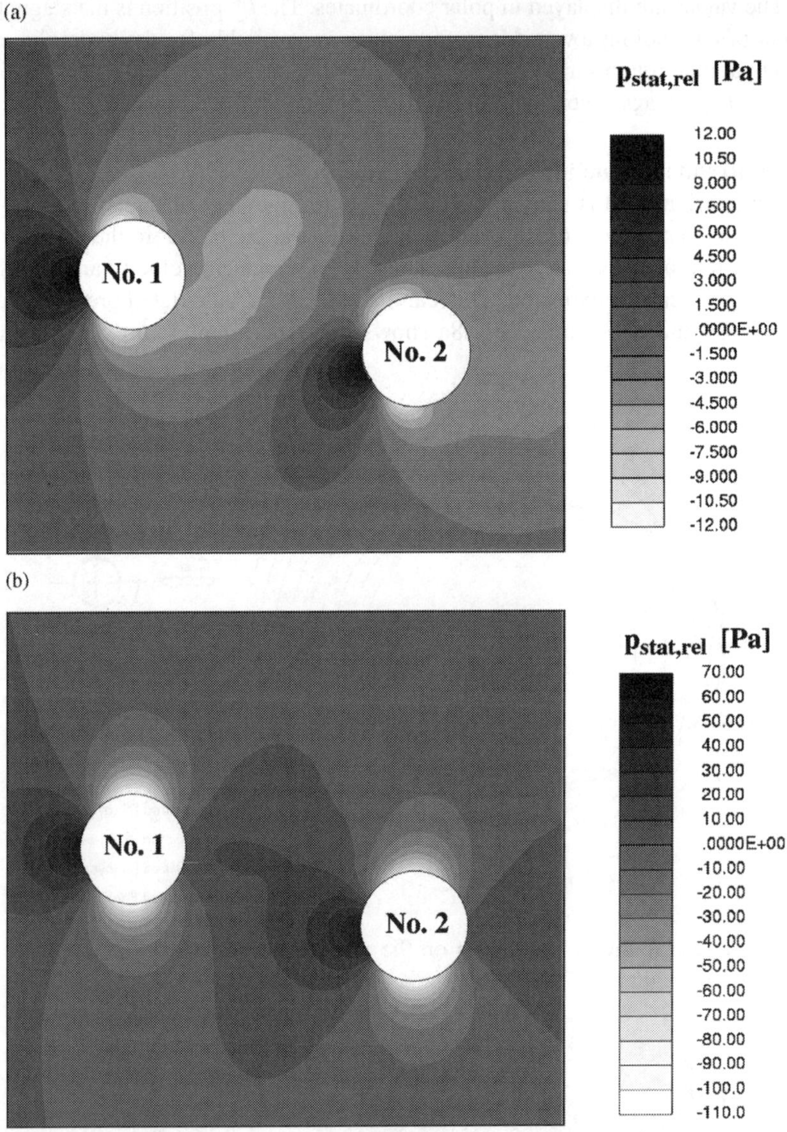


Figure 8: Calculated static pressure fields of the group arrangement at a height of $0.79H_{tot}$: (a) hydraulic rough surface, scale $M$ 1:275, Re = $6.3 \times 10^5$; (b) hydraulic even surface, scale $M$ 1:1, Re = $4.327 \times 10^7$.

surface, Fig. 8b shows the pressure field by taking into account the original cooling tower size and a hydraulic even surface. From the significant differences in the pressure field, it can be concluded that the experimental results cannot be transferred to the real flow situation. Due to the limitation of the established Re number, aerodynamic interaction phenomena are overestimated in the experimental case.

With respect to the aerodynamic validation purpose, the comparison of the measured and calculated pressure coefficients for the second cooling tower in Fig. 9 shows a good agreement. It can also be seen that for the realistic very high Re number situation, there are significant differences in the distributions that are showing the reduced effect of the front cooling tower on the second one.

## 3.2 Validation of the coupled aerothermodynamic and heat transfer model

As the major purpose of the numerical investigation has been to get detailed results on the interaction of the cooling tower aerodynamics, thermodynamics and heat transfer, the basic models as described in Section 2 have been implemented to the code. Thus, it has been necessary to validate the code and models by comparison of the calculated data with results of a counterflow cooling tower that have been available from the operator of a power plant. Initial calculations for the validation purpose are without cross wind effects.

### 3.2.1 Numerical model and boundary conditions

Basic geometry parameters of the investigated cooling tower are a total height of $H_{tot}$ = 117 m, a waist diameter of $D_w$ = 61.3 m at a height of $H_w$ = 90.7 m. Base diameter of the cooling tower is $D_b$ = 92.8 m. A three-dimensional half-grid with symmetry condition has been generated for the cooling tower (Fig. 10). The half-grid contains approximately 500,000 cells. In the internal part of the cooling tower,

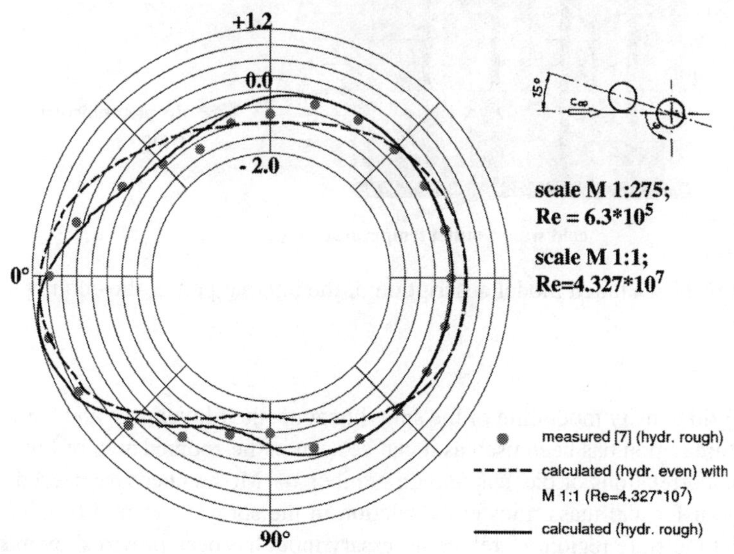


Figure 9: Distribution of $c_p$-values on the second cooling tower surface (group arrangement) at a height of $0.79H_{tot}$.

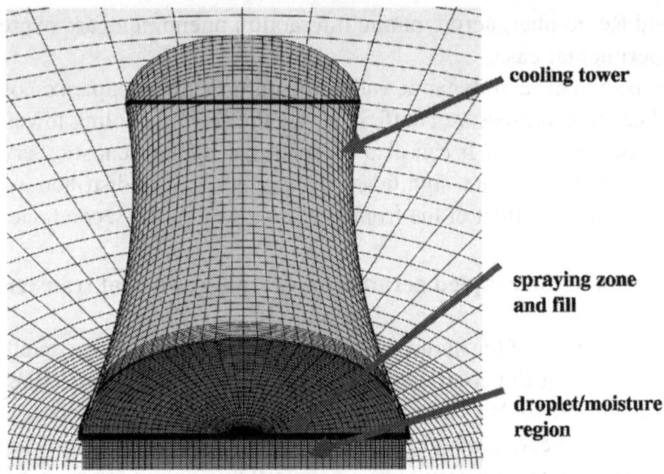


Figure 10: Numerical grid for coupled aerothermodynamic and heat transfer calculation (3D view into the tower).

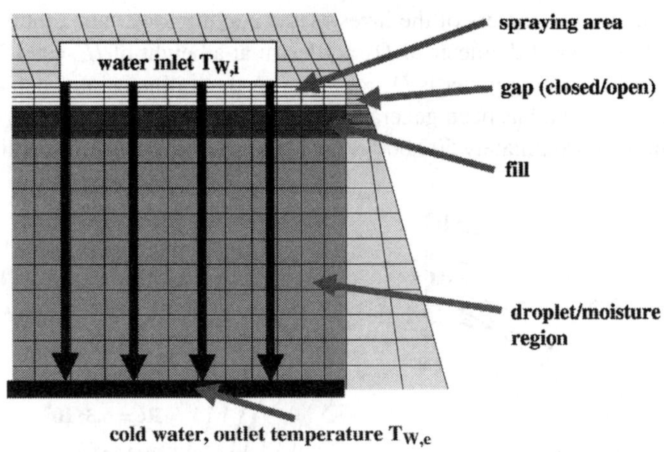


Figure 11: Detailed model description of the internal part of the cooling tower.

high-grid density modelling of the spraying zone, the fill volume, and the droplet/moisture region has been used as it can be seen in the zoomed area of Fig. 11.

The correlations of the evaporation number $k_V$ (Me number) are needed for the heat transfer and mass transfer calculation in the spraying zone, fill volume and droplet/moisture region. Here, the necessary input has been provided by manufactures of the fill and by the power plant operators. In the investigated case conditions, the fill type of the cooling tower is fibrated concrete that has been replaced by a plastic fill type (see Section 4.1). For the fill type, the following correlation in dependency of the local mass flow ratio λ can be given:

$$k_{V,fill} = 1.22 \cdot \lambda^{0.67} \tag{11}$$

For the spraying zone and droplet/moisture area, the following correlations have been used:

$$k_{V,spray} = 0.15 \cdot \lambda^{0.67} \tag{12}$$

$$k_{V,drop} = 0.03 \cdot \lambda^{0.67} \tag{13}$$

Based on the work provided by Poppe [6] the value of the Le number is $Le$ = 0.865 and the exponent $n$ in the Lewis analogy has a value of $n = 0.33$. Whereas an exponent of $n = 1.0$ is valid for a fully turbulent flow, a value of $n = 0.33$ indicates a flow with beginning thermal and hydrodynamic boundary layer formation. For the Lewis number, a value of $Le = 0.865$ is valid in a wide temperature range.

Furthermore, the pressure loss coefficients for the different regions have to be provided. These values have been provided by the designers of the cooling tower and the manufactures of the fill elements. Thus, these values depend on the specific design of the cooling tower. As precise values are part of the companies' proprietary, it is necessary to contact the manufactures so that such values can be provided. Here, the pressure loss coefficient for the fill elements itself is also dependent on the local flow velocity of the humid air. For velocities from $v = 1$ m/s to $v = 2.5$ m/s the loss coefficient is decreasing from $\zeta = 18$ to $\zeta = 10.8$. For the moisture/droplet region loss coefficients from $\zeta = 21.1$ for $v = 1$ m/s to $\zeta = 12.6$ for $v = 2.5$ m/s can be provided. For the spraying zone the values for the loss coefficients are $\zeta = 7.03$ for $v = 1$ m/s to $\zeta = 4.2$ for $v = 2.5$ m/s.

Finally, some basic thermodynamic data for the cooling tower operation are needed as input data (boundary conditions). These values are the air pressure ($p_0$ = 1013.0 hPa), the air temperature ($T_A$ = 282.65 K), humidity ($\varphi$ = 77.17%), water inlet temperature ($T_{W,i}$ = 307.35 K), and the water mass flow ($m_{W,i}$ = 58500.0 t/h). Due to the huge size of the cooling tower, the atmospheric layering of pressure and temperature as a function of the height has to be taken into account at the far field boundaries of the grid.

The dry air mass flow ($m_{A,dry}$) and the cold water temperature ($T_{W,e}$) are results of the numerical simulation. Thus, the local mass flow ratio Λ distribution will be also a calculation result at the end of the numerical iteration process. Of course, the local velocity field, local heat fluxes, and local mass fluxes are also a result of the CFD analysis. Thus, the great advantage of this comprehensive CFD approach is not only the precise calculation of the integral result values of the cooling tower but also the inclusion of several influences on these values for a systematic investigation on the improvement potential.

### 3.2.2 Aerothermodynamic calculation results

The calculated cold water temperature is the most critical parameter with respect to the calculated performance of the cooling tower. Initially, the temperature values for the water-side of the simulation have been set equally to $T_{W,i} = 307.35$ K, that is the inlet warm water temperature. During the iteration process, the calculated cold water temperature decreases with ongoing calculation with only few numerical instability. This has been documented by the graph in Fig. 12, which shows the calculated cold water temperature during the iteration process. Finally, a converged value of $T_{W,e} = 295$ K has been reached after 1500 steps with only minor further change. The final calculation value $T_{W,e} = 294.96$ K is very close to the target design value of 294.75 K and clearly within a confidence level of ±0.5 K requested by the cooling tower designers.

Table 1 lists the most important integral values of the numerical simulation. The comparison of the total transferred heat shows that there is only a 1.8% difference

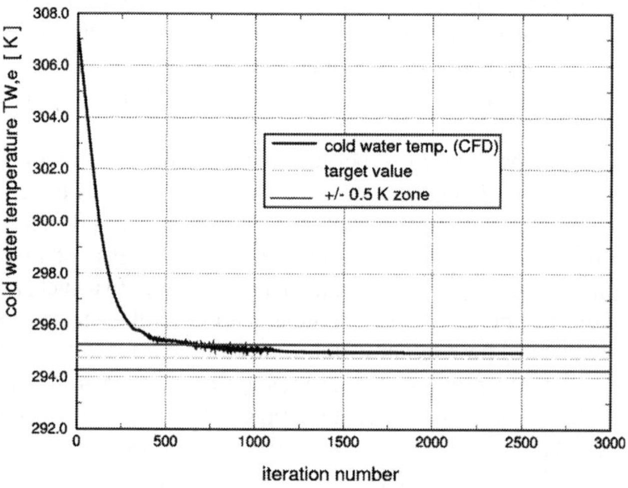


Figure 12: Cold water temperature calculation.

Table 1: Integral results on transferred heat and mass flow.

| | Calculation | | Design point |
|---|---|---|---|
| Cold water temperature | 294.960 K | | 294.750 K |
| Total heat (transferred from water) | 835.87 MW | | 850.0 MW |
| Convective heat | 226.74 MW | 27.13% | |
| Evaporation heat | 609.13 MW | 72.87% | |
| Total transferred heat in spraying area | 76.75 MW | 9.18% | |
| Total transferred heat in fill region | 577.51 MW | 69.09% | |
| Total transferred heat in droplet region | 181.61 MW | 21.73% | |
| Evaporated water mass flow | 246.98 kg/s | 1.52% | |
| Dry air mass flow | 13779 kg/s | | 13390 kg/s |
| Averaged cooling tower outlet temperature | 298.93 K | | |

between the value obtained from calculation and the cooling tower design value of 850 MW. The dry air mass flow is only 2.9% higher than the design value, which is also an acceptable deviation as the dry air mass flow is a calculated value based on the full aerothermodynamics and not a boundary condition.

Based on the local values in the flow field, one can distinguish precisely between the transferred convective heat and the evaporation heat, which is the major part (72.87%) of the heat transferred from the water side. Furthermore, it can be distinguished between the heat transferred in the spraying area, the fill, and the moisture/droplet region (Table 1). The evaporated water mass flow is 246.98 kg/s (1.52% of inlet water mass flow).

The flow visualization by flow vectors in a two-dimensional sectional cut is presented in Fig. 13. The dry air is sucked into the tower at the bottom and then turned into vertical direction so that only a minor part of the air streams through the complete droplet/moisture region until the centre of the tower. Then, the flow direction is homogenized in the fill area by the fill installation. Maximum flow velocities in the centre region of the tower are approximately 8 m/s. Furthermore, it can be seen in the zoomed region (Fig. 13b) that for the stationary case and without side wind effects no cold air entry in the tower exit area close to the tower rim can be observed.

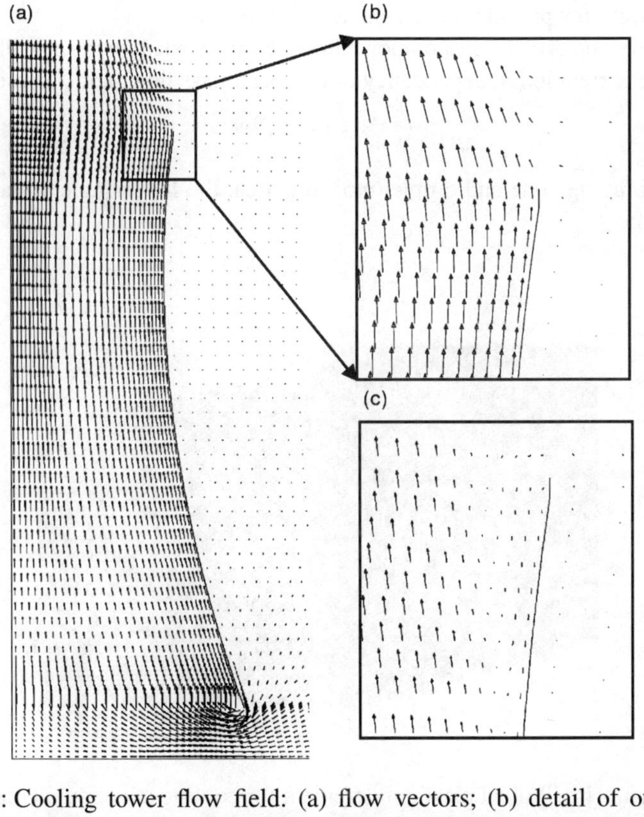


Figure 13: Cooling tower flow field: (a) flow vectors; (b) detail of outlet flow (closed gap); (c) detail of outlet flow (open gap).

The situation is different in the case, if the circular gap, as indicated in Fig. 11, is opened in the calculation model. In that case (Fig. 13c), a part of the inlet air is able to flow through the gap and leads to a low kinetic boundary layer flow. As a result, some part of the environmental cold air is able to generate a partly blockage of the tower outlet area close to the tower rim. A recirculating area of cold air can be observed. Overall performance of the cooling tower is significantly reduced. The simulation underlines the importance of the typical design feature to close gaps in the fill area near to the cooling tower wall.

## 4 Influences on the cooling tower performance

Based on the validation calculation of the cooling tower as shown in Section 3.2, the influence of several design features on the cooling tower performance is investigated by application of the numerical method.

### 4.1 Different fill types

With respect to the calculations for validation (reference case), the fill type has been fibrated concrete. It has been replaced by a modern plastic fill type as shown in Fig. 14 for an example. Such modern fill types are characterized by an improved heat transfer, low pressure losses and a high life-span.

The correlations of the evaporation number $k_V$ are needed for the new fill type. The following correlation in dependency of the local mass flow ratio $\lambda$ can be given:

$$k_{V,fill} = 1.92 \cdot \lambda^{0.633} \tag{14}$$

For the spraying zone and droplet/moisture area, the following correlations have been used:

$$k_{V,spray} = 0.15 \cdot \lambda^{0.633} \tag{15}$$

Figure 14: Typical plastic fill element for cooling tower application.

$$k_{\text{V,drop}} = 0.03 \cdot \lambda^{0.633} \tag{16}$$

Furthermore, the pressure loss coefficients for the different regions have to be provided. The pressure loss coefficient for the fill elements itself is dependent on the local flow velocity of the humid air. For velocities from $v = 1$ m/s to $v = 2.5$ m/s the loss coefficient is decreasing from $\zeta = 19.25$ to $\zeta = 11.52$. For the moisture/droplet region loss coefficients and the spraying zone, the same pressure loss coefficients have been applied as for the reference calculation (see Section 3.2.1).

Whereas the environmental conditions in the calculation are the same as in the reference case, some operating parameters have been changed (Table 2). The water inlet temperature is $T_{\text{W,i}} = 306.05$ K, and the inlet water mass flow $m_{\text{W}} = 63{,}000.0$ t/h (17500 kg/s). Initially, the temperature values for the water-side of the simulation have been set equally to $T_{\text{W,i}} = 306.05$ K, that is the inlet warm water temperature. During the iteration process, the calculated cold water temperature decreases with ongoing calculation and shows again an excellent convergence behaviour. That has been documented by the graph in Fig. 15. Finally, a converged value of

Table 2: Comparison of the operating conditions.

| | Plastic fill type | Fibrated concrete fill type |
|---|---|---|
| Environmental temperature | 282.65 K | 282.65 K |
| Humidity | 77.17% | 77.17% |
| Environmental air pressure | 1.013 bar | 1.013 bar |
| Water inlet temperature | 306.05 K | 307.35 K |
| Inlet water mass flow | 17500 kg/s | 16250 kg/s |

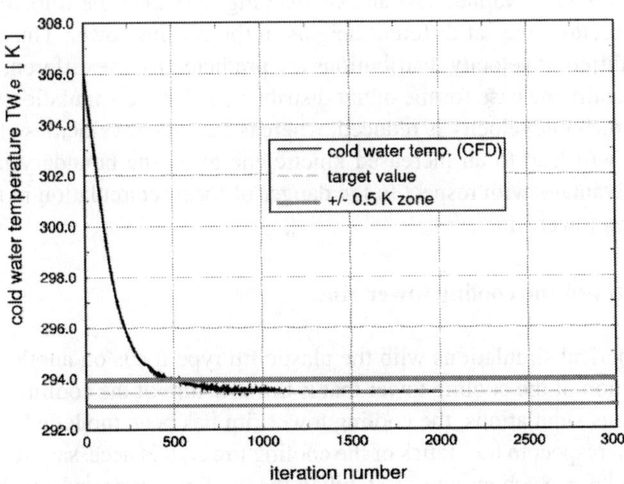


Figure 15: Cold water temperature calculation.

Table 3: Integral results on transferred heat and mass flow.

| | Calculation | | Design point |
|---|---|---|---|
| Cold water temperature | 293.455 K | | 293.450 K |
| Total heat (transferred from water) | 914.35 MW | | 915.0 MW |
| Convective heat | 252.63 MW | 27.63% | |
| Evaporation heat | 661.72 MW | 72.37% | |
| Total transferred heat in spraying area | 55.60 MW | 6.08% | |
| Total transferred heat in fill region | 699.51 MW | 76.50% | |
| Total transferred heat in droplet region | 159.24 MW | 17.42% | |
| Evaporated water mass flow | 267.90 kg/s | 1.53% | |
| Dry air mass flow | 14137 kg/s | | 13230 kg/s |
| Averaged cooling tower outlet temperature | 300.28 K | | |

$T_{W,e}$ = 293.455 K has been reached. That is almost identical to the design value (Table 3) and, therefore, clearly within a confidence level of ±0.5 K requested by the cooling tower designers.

Table 3 also lists the most important integral values of the numerical simulation. The comparison of the total transferred heat shows that this value is also almost identical to the design value of 915 MW. Compared to the reference case this means a significant increase in total heat transferred within the cooling tower (+7.6%). Thus, this improvement has a significant effect on the thermal efficiency of the steam cycle in the power plant. However, the percentage distribution of the heat transferred in the different regions of the cooling tower has only minor changes. The calculated dry air mass flow is approximately 6.8% higher than the design value.

One of the great advantages of the CFD approach is that detailed information on the local flow field is available. As an example, Fig. 16 shows the velocity distributions along section lines at different heights in the cooling tower. Thus, it can be shown that different velocity distributions are predicted for the different fill types. This is especially the case for the outlet distribution. For the simulation with plastic fill type, the bulk velocity is reduced, whereas the higher velocities close to the tower walls will lead to an increased kinetic energy in the boundary layer flow. That is of advantage with respect to the danger of local recirculation in the region of the cooling tower rim.

## 4.2 Geometry of the cooling tower rim

Further numerical simulations with the plastic fill type focus on another detail of the construction of the cooling tower that is the structure of the cooling tower rim. In the previous simulations, the cooling tower rim has been modelled in a simple way, but with respect to the statics of the cooling tower, it is necessary to add a special rim structure. Such structures stabilize the cooling tower and can be located at the outer or inner side of the cooling tower rim. Figure 17 shows a stabilizing structure at the inner rim side.

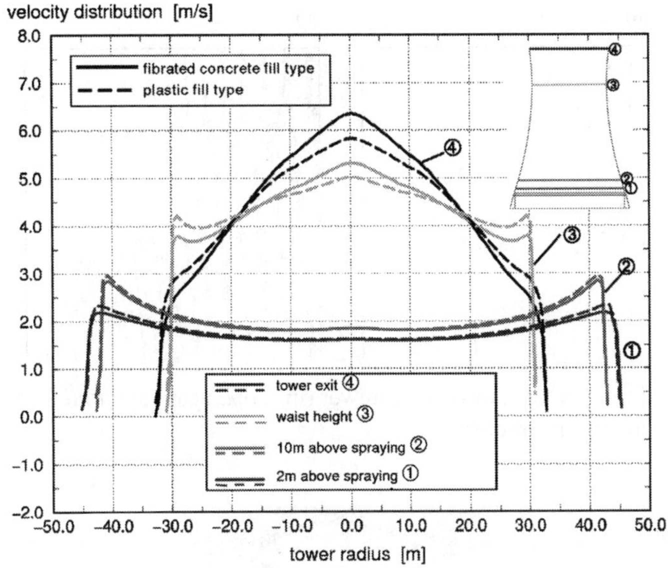


Figure 16: Velocity distribution in the cooling tower for different fill types.

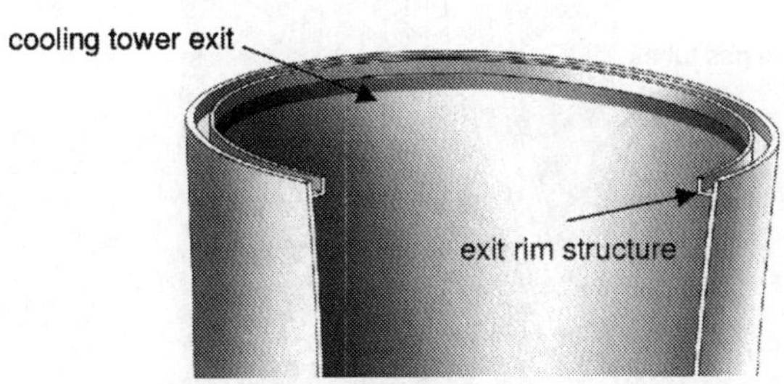


Figure 17: Stabilizing structure at the inner rim side.

Due to the fact, that the inner rim structure reduces the open exit surface of the cooling tower, an increase in the calculated cold water temperature of 0.175 K ($T_{W,e} = 293.63$) is predicted, that is still by far within the confidence level of ±0.5 K requested by the cooling tower designers. The calculated dry air mass flow is reduced to 13830 kg/s compared to the 14137 kg/s for the calculation without the inner rim structure. Thus, the calculated air mass flow is now 4.5% higher than the design value. Based on these results, it might be of advantage to install a rim structure at the outer side. This might be correct with respect to the viewpoint of the aerothermodynamics without flow instabilities at the cooling tower exit, but a simulation under the same conditions but with opened gap between the fill and the tower wall disclose the great advantage of the inner rim structure for stabilizing the tower exit flow. Whereas in the case without the inner rim structure an unstable flow with local recirculation near the cooling tower rim can be observed in Fig. 18a, the situation is significantly improved for the case with inner rim structure

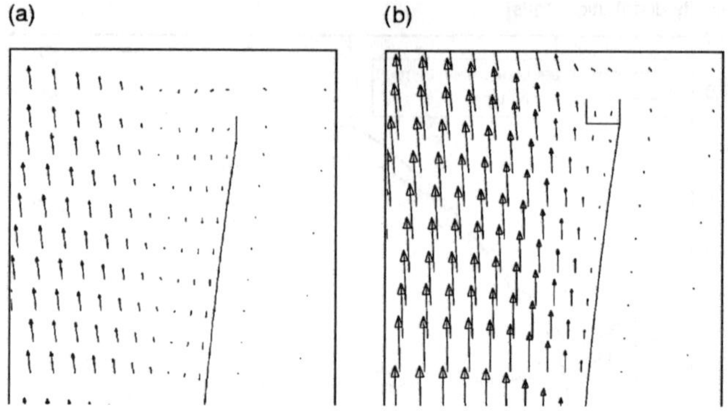


Figure 18: Flow vector at tower exit (tower rim): open gap (a) without and (b) with inner rim structure.

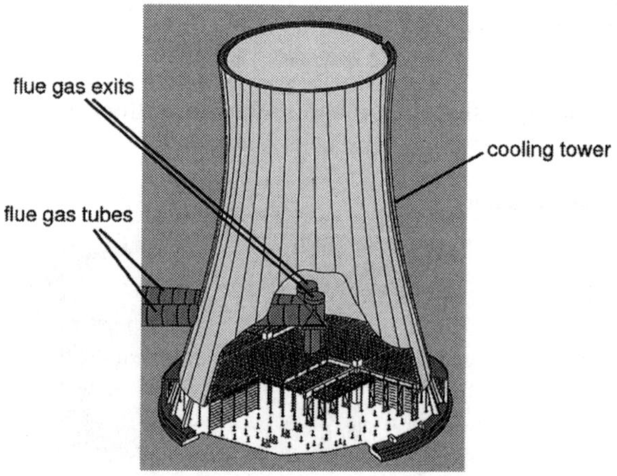


Figure 19: Natural draft cooling tower with additional flue gas operation.

(Fig. 18b). In that case, the velocity vectors show not only increased velocities at the exit but also that the recirculating flow area has vanished.

### 4.3 Additional flue gas discharge operation

Additional flue gas discharge operation for cooling towers has been established in recent years. In that case, the additional construction of a high chimney for the flue gas is not necessary. Thus, the costs for the construction of the power plant are reduced. Another advantage might be seen in the optical appearance of the power plant. Figure 19 shows a drawing for a typical construction of a cooling tower with additional flue gas discharge operation.

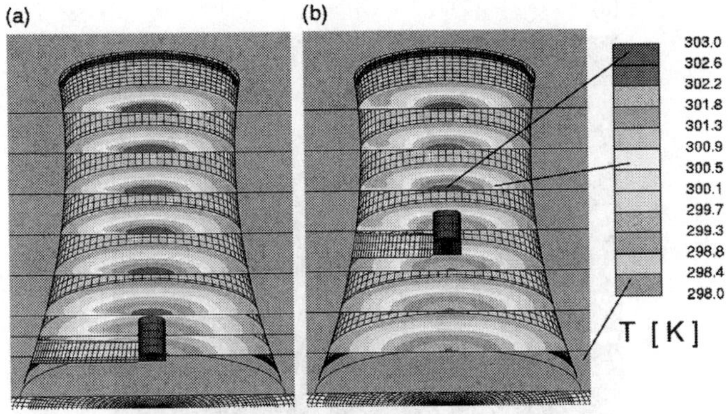


Figure 20: Temperature distributions for the cooling tower (plastic fill type, inner rim structure, closed gap) with additional flue gas operation: flue gas outlet at (a) $H_{flue}$ = 13 m and (b) $H_{flue}$ = 51 m.

However, several questions are of high interest in case an additional flue gas operation is implemented. At first, there is the question on the changed aerothermodynamics of the cooling tower. Furthermore, it is of significant importance to avoid any contact of the aggressive gases with the tower walls. Thus, numerical simulations with additional flue gas outlets in two different height, $H_{flue}$ = 13 m and $H_{flue}$ = 51 m, have been performed. The calculations have been executed with plastic fill type, inner rim structure and a closed gap between the fill and the tower wall.

Figure 20a and b show the three-dimensional temperature distribution in cutting planes at different tower heights for both cases. The high temperatures are in the centre region of the cooling tower. There is no hint that in case of stable operation of the cooling tower there is a danger of flue gas contact with the tower walls. Furthermore, the integral evaluation of the results shows that there is only a minor effect of the additional flue gas discharge operation on the aerothermodynamics of the cooling tower. For a height of $H_{flue}$ = 13 m, there is a very slight improvement of the cold water temperature with a reduction of just –0.004 K. Some small blockage effects of the flue gas piping are compensated by a small improvement of the draught in the tower due to the warmer flue gases. In the case of an additional flue gas discharge operation with an outlet height of $H_{flue}$ = 51 m, there is a very slight increase in the cold water temperature due to the blockage effect and only very minor improvement of the draught in the tower. However, to reduce the danger of a flue gas contact with tower walls, it might be even more secure to install the flue gas outlet at a high position as all negative effects on the aerothermodynamic performance of the cooling tower can be stated as negligible.

## 5 Cross wind effects on the cooling tower performance

A final set of numerical simulations for the cooling tower application deals with the impact of different cross wind situation, i.e. different wind velocities. It is

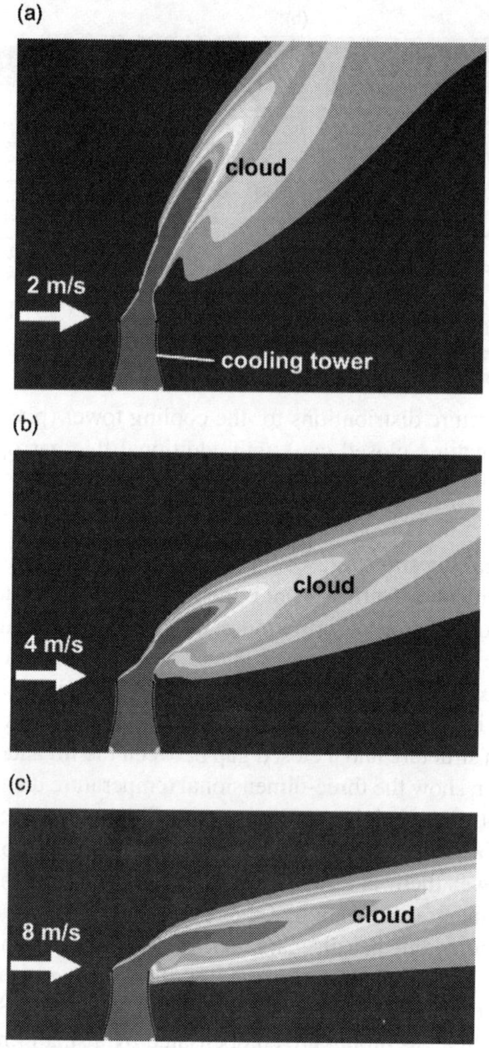


Figure 21: Temperature distributions for visualization of the cooling tower cloud propagation under cross wind effects: cross wind velocity of (a) 2 m/s, (b) 4 m/s and (c) 8 m/s.

well known that the interaction of the cooling tower flow and the cross wind can lead to a situation with a large region of cold air entering the tower exit. This phenomenon leads to a reduction in the effective cooling tower outlet area and, thus, has also a significant negative impact on the cooling tower performance and, furthermore, on the steam cycle performance. Thus, such operating situations have to be avoided or at least to be reduced in size by a modern design of the cooling tower.

## 5.1 Stable configurations

The modelling of the cross wind is done in the same way as described in the Section 3.1 for the aerodynamic validation. The cooling tower model that has been used contains the plastic fill type, the inner rim structure and a closed gap between the fill and the tower wall. Calculations (three-dimensional) have been performed with different wind velocities of $v_w = 2$ m/s, $v_w = 4$ m/s, $v_w = 8$ m/s, and $v_w = 14$ m/s.

Figure 21a–c shows the propagation of the cooling tower cloud due to the impact of different wind velocities. For all velocities, a stationary solution has been reached as shown in Fig. 21a–c. Therefore, the numerical simulations do not show any hint for a cold air ingestion at the tower exit. Nevertheless, an impact on the cooling tower performance, i.e. the cold water temperature, can be found as described by Fig. 22. For very small cross wind velocities of approximately $v_w = 2$ m/s a slight improvement, i.e. a reduction of the cold water temperature can be observed. With further increasing cross wind velocities, there is a significant negative effect on the cooling tower performance and the cold water temperature increases. As it is shown by the curve in the diagram of Fig. 22, the increase in the cold water temperature is not linear but follows a parabolic shape and the situation worsens for high cross wind velocities. However, cold air ingestion is even not observed for the highest investigated cross wind velocity of $v_w = 14$ m/s.

The reasons for the stable behaviour of the cooling tower can be found in the advantageous tower design. As it has been shown in Section 4.2, the inner rim structure has a stabilizing effect on the flow regime near to the tower rim. Furthermore, the plastic fill type leads to a boundary layer flow with a higher kinetic energy (see Section 4.1). Altogether, these two design features lead to a stable flow situation at the tower outlet even in a cross wind situation of up to $v_w = 14$ m/s.

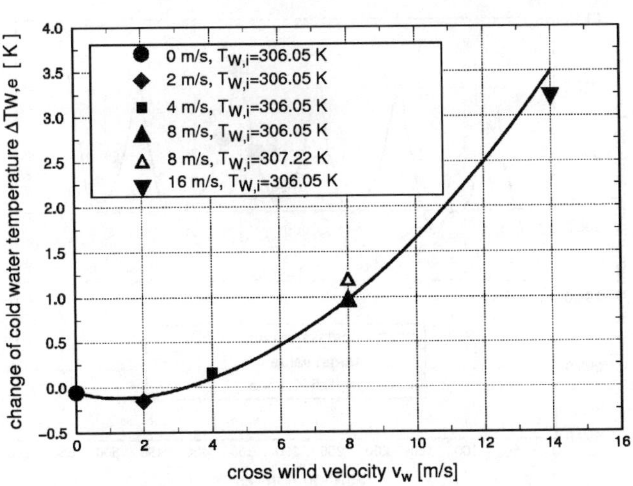


Figure 22: Calculated change of cold water temperatures for different cross wind velocities.

## 5.2 Unstable configuration with cold air ingestion

Based on the understanding of the positive effects of the inner rim structure and the plastic fill type on the exit flow of the cooling tower, a more unstable situation has been expected for a cooling tower simulation without the inner rim structure and the fibrated concrete fill type. Furthermore, the gap between the fill and the tower wall has been opened in the numerical model as investigations presented in Section 3.2.2 have shown that an open gap leads to a recirculating flow area at the tower rim even in the case without cross wind. The three-dimensional simulation has been started based on the case without cross wind as the initial flow and temperature field and with a low cross wind velocity of $v_w = 2$ m/s. Figure. 23 shows the calculated cold water temperature during the iteration process. Whereas in other simulation a good convergence has been observed, the situation is now different. A convergence of the cold water temperature has not been reached. Instead, there is a periodic fluctuation of the cold water temperature with the time steps.

Figure 24 shows the visualization of the temperature field at four different time steps as indicated in Fig. 23. At position 1 (Fig. 24a), there is a low cold water temperature and there is only a minor area of cold air ingestion near to the tower rim. Then, with increased cold air ingestion as shown in Fig. 24b for position 2, the cold water temperature increases rapidly. It reaches its maximum at position 3. Figure 24c shows for that case that a major part of the tower outlet is blocked by cold air. However, warm air in the lower part of the tower has started to push the cold air towards the outlet. Figure 25 shows a three-dimensional view of the temperature distributions for position 2 and position 3. Thus, it becomes even more obvious how deep the cold air can penetrate into the interior of the cooling tower.

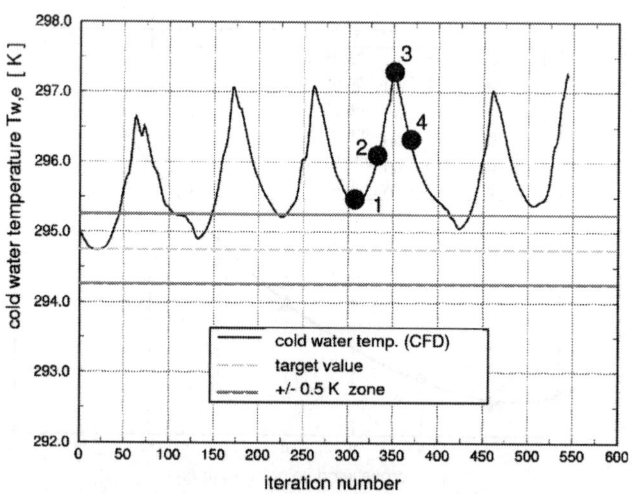


Figure 23: Cold water temperature calculation, cross wind velocity 2 m/s, cooling tower *without* inner rim structure.

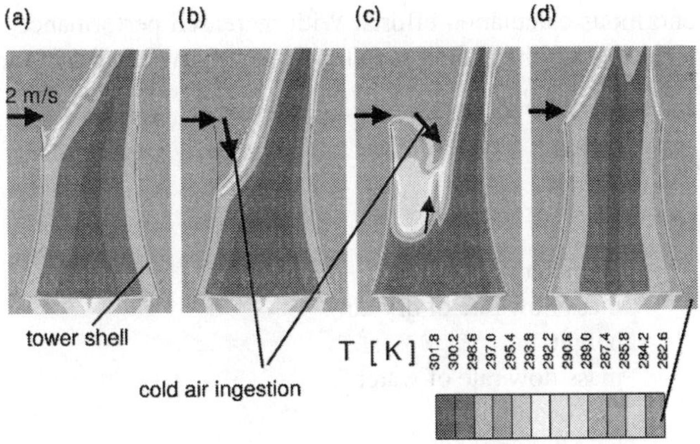


Figure 24: Temperature distribution at indicated iteration steps, cross wind velocity 2 m/s, cooling tower *without* inner rim structure: (a) position 1, (b) position 2, (c) position 3 and (d) position 4.

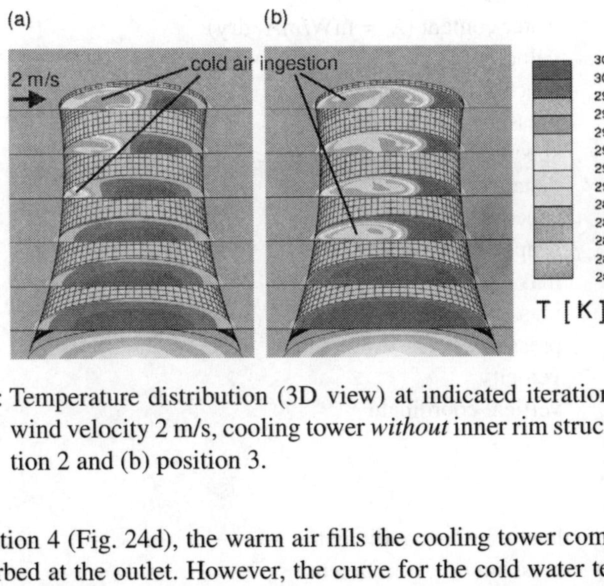


Figure 25: Temperature distribution (3D view) at indicated iteration steps, cross wind velocity 2 m/s, cooling tower *without* inner rim structure: (a) position 2 and (b) position 3.

At position 4 (Fig. 24d), the warm air fills the cooling tower completely but is still disturbed at the outlet. However, the curve for the cold water temperature in Fig. 23 shows that it has decreased now rapidly and will further decrease until a similar situation is reached as it has been at position 1. Then, the next cycle starts. Thus, the effect of cross wind for cooling tower design with significant deficiencies might lead to a timely periodic cold air ingestion as it has been shown by Figs. 23–25. However, it has to be taken into account that the results shown in Section 5.2 are only of limited numerical value as the calculation is not unsteady, but results are obtained of a periodic numeric instability in a steady calculation using a time-marching scheme. Unsteady calculations of the cold air ingestion phenomenon in case of a cross wind situation have not been performed now by the authors due to the enormous calculation efforts. With increased performance of modern PC-Clusters, this possibility should soon be in reach for validation of the observed and described phenomenon.

## Nomenclature

| | |
|---|---|
| $A$ (m$^2$) | area |
| $D$ (m) | diameter |
| $G$ (kg/s) | mass flow rate of dry air |
| $H$ (m) | height |
| $L$ (kg/s) | mass flow rate of water |
| $Le$ (–) | Lewis number |
| $\dot{Q}$ (J/s) | heat flow rate |
| $Ka$ (kg/(m$^3$ s)) | mass transfer coefficient |
| $Me$ (–) | Merkel number |
| $Re$ (–) | Reynolds number |
| $T$ (K) | temperature |
| $V$ (m$^3$) | volume |
| $X$ (–) | water content ($X$ := mW/mA, dry) |
| $c$ (m/s) | velocity |
| $c_p$ (J/(kg K)) | specific heat capacity |
| $cp$ (-) | pressure coefficient |
| $c_W$ (J/(kg K)) | specific heat capacity of water |
| d (m) | distance |
| $h$ (J/kg) | specific enthalpy |
| $k_V$ (–) | evaporation number |
| $m$ (kg) | mass |
| $\dot{m}$ (kg/s) | mass flow rate |
| $p$ (N/m$^2$) | pressure |
| $v$ (m/s) | velocity |
| $z$ (m) | vertical coordinate |

### Greek letters

| | |
|---|---|
| $\alpha$ (W/(m$^2$ K)) | heat transfer coefficient |
| $\beta_x$ (kg/(m$^2$ s)) | mass transfer coefficient |
| $\lambda$ (–) | ratio of dry air mass flow and cooling water |
| $\rho$ (kg/m$^3$) | density |
| $\zeta$ (–) | pressure loss coefficient |

### Subscripts

| | |
|---|---|
| A | air, air-side |
| S | steam, vapour |
| W | water, water-side |

| | |
|---|---|
| b | base |
| e | exit, outlet |
| ev | evaporation |
| drop | droplet/moisture region |
| dry | dry air |
| fill | cooling fill |
| flue | flue gas |
| i | inflow, inlet |
| rel | relative (with respect to reference value) |
| spray | spraying zone |
| stat | static |
| tot | total |
| w | waist |
| w | wind (velocity) |
| $\infty$ | infinity |

**Superscripts**

| | |
|---|---|
| $''$ | condition at phase liquid/gas interface |

## References


[1] *Cooling Tower Performance Curves*, Cooling Tower Institute: Houston, TX, 1967.

[2] Kelly, N.W., *Kelly's Handbook of Crossflow Cooling Tower Performance*, Neil W. Kelly and Associates: Kansas City, MO, 1976.

[3] Majumdar, A.K., Singhal, A.K. & Spalding, D.B., Numerical modeling of wet cooling towers – Part 1: Mathematical and physical models. *Journal of Heat Transfer*, **105**, pp. 728–735, 1983.

[4] Majumdar, A.K., Singhal, A.K., Reilly, H.E. & Spalding, D.B., Numerical modeling of wet cooling towers – Part 2: Application to natural and mechanical draft towers. *Journal of Heat Transfer*, **105**, pp. 736–743, 1983.

[5] Star-CD Version 3.20, CD adapco Group, 2004.

[6] Poppe, M., *Wärme- und Stoffübertragung bei der Verdunstungskühlung im Gegen- und Kreuzstrom* (in German), PhD Thesis, Technical University of Hannover, 1972.

[7] Beger, G., Modellversuche zur Windlastbestimmung an Kühltürmen in Gruppenaufstellung (in German). *Energietechnik*, **42(2)**, pp. 45–48, 1992.

[8] Beger, G., Modelluntersuchungen der Strömungsverhältnisse im Kopfbereich von Naturzugkühltürmen (in German). *Energietechnik*, **40(4)**, pp. 135–139, 1990.

# CHAPTER 7

# Innovative gas turbine cooling techniques

R.S. Bunker
*GE Global Research Center, USA.*

## Abstract

Advanced heat transfer and cooling techniques form one of the major pillars supporting the continuing development of high efficiency, high power output gas turbine engines. Conventional gas turbine thermal management technology is composed of five main elements including internal convective cooling, external surface film cooling, materials selection, thermal-mechanical design at the component and system levels, and selection and/or pre-treatment of the coolant fluid. The present summary will examine specific cooling technologies representing cutting edge, innovative methods expected to further enhance the aero-thermal-mechanical performance of turbine engines. The techniques discussed will include forced convective cooling with unconventional turbulators and concavity surface arrays, swirl-cooling chambers, latticework cooling networks, augmentations of impingement heat transfer, synergistic approaches using mesh networks, and film cooling.

## 1 Introduction

The technology of cooling gas turbine components, primarily via internal convective flows of single-phase gases and external surface film cooling with air, has developed over the years into very complex geometries involving many differing surfaces, architectures, and fluid-surface interactions. The fundamental aim of this technology area is to obtain the highest overall cooling effectiveness with the lowest possible penalty on the thermodynamic cycle performance. As a thermodynamic Brayton cycle, the efficiency of the gas turbine engine can be raised substantially by increasing the firing temperature of the turbine. Modern gas turbine systems are fired at temperatures far in excess of the material melting temperature limits. This is made possible by the aggressive cooling of the hot gas path components using a portion of the compressor discharge air, as depicted in Fig. 1. The use of 15–25%

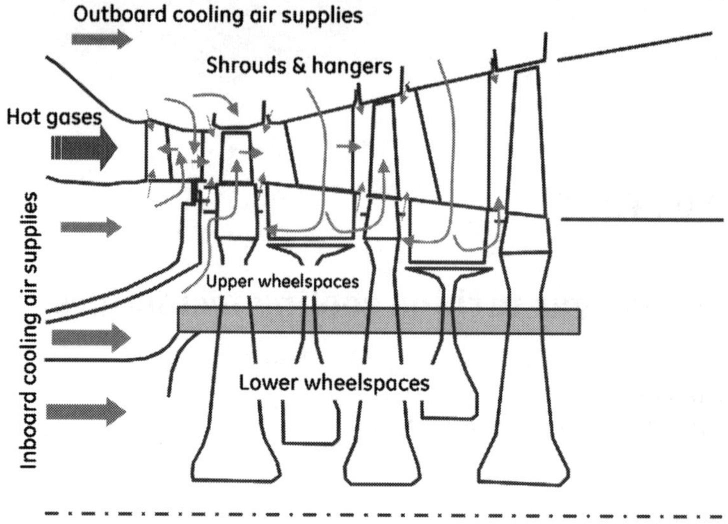


Figure 1: Schematic of turbine with cooling flows.

of this compressed air to cool the high-pressure portions of the turbine presents a severe penalty on the thermodynamic efficiency unless the firing temperature is sufficiently high for the gains to outweigh the losses. In all properly operating cooled turbine systems, the efficiency gain is significant enough to justify the added complexity and cost of the cooling technologies employed. Actively or passively cooled regions in power generating gas turbines include the stationary vanes or nozzles and the rotating blades or buckets of the high-pressure stages, the shrouds bounding the rotating blades, and the combustor liners and flame holding segments (fuel nozzles, splash plates). All such engines additionally cool the interfaces and secondary flow regions around the immediate hot gas path. A more detailed schematic of the cooling for an aircraft engine combustor and turbine first stage, which may be thought of also as an aero-derivative power turbine, is shown in Fig. 2.

Cooling technology, as applied to gas turbine components is composed of five main elements, (1) internal convective cooling, (2) external surface film cooling, (3) materials selection, (4) thermal-mechanical design, and (5) selection and/or conditioning of the coolant fluid. Cooled turbine components are merely highly specialized and complex heat exchangers that release the cold side fluid in a controlled fashion to maximize work extraction. The enhancement of internal convective flow surfaces for the augmentation of heat transfer was initially improved some 25–30 years ago through the introduction of rib-rougheners or turbulators, and also pin-banks or pin-fins. Figure 3 shows an example schematic of a blade cooling circuit that utilizes many turbulated passages, a pin bank in the trailing edge, and impingement in the leading edge (coolant is released via film holes, tip holes, and trailing edge). These surface enhancement methods continue to play a large role in today's turbine cooling designs. Film cooling is the practice of bleeding

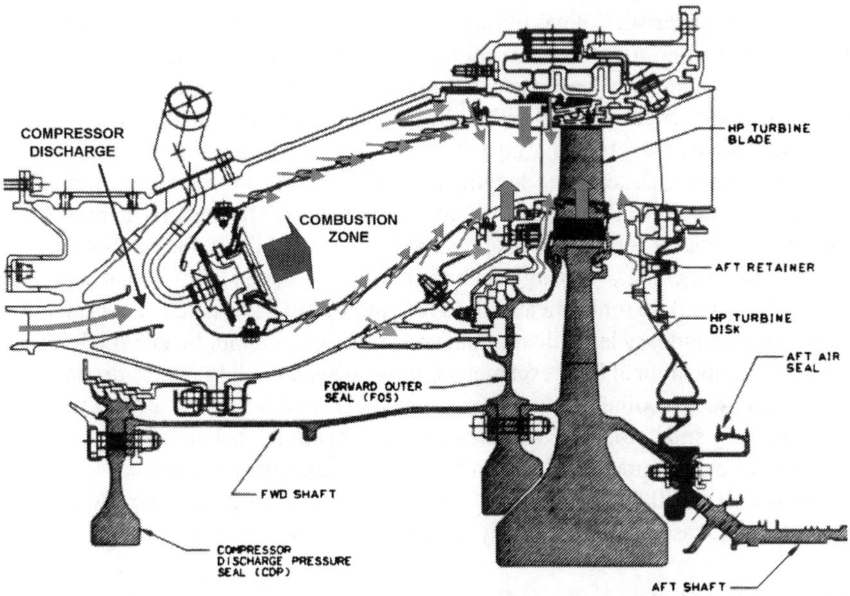


Figure 2: Cooling flows for a combustor and high-pressure turbine.

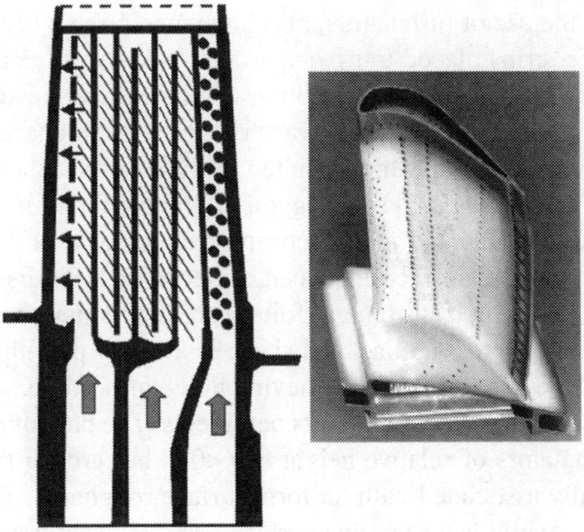

Figure 3: Schematic of a blade cooling circuit.

internal cooling flows onto the exterior skin of the components to provide a heat flux reducing cooling layer, as shown by the many holes placed over the airfoil in Fig. 3. Film cooling is intimately tied to the internal cooling technique used in that the local internal flow details will influence the flow characteristics of the film jets injected on the surface.

Several characteristics of gas turbine cooling are worth noting prior to describing any specific technologies. Almost all highly cooled regions of the high-pressure turbine components involve the use of turbulent convective flows and heat transfer. Very few, if any, cooling flows within the primary hot section are laminar or transitional. Moreover, the typical range of Reynolds numbers for cooling techniques, using traditional characteristic lengths and velocities, is from 10,000 to 60,000. This is true of both stationary and rotating components. The enhancement of heat transfer coefficients for turbine cooling makes full use of the turbulent flow nature by seeking to generate mixing mechanisms in the coolant flows that actively exchange cooler fluid for the heated fluid near the walls. These mechanisms include shear layers, boundary layer disruption, and vortex generation. In a marked difference from conventional heat exchangers, most turbine cooling means do not rely on an increase in cooling surface area, since the available surface area to volume ratios are very small. Surface area increases are beneficial, but are not the primary objective of enhancements. The use of various enhancement techniques typically results in at least 50% and as much as 300% increase in local heat transfer coefficients over that associated with fully developed turbulent flow in a smooth duct.

## 2 Turbulated channel cooling

One of the most common means for enhancing heat transfer coefficients within internal cooling passages, and especially the serpentine passages of many turbine blade designs, is the use of turbulators, also known as rib rougheners. Turbulators in the form of trip strips placed transverse to the bulk flow direction were one of the first improvements made to the cooling of blades, and hence many investigations have been made into the heat transfer and friction characteristics. Basic transverse turbulator research, which resulted in widely used data and correlations, was performed by Webb *et al.* [1], Burggraf [2], and Han *et al.* [3, 4]. With the advancements in materials and manufacturing technologies of the last decade, a drastically larger realm of surface enhancement techniques has become cost effective for use in the cooling of turbine airfoils. Turbulators may now be of varying shapes, orientations, segmentations, and sizes, essentially providing a continuous spectrum of possible geometries for achieving flow-surface interactions that serve to enhance local and global heat transfer coefficients. The bulk of researchers concentrated on turbulators of relative height $e/D < 0.2$, but greater than that relative roughness typically associated with uniform surface roughness. The general findings of all such research has been that surface averaged heat transfer coefficients within stationary, turbulated passages may be enhanced by factors from 1.8 to 2.8, while the friction factors or required pumping power are increased by factors of 3–10. Many parameters have been investigated in turbulated passages (angle to the bulk flow, $P/e$, $e/D$, shaping, taper, etc.), and many more in serpentine circuits, but the range of effects has remained much the same over the years.

Improvements over the use of transverse turbulators within channels have in the bulk of the literature focused on angled turbulators, segmented turbulators, and many other combinations of these in such forms as chevrons and segmented discrete strips. Many researchers have noted significant heat transfer coefficient enhancements for surfaces with upstream pointing chevrons and staggered segments of chevrons. For example, Han and Zhang [5] investigated four cases of various segmented chevron geometry, as well as chevrons, segmented angled strips, and conventional turbulators in a simple duct flow. They found heat transfer coefficient enhancements as high as 3–4 compared to smooth surfaces for Re from 15,000 to 80,000. The friction coefficient enhancements were also generally higher than those for more conventional turbulators. Similar results at lower Re conditions were obtained by Taslim *et al.* [6] and Kiml *et al.* [7]. Cho *et al.* [8, 9] tested variations on segmented angled or transverse turbulators, finding average mass transfer coefficient augmentations in the range of 2.5–3 for Re up to about 70,000. While beneficial effects of turbulator angle were maintained, the number of segmentations did not show a consistent improvement trend. Chyu and Natarajan [10] presented detailed mass transfer data for transverse segmented turbulators where the segments were aligned or staggered. Alignment of the segments actually lowered the average mass transfer. While the segmentation gaps served to increase vertical mixing locally, these also disrupted the benefits of reattaching flows. Staggered segments, on the other hand, did produce higher average mass transfer coefficients by providing a more periodic nature to the disturbances. All of the foregoing studies examined enhancement features inside square or nearly square channels, so the features were not repeated laterally as would be the case between parallel plates or in ducts of higher aspect ratio.

Recent research into further turbulated channel or turbulated surface cooling is taking the direction of even more complex geometries such as perforated turbulators, turbulators detached from the wall, and turbulators with other vortex generators added on top. Care must be taken, however, to maintain robustness and high yield manufacturing also. Amongst the most promising geometries are variations on chevron turbulators, including broken or segmented chevrons, as well as the staggering of these features to increase surface flow interactions. Figure 4 shows an example of heat transfer coefficient data from several surfaces placed in a parallel plate channel flow at a Reynolds number of 100,000. Among the geometries tested are a smooth surface, transverse turbulators, unbroken chevron arrays, and broken/staggered counter-angled strips formed by removing the apex of each chevron, as shown in Bunker *et al.* [11]. The parallel plate channel is large enough in width to have about a dozen such pairs of features in a full adjacent surface array. The laterally averaged smooth surface Nusselt number distribution agrees with literature values for a developing flow between parallel plates. The Nusselt number distribution for the transverse turbulators is elevated by a factor of about 1.75, and exhibits the characteristic trait of decay with axial distance from the channel entry. The somewhat surprising behavior of the chevrons is that no decay characteristic is present; the heat transfer coefficient is enhanced to a level of more than two times the developed smooth surface value and remains there throughout. The broken and staggered chevron strip heat transfer coefficients improve a bit more, which might be attributed to the increased free edges generating additional mixing. Friction coefficients also increase with these changes.

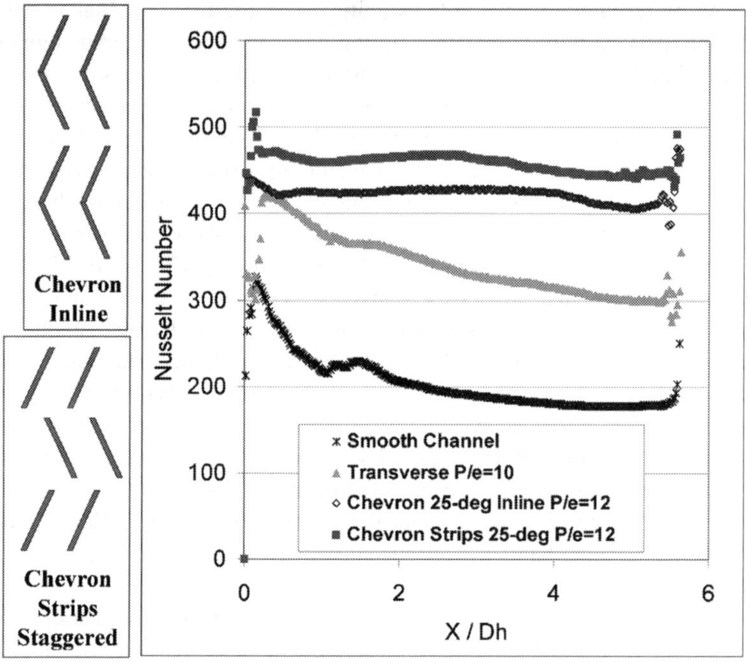


Figure 4: Turbulated heat transfer with forms of chevrons.

In a second example of the potential for such new turbulated surfaces, Fig. 5 presents laterally averaged heat transfer coefficients for staggered and counter-angled strips inside a duct of curved cross sectional geometry at Reynolds number of 100,000. The smooth surface behavior is as expected, with characteristic entry region decay. The transverse turbulators (90°) also exhibit this decay, but clearly show the periodic trend associated with flow separation and reattachment at each turbulator; the heat transfer coefficients are quite non-uniform. The angled 45° turbulators serve to eliminate this periodic behavior, but at the price of a somewhat lower average heat transfer coefficient, and with no benefit to friction coefficient (not shown). The counter-angled strips result in the best of both parameters by not only maintaining an elevated and uniform heat transfer coefficient, but also leading to almost a 50% lower friction coefficient compared with the other turbulated cases. It is speculated that the interaction of surface curvature effects with the counter-angled turbulator strips results in this very beneficial behavior.

## 3 Mesh network and micro cooling

As turbine blade cooling advances beyond the conventional methods, cooling will migrate from a largely internal bulk flow medium within hollow airfoils (e.g. Fig. 3) to one that may be built into or "within" the exterior walls of an airfoil.

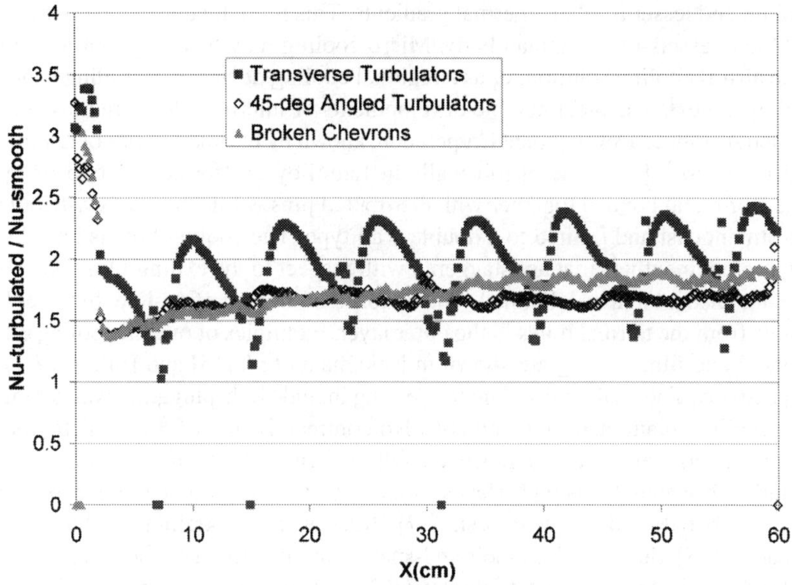


Figure 5: Heat transfer enhancement distribution comparison.

In essence, this change seeks to move the heat sink (the coolant and its wetted surface area) closer to the heat source (the hot gas path surfaces), thereby producing a more effective heat exchanger. This approach has the benefit of retaining a cooler internal bulk structure for the blade, which results in a longer rupture life. In the extreme, this type of design might be conceived as a micro cooling method. The use of many small cooling passages near the surface has potential benefits, but must also consider the risks associated with dirt plugging, manufacturing tolerances, and cost. The nearer term version of in-wall cooling is sometimes referred to as double-walled cooling, or here as mesh network cooling, in which the passages are of sizes amenable to investment casting. A greater degree of uniformity in internal cooling is key to many current thermal stress–strain limitations. Turbine airfoils can generally withstand higher bulk temperatures as the local thermal gradients are reduced. Zhang *et al.* [12] addressed this type of cooling network for different open flow areas, or alternately different solidity of channel meshes, using straight channels that intersect with included angle of 60 degrees, showing 2–2.5 times local heat transfer enhancement compared to a smooth channel. Heat transfer in small pre-film impingement chambers within airfoil wall sections was studied by Gillespie *et al.* [13] showing the contributions to cooling of both wall chamber interior surfaces.

The concept of micro cooling for airfoils is the natural extension of the more macroscopic mesh network cooling to reach the limits of thermodynamics and heat transfer. Micro cooling spreads out the cooling network in a series of smaller and highly distributed channels, or sub-channels, providing better uniformity of cooling and lesser in-plane thermal gradients. This is analogous to the distribution of blood vessels in the human body. Micro cooling may be thought of as a complete airfoil cooling solution, or as a regional cooling device (e.g. leading edges or trailing edges). The most notable developments of micro cooling are those of the Allison Advanced Development Corporation, known by the trade names of Lamilloy® and CastCool® [14]. The airfoil walls in Lamilloy are formed as two or more sheets of metal bonded together with distributed pins, while those of CastCool are investment cast and limited to a double-wall type. The cooling flow is introduced via impingement jets that are staggered with respect to the exiting film holes. Full cooling distribution is obtained, and a form of transpiration-like film cooling results from the normal holes in the outer layer. Examples of overall cooling effectiveness and film cooling are shown in Nakamata *et al.* [15] and Hale *et al.* [16], respectively. The challenges of micro cooling include hole plugging, wall strength, film cooling, manufacturing, and cost. No commercial use of these micro cooled solutions has yet appeared however. Other forms of proposed micro cooling include fabricated diffusion surfaces formed by electro-galvanic metal deposition, such as that described by Battisti [17]. Russian and Ukrainian turbine airfoil research [18] developed a shell-and-spar approach to manufacturing. In this method, the interior portion of the airfoil is cast in a simple format, and cooling channels machined in the exposed surface. The channels are then filled with a leachable material, a thin outer airfoil skin is bonded to the non-channel metal regions, and the filler is leached out. The result is a distribution of small cooling channels close to the outer surface. Again, these micro cooling forms face significant material and processing challenges, as well as unexplored thermal and fluid design domains.

A demonstration of the capability of mesh network cooling is provided in the research of Bunker *et al.* [19]. Cooling meshes were tested using round pins and rounded diamond shaped pins with height-to-diameter ($H/D$) ratios of 0.2 and center spacing ($S/D$) ratios of 1.5, as well as less dense rounded diamond pins of smaller $H/D$ of 0.3 and $S/D$ of 2.14. These geometries differ substantially from conventional pin fin arrays in which $H/D > 1$ and $S/D$ ratios are typically about 2.5. Special attention was paid to the combination of techniques including pin meshes, turbulators, and dimple/concavity arrays that are shown to provide multiple design solutions for heat transfer and pressure loss objectives. Figure 6 shows one such combined geometry using pins, concavities, and turbulators. Figure 6 also shows the potential use for these new in-wall mesh networks as improvements in cooling designs for turbine blades per Lee and Bunker [20]. Figure 7 shows that average channel area-corrected heat transfer capabilities exceeding three times that of smooth channels have been demonstrated using actual surface wetted area enhancements of no more than 20%. This cooling capability increase is also realized with a large decrease in channel material solidity, up to 30%, compared to conventional pin fin arrays. For both heat transfer and friction coefficients, some effects are additive, while others are synergistic in exceeding strictly additive behavior. The use of concavity surfaces within other structural geometries is particularly important for minimizing increases in friction.

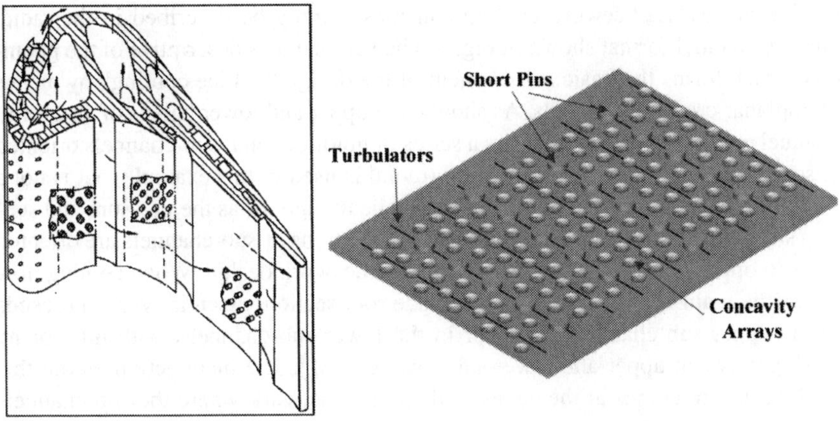


Figure 6: Mesh network geometry and turbine blade design.

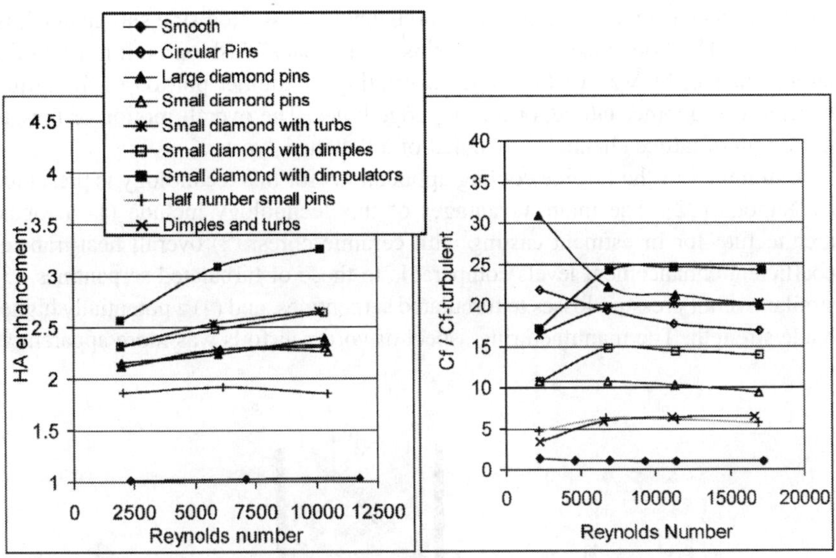


Figure 7: Mesh network heat transfer and friction enhancements (reproduced with permission from ASME).

## 4 Latticework (vortex) cooling

Latticework cooling, also known as vortex cooling or bounded vortical duct cooling, in its application to high-temperature gas turbine components originated within the former Soviet design bureau engineering system 25 years ago [21]. This method of vane and blade structure and cooling developed into the most recognizable standard within the many Soviet designs (Russian and Ukrainian designs today). Its application history parallels the Western use of serpentine cooling for turbine blades. Latticework cooling can most simply be described in the radial cooling channel format shown in Fig. 8. The very name is descriptive of the geometry which forms the basic sub-element of the design, "lattice cooling" by means of coplanar crossing channels. As shown, the upper and lower halves of an overall channel or cooling region each has a series of unidirectional sub-channels oriented at some angle $\beta$ to the radial direction (radial is used in this example, but is not a limitation). The upper and lower walls may be thought of as the pressure and suction sides of an airfoil, for example. The two portions of sub-channels are oriented so as to oppose each other, or cross as in a latticework design. When cooling flow enters the main network, such as at a blade root section, essentially half proceeds in the upper sub-channels, and half in the lower sub-channels, with little or no mixing between upper and lower sub-channels [21]. The main action within this cooling design comes at the edges of the lattice network where the sub-channels encounter seemingly "dead ends" or "bounding" walls, these would be the interior rib or the external wall of an airfoil design. Upon encountering the side wall, the flow must turn by the angle $2\beta$ as it enters the upper (or lower) crossing channel, i.e. it switches from a pressure side sub-channel to a suction side sub-channel, or vice versa. The flow makes its way in this "switchback" fashion until it leaves the lattice channel by way of film holes or routing to another portion of the airfoil interior, e.g. another lattice, or trailing edge holes. The overall motion of flow in the length of lattice channel is then that of a flattened "vortex".

A summary of the turbine cooling applications for this technology is presented by Nagoga [22]. The main advantages of this technology include (1) a robust architecture for investment casting with ceramic cores, (2) overall heat transfer coefficient enhancement levels comparable to those of turbulated serpentines, (3) similar overall pressure losses to turbulated serpentines, and (4) a potentially higher blade strength. The manufacturing aspect of vortex airfoils was more apparent as

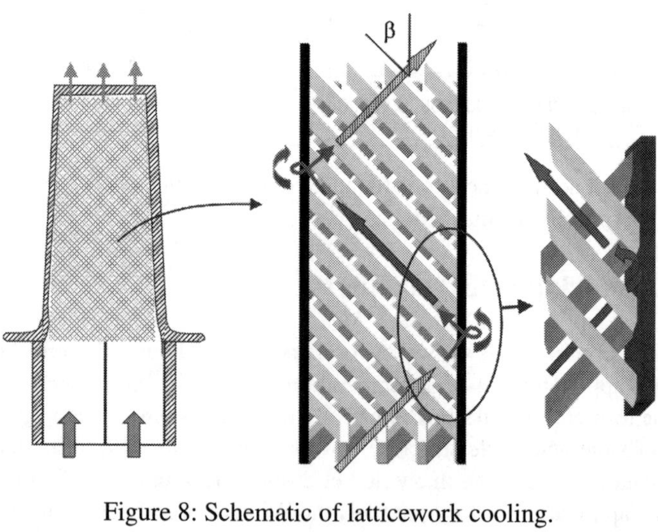


Figure 8: Schematic of latticework cooling.

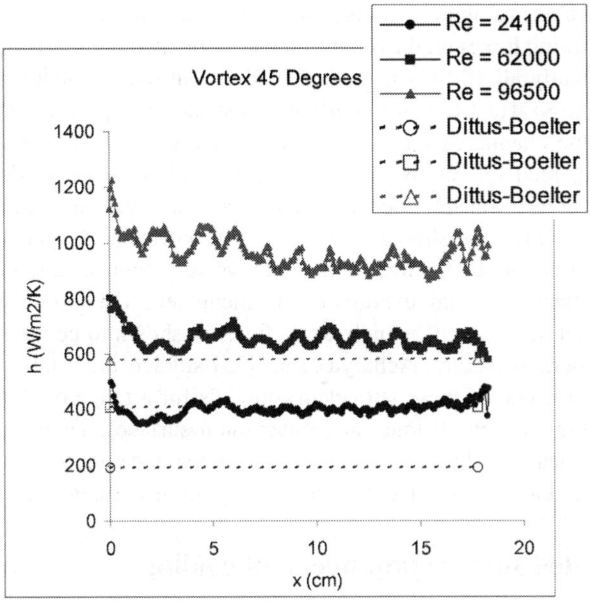


Figure 9: Latticework laterally averaged heat transfer coefficients (reproduced with permission from ASME).

an advantage in the initial years of development when ceramic core material strength required higher core connectivity, but may be diminished today with the advances of investment casting technology. The blade strength advantages are due to the use of distributed internal ribs of full or partial extent with substantial radial orientation components, but may be countered by other designs and materials. The overall cooling effectiveness of a turbine blade with leading and trailing edge latticework cooling is shown in Goreloff *et al.* [23].

The study of Bunker [24] provides detailed information concerning the heat transfer coefficients and pressures in latticework cooling channels. This study used two test methods to determine the local and overall heat transfer coefficients for a vortex channel with $\beta$ of 40–45°. Both liquid crystal and infrared thermographic methods were used on acrylic and metallic models, respectively, to discern the heat transfer coefficients without and with the important effects of internal rib fin effectiveness. Tests with insulating ribs determined the heat transfer on the primary surfaces representing the pressure and suction walls of an airfoil. Tests with integral metal ribs determined the additional impact of the fin effectiveness provided by the lattice ribs. A simple radial vortex channel design was employed throughout with sub-channel aspect ratios near unity and sub-channel Reynolds numbers from 20,000 to 100,000. Figure 9 shows an example of the laterally averaged primary surface heat transfer coefficients, varying in the radial ($x$) direction for a narrow, 45° channel latticework model. Primary surface enhancements average about 1.5 over fully developed, smooth duct behavior (Dittus–Boelter), but reach local values of about 3 immediately after each turn. Pressure distributions show high turning losses on the order of those associated with serpentine 180° turn circuits (virtually all 180° turns in real blades contribute a major portion of the circuit total loss). Heat transfer coefficient distributions are remarkably uniform throughout the channels excepting the turns themselves, because turn enhancements are retained for relatively long distances. Figure 10 provides the overall vortex channel heat transfer coefficient enhancement levels, including internal rib effectiveness, which are shown to be 2.5–3. In narrow vortex channels, overall enhancements are about 3, while in the wider vortex channels, turn effects are less of the total effect producing overall enhancements levels of about 2.8. The effects of sub-channel internal ribs, which act as fins, are shown to be very important in the overall thermal picture. Acharya *et al.* [25] studied the rotational effects on heat transfer for one of these latticework models for a range of Rotational numbers; obtaining data on all inner and outer main surfaces. Their results indicate very uniform heat transfer, but more importantly that the effects of rotation are not at all significant as they are in conventional serpentine cooling channels.

## 5 Augmented surface impingement cooling

The use of impingement jets for the cooling of various regions of modern gas turbine engines is widespread, most especially within the high-pressure turbine. Since the cooling effectiveness of impingement jets is very high, this method of cooling provides an efficient means of component heat load management given sufficient available pressure head and geometrical space for implementation. Regular arrays of impingement jets are used within turbine airfoils and endwalls to

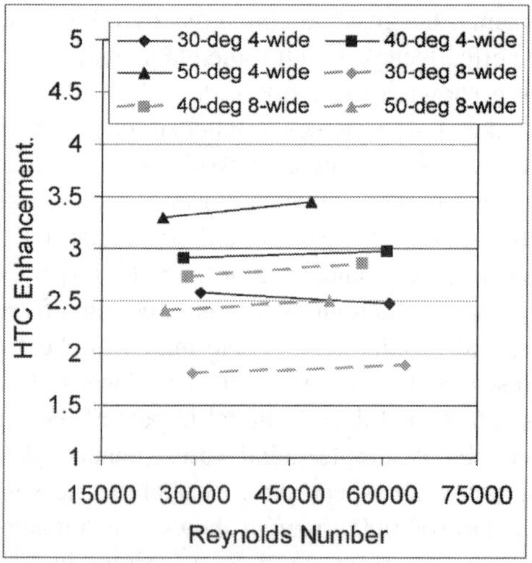


Figure 10: Latticework overall heat transfer enhancements (reproduced with permission from ASME).

provide relatively uniform and controlled cooling of fairly open internal surface regions. Such regular impingement arrays are generally directed against the target surfaces by using sheet metal baffle plates, inserts, or covers which are fixed in position relative to the target surface. These arrangements allow for the design of a wide range of impingement geometries, including in-line, staggered, or arbitrary patterns of jets. In more confined regions of airfoils such as the leading edge or trailing edge, spanwise lines of impingement jets are sometimes used to focus cooling on one primary location of high external heat load like the airfoil aerodynamic stagnation region. There also exist many other applications for individual impingement jets on selected stationary and rotating surfaces. Vane endwalls, blade platforms, unattached shrouds, and combustor liners may all have specific local cooling requirements well suited to the use of individual jet cooling. Impingement jets are also used on rotor disk cavity faces and in some applications may provide additional functions of sealing. Summaries of applicable impingement heat transfer research may be found in Martin [26] and Han and Goldstein [27].

A wealth of information exists on the basic cases of individual and array jet impingement heat transfer. The heat transfer coefficient distributions due to a single axisymmetric jet impinging normally on a smooth flat plate with free spent air discharge were investigated by Gardon and Cobonpue [28]. This study determined the basic effects of jet-to-target spacing and jet Reynolds number on stagnation region and radial heat transfer. Perry [29] measured single air jet impingement heat transfer characteristics on a free surface for normal and oblique impingement angles, determining the local and averaged effects of jet angling. Mass transfer normal jet impingement experiments of Schluender and Gnielinski [30] and Petzold [31], as summarized in Martin [26], show similar stagnation and non-stagnation region Sherwood number behavior for very high jet Re conditions of as much as 375,000.

While jet Reynolds number, target distance, and impingement angle all have some effect on stagnation region heat transfer coefficient, as well as the overall surface heat transfer distribution, target surface roughness and texturing can have a major impact on the impingement dominated heat transfer portion of the surface. Moreover, roughness can frequently be tailored to address cooling needs in specific localities. Effects that tend to thin the boundary layer relative to the roughness element heights, such as increased Reynolds number or decreased jet diameter, lead to increased heat transfer. Chakroun *et al.* [32] examined impingement heat transfer for a single normal air jet on a deterministic patterned rough surface. Heat transfer augmentations of up to 28% were noted, with a measured increase in the turbulence intensity of the flow around the roughness elements. El-Gabry and Kaminski [33] measured peak and average heat transfer coefficients for an array of normal or angled jets impinging on a randomly rough surface. Figure 11 shows a photo of the close-packed particles forming the rough surface (average roughness Ra value of 33 μm), as well as the basic heat transfer data for normal and angled jet arrays spaced 2 jet diameters from the surface. Heat transfer coefficients were observed to increase by as much as 25% in this study, much more than the roughly 8% surface area increase. Also of note, due to the use of a rough surface, the ratio

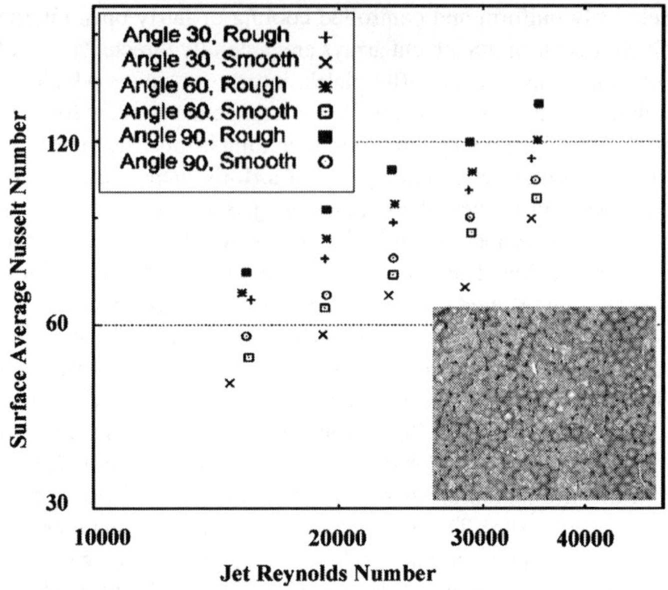


Figure 11: Rough surface impingement jet array heat transfer (reproduced with permission from ASME).

of peak-to-average heat transfer coefficient was reduced to half that of the smooth surface. As an example of the very localized enhancement due to roughness, Fig. 12 provides the smooth and rough surface heat transfer coefficient distributions for the turbine blade platform cooling model study of Bunker and Bailey [34]. Also shown is a photo of the random roughness surface, which, in this case, is not close-packed (average roughness Ra value of 30 μm). In this study, a single nearly normal impingement jet is used with Reynolds number of 130,000, spaced 2.5 jet diameters from the surface. The immediate stagnation region of the jet exhibits a 50% increase in heat transfer coefficient, while this benefit is very quickly lost as flow transitions to a convective condition over the surface.

Several forms of patterned surface augmentation, more amenable to investment casting, have also received increasing attention for use under impingement jets. These include pin arrays in the form of discrete shaped bumps, regularly spaced turbulators, and even arrays of concavities. The intent of these surface augmentation methods is to derive a greater benefit than simple wetted surface area increase by also creating additional vortices and turbulence. The study of Son *et al.* [35] examined the use of cylindrical surface pins under impingement jets for a flat surface, while that of Taslim *et al.* [36] investigated conical bumps to augment airfoil leading edge impingement cooling. Both studies found heat transfer to be augmented by roughly an amount equal to the increased surface area ratio factor. Kanokjaruvijit and Martinez-Botas [37] tested several configurations of impingement jet array parameters with dimpled target surfaces of various forms. They found that shallow dimples could provide as much as 50% improved heat flux

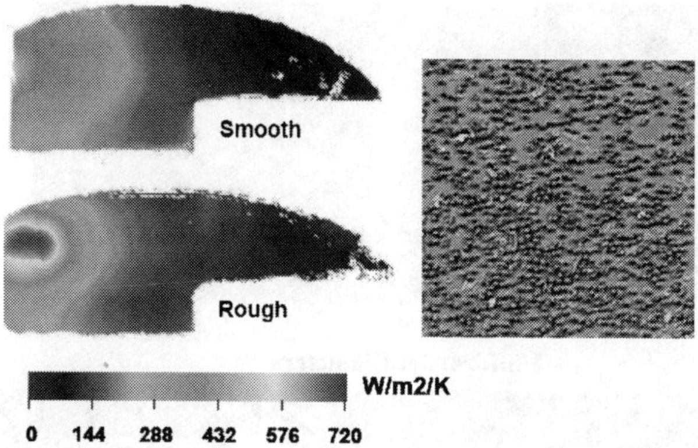


Figure 12: Rough surface single impingement jet heat transfer augmentation (reproduced with permission from ASME).

(including area factor), while deeper dimples could actually serve to create poor recirculation zones. Gau and Lee [38] studied various forms of two-dimensional jet impingement on rib-roughened surfaces, also showing that with proper relative geometry control heat flux can be significantly increased, but creation of trapped flows can decrease heat flux below the non-ribbed condition. Of great significance throughout these studies was the finding that pressure losses were negligibly affected by the use of such surface methods with impingement flows.

As noted above, the correct relative size and location of impingement jets and surface augmentations is very important for obtaining desired results. Most research has focused on the modification of the surfaces, but the jet arrays may also be modified from the conventional uniform in-line or staggered spacing. Bailey and Bunker [39] examined heat transfer with jet array spacings outside that of prior literature, both smaller and larger. Significant deviations from past data were noted, most especially in cases where jet arrays become very dense, leading to large variations in jet Reynolds numbers. As an extension to this work, Gao *et al.* [40] studied the effects of linearly stretched jet arrays in which the row-to-row spacing is increased with distance inside the channel. Fig. 13 shows an example of the Nusselt number distribution for average jet Reynolds number of 10,000. The streamwise averaged heat transfer coefficients decline as the jets become less numerous. Fig. 13 also shows a second modification to compensate for this decline by increasing the jet diameter on each row. The combined effect serves to make the streamwise distribution much more uniform.

## 6 Concavity surfaces cooling

Another class of surface enhancements results from the depression of features into the cooling channel or surface walls, forming recesses rather than projections. Generically, such features are known as concavities (or dimples), and may

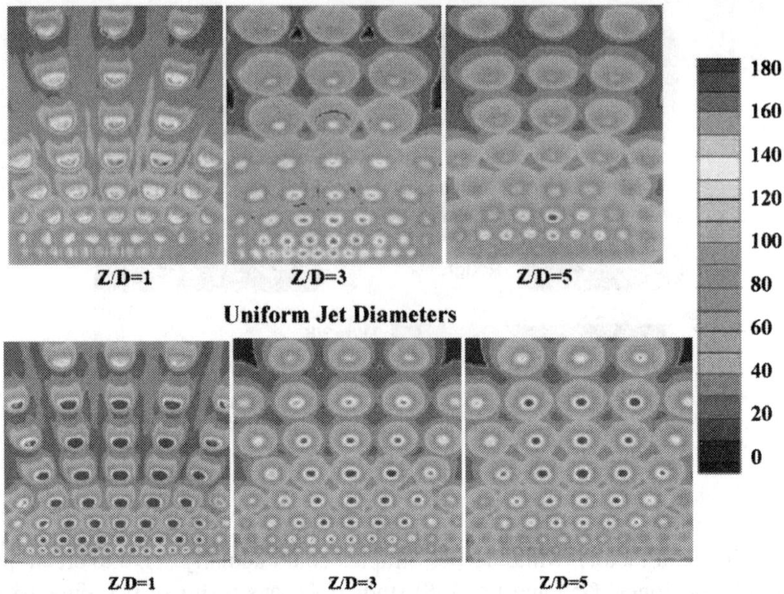


Figure 13: Stretched and variable jet diameter impingement Nusselt distributions (reproduced with permission from AIAA).

be formed in an infinite variation of geometries with various resulting heat transfer and friction characteristics. Concavity surfaces are commonly known for their drag reduction characteristics in external flows over bodies. The most famous example being golf balls, where they serve to delay the point of boundary layer separation, thereby reducing overall drag for the sphere, as shown by Bearman and Harvey [41]. Application of such concavity surfaces to external flows for marine vessels and airframes has been suggested by Kiknadze *et al.* [42], in which defined arrays of shallow concavities may cover the main hull or fuselage surfaces to prevent a thick boundary layer from forming, thereby reducing the drag.

Applications of concavity surfaces involving heat transfer remained largely unknown until the recent dissolution of the Soviet Union, at which time Russian research from the 1980s began to surface. The basic fluid dynamic condition for flow over concavities of spherical or cylindrical shape is well described in the study of Afanas'yev *et al.* [43]. A flow with boundary layer thickness less than the concavity surface diameter reacts with the cavity by flowing into the "bowl", experiencing a separated region of some extent on the entry side as depicted in Fig. 14. The spherical shape, or one nearly approximating it, creates a pressure field within the bowl acting to collapse or concentrate the flow in the downstream portion of the recess, creating a vortex structure. In a steady flow over a symmetric spherical dimple, a pair of symmetric, counter-rotating vortices is ideally formed, as shown by the computational work of Isaev *et al.* [44] in Fig. 15. In most real flow cases,

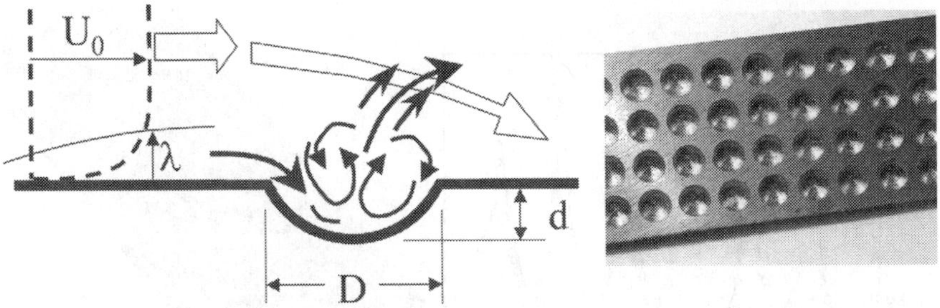


Figure 14: Schematic of concavity flow and sample surface array.

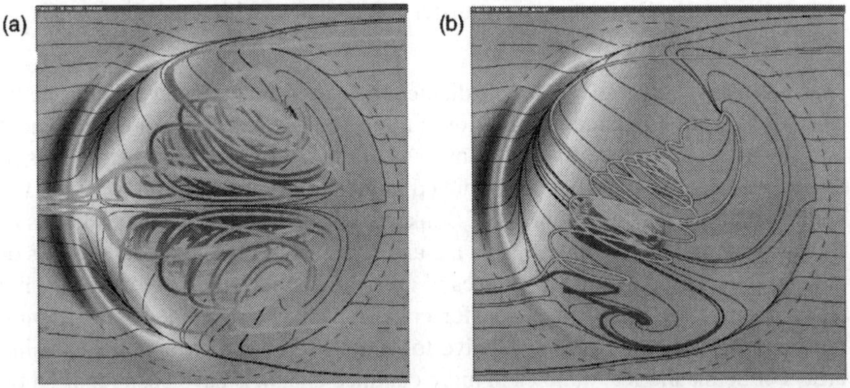


Figure 15: CFD showing jet-vortical structures in a deep spherical dimple in turbulent flow: (a) symmetric vortex pair; (b) asymmetrical single vortex (graphic courtesy of S. Isaev).

only one vortex is created, and this vortex may move side-to-side with some frequency. This condition is said to "expel" flow from the cavity as an organized vortical structure. As the vortex penetrates into and interacts with the mainstream flow, it provides a scrubbing action, which brings fresh core fluid to the surface for enhanced heat transfer. Because the motion is organized, rather than the more dissipative effect of shearing layers, the pressure loss is less than that observed with projecting obstructions such as turbulators. In fact, some cases of concavity surfaces have friction nearly the same as smooth surfaces. Studies focusing on the heat transfer coefficients on a surface with a single hemispherical concavity show the fundamental potential of this method for thermal enhancement. Kesarev and Kozlov [45] show the detailed shear stress and heat transfer coefficient distributions relative to a flat surface for a single dimple, observing overall heat transfer augmentations of up to 1.5.

The full potential for concavity surface heat transfer comes in the application of full-surface arrays of ordered concavities. Wind tunnel surface studies of Afanas'yev and Chudnovskiy [46] and also Afanas'yev *et al.* [47] for arrays of

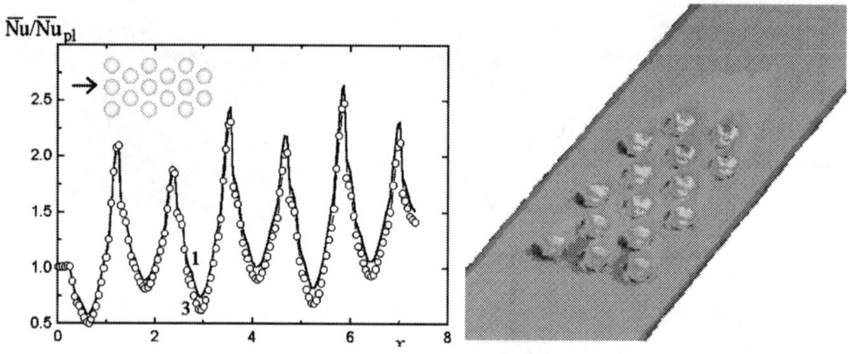


Figure 16: Computed heat transfer enhancement with dimple array.

spacings from about 1 to 1.7 dimple diameters showed heat transfer enhancements of 30–40% with no increase in friction. Figure 16 shows the typical development and streamwise increase in heat transfer coefficients for an array of dimples at fairly low Reynolds number, as predicted numerically by Bunker *et al.* [48], and confirmed in parallel experiments of Bunker *et al.* [49]. Heat exchanger studies of Belen'kiy *et al.* [50, 51] investigated the use of dimple arrays on the inner walls of annular passages and on tube surfaces of crossflow bundles, respectively. In some cases, as much as a 2.4 heat transfer enhancement was obtained with friction increases on the order of 2–4, relative to smooth surfaces. Recent studies using arrays of hemispherical dimples in more confined channels have been reported by Chyu *et al.* [52], Moon *et al.* [53], and Mahmood *et al.* [54]. In these studies, heat transfer enhancements of 2 to 2.5 have been demonstrated with friction factor increases from about 1.5 to 4. For application in turbine cooling, the summary by Nagoga [22] provides some insights into the effects of concavity array geometric parameters, namely the dimple depth-to-diameter ratio, the channel height-to-dimple diameter ratio, and the dimple spacing or surface density "f". An important feature to note about dimpled surfaces is the analogous form of the friction coefficient to that seen for rough surfaces, i.e. a decreasing magnitude of coefficient as Reynolds number increases, and a limiting value for a "fully rough" zone. A critical Reynolds number is demonstrated above which the friction coefficient does not change.

As with many other cooling technologies, care must be taken to examine the specific use and application of the augmentation means. For example, Bunker and Donnellan [55] investigated the use of dimpled surface arrays in circular cooling passages with Reynolds numbers up to 90,000. The resulting behavior of Nusselt number enhancement versus friction factor enhancement relative to a smooth passage is shown in Fig. 17, along with a photo of a sample dimpled tube (split for viewing). In this instance, the dimpled surfaces obtained as much as two times smooth surface heat transfer with moderate friction factor increases, though not matching a Reynolds analogy ideal. Still, the performance is much better than many conventional forms of turbulated channels.

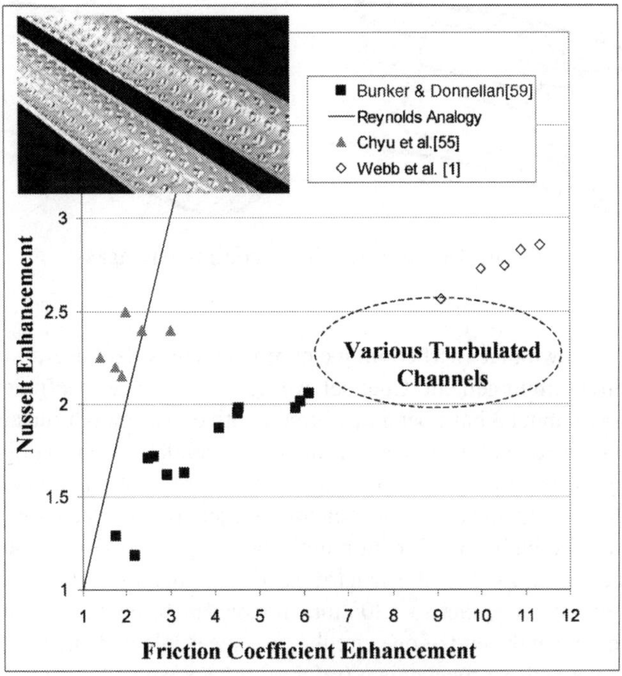


Figure 17: Concavity array heat transfer inside circular passage (reproduced with permission from ASME).

## 7 Swirl (cyclone) cooling

In the broadest sense, the concavity surface flows are one of a larger category known as "vortex" technologies, as described in the summary of Khalatov [56], which include various means of the formation of organized vortical or swirling flows in turbines. The latticework cooling presented earlier is one of these vortex technologies, utilizing the angled turning regions to generate bulk swirl. Another emerging vortex technology is the use of discrete wall jets injected into concave cooling passages, or along concave internal wall sections, to induce a bulk swirl motion. This cooling technique is generally known as swirl cooling, and also cyclone cooling from the original Russian studies. The studies of Glezer *et al.* [57], Hedlund *et al.* [58], Ligrani *et al.* [59], and Glezer *et al.* [60] provide the fundamentals of swirl cooling for blade leading edge passages.

Figure 18 shows two sketches of swirl cooling chambers implemented inside a blade leading edge, one without film extraction and the other with film extraction. Rather than using direct impingement jets aimed at the apex of the concave region, racetrack shaped wall jets are injected tangential to the surface at intervals along the blade height. The jets provide high heat transfer locally, but also serve to refresh the local coolant nearest to the surface at each new injection site. These

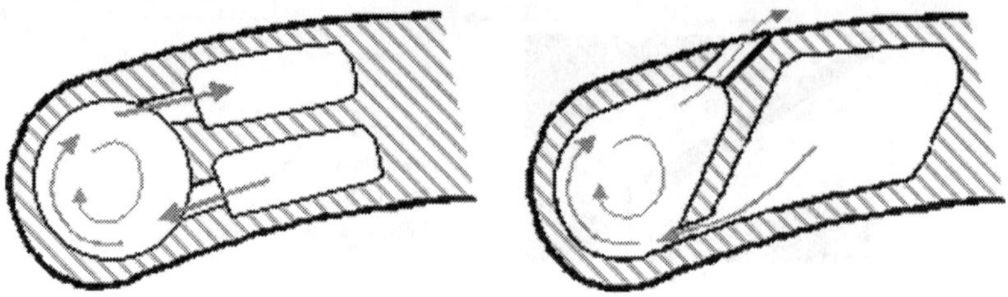

Figure 18: Swirl (cyclone) cooling passages.

studies have shown that swirl cooling can provide equivalent overall heat transfer to that of direct impingement. Channel average heat transfer coefficient enhancement factors of about 3 have been measured relative to a smooth surface, turbulent channel flow for Reynolds numbers from 5000 to 80,000. In an analogous manner to the decay of heat transfer coefficients downstream of abrupt entry regions, the swirl heat transfer is highest at the location of injection and decreases in a regular fashion along the bulk flow direction until the next injection location is reached. Other studies, such as that of Khalatov *et al.* [61] demonstrate the use of swirl cooling as an enhancement to 180° turn regions in serpentine channels. In this form, the coolant at the end of one circular passage is injected into the next circular passage via the wall jet mechanism (referred to as a tangential swirl generator in this case), thereby forming an initial cyclone effect that is allowed to decay along the remainder of the channel with no addition injections.

## 8 Film cooling

Film cooling is one of the major technologies allowing today's gas turbines to obtain extremely high turbine firing temperatures, subsequent high efficiencies, and longer life parts. The art and science of film cooling concerns the bleeding of internal component cooling air through the external walls to form a protective layer of cooling between the hot gases and the component external surfaces. The application of effective film cooling techniques provides the first and best line of defense for hot gas path surfaces against the onslaught of extreme heat fluxes, serving to directly reduce the incident convective heat flux on the surface. Fundamental early research concentrated on nearly ideal film cooling formed by a two-dimensional layer as depicted in Fig. 19a. Goldstein [62] provides a thorough summary of the theory, modeling, and experimental studies surrounding such idealized film cooling layers. Because of its high importance and widespread application, research into the many aspects of film cooling has seen a tremendous increase in the last ten to fifteen years. The publications relating directly or indirectly to film cooling deal with the major effects of film hole internal fluid dynamics, interactions with the mainstream gas flow, turbulence and vorticity production,

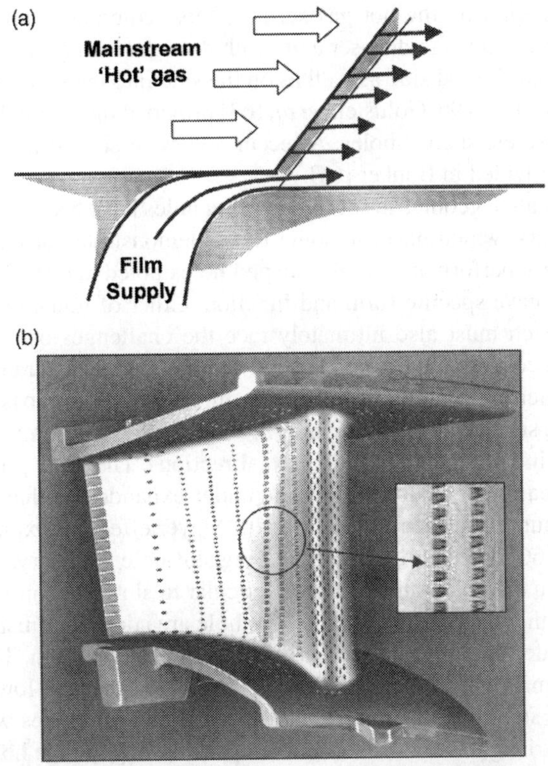


Figure 19: (a) Idealized two-dimensional film cooling; (b) typical film cooled turbine inlet guide vane.

effects of approach flows prior to the hole entry, hole shaping, orientation, and spacing, hole length-to-diameter ratio, density ratio, blowing strength, momentum flux ratio, effects of mainstream turbulence intensity, mainstream acceleration, external surface curvature, and external surface roughness.

Over the past 30+ years, investigations have been performed by a broad spectrum of researchers to understand the fundamental physics of film cooling, and to improve the state-of-the-art. The primary focus of most research has been on the use of discrete film holes, or rows of film holes, on the hot gas path surfaces of the turbine, since mechanical constraints dictate this format. Fig. 19b shows an example of a typical high-pressure turbine inlet guide vane with many film cooling rows of varying spacing and orientation. Only one primary advance in this technology has been put into widespread practice over these many years. That single improvement has been the change from round film holes to shaped film holes. Furthermore, the use of the term "shaped", while allowing a potentially vast number of geometries, is actually limited again to a single general class of geometry. Shaped holes are composed of round metering or throat sections with a uniform and symmetric expanded exit region on the hot gas surface. Most commonly, all shaped holes applied in practice have fan diffuser exits with divergence angles between 10 and 15 degrees on each lateral side as well as on the side into the surface, as shown in the inset photo of Fig. 19b. Goldstein *et al.* [63] provided the seminal study on film cooling with discrete shaped holes. A recent review of shaped film hole cooling technology is provided in Bunker [64].

Several alternative geometries of film cooling holes have been proposed within the last few years, which have, in some form, demonstrated at least equivalent film effectiveness performance to the shaped holes noted above. These differing film holes may have specific form and function, either of limited or widespread potential, but each must also ultimately face the challenges of manufacturing, operability, and cost effectiveness. Moser *et al.* [65] analyzed a transonic wall jet geometry in which the film hole transitions from some interior cross section, such as circular, to a surface slit, as shown in Fig. 20. This transition is such that the hole converges in both the axial and lateral directions. The basic principle of this design is the creation of a choked flow and under-expanded jet that will conform to the exterior curved surface via the Prandtl–Meyer effect for expansion waves. Sargison *et al.* [66] demonstrated a converging slot-hole geometry, also shown in Fig. 20, in which the hole transitions from circular to slot with convergence in the axial direction and divergence laterally. The hole area does diminish to cause the flow to accelerate, though not necessarily to a choked condition. The exit is the metering section. This accelerated flow is speculated to have lower jet turbulence and more stability. Nasir *et al.* [67] tested round film holes with the addition of various triangular tabs covering the upstream edge of the holes, as drawn in Fig. 20. The tabs in some cases modified the exit flow and vortex structure in a manner that kept the coolant from lifting off. Fric and Campbell [68] investigated a so-called cratered film hole in which the circular hole exits into a shallow right circular surface cup or depression. The flow actually impinges on the edge of this depression causing it to deflect and fill the depression prior to issuing onto the external surface. Figure 21 shows water tunnel planar laser induced

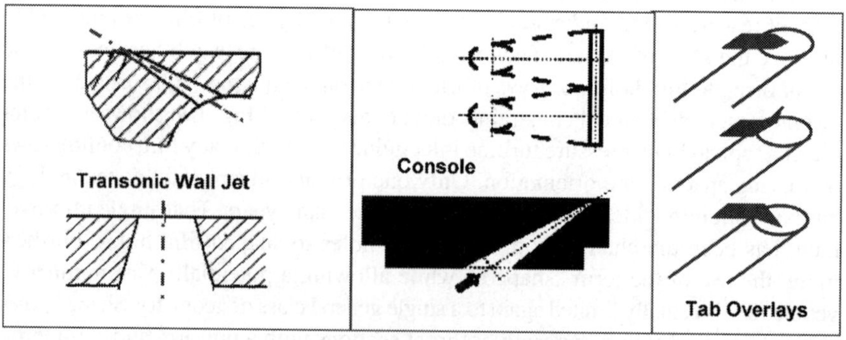


Figure 20: Innovations in film cooling holes.

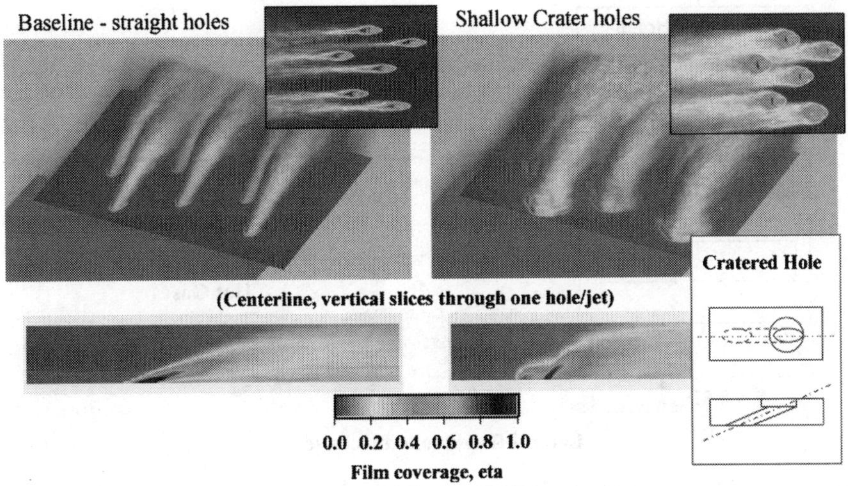


Figure 21: Crater film hole modification of injected coolant.

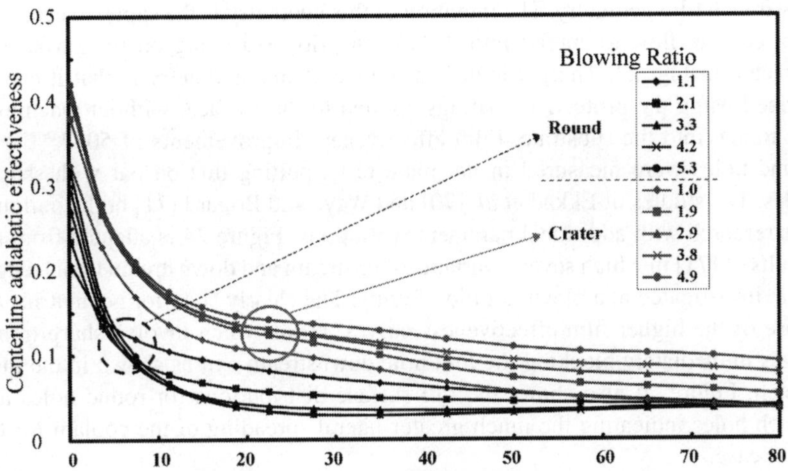


Figure 22: Improved centerline adiabatic effectiveness using crater film holes.

fluorescence data comparing the crater hole effect to that of a round film hole at a blowing ratio of 0.8. Three-dimensional, near-surface, and centerline cross-sectional views all show the effect of film jet lateral spreading due to the crater. Flat plate high-speed wind tunnel data shown in Fig. 22 result in 50–100% improved effectiveness over round holes, with better performance as blowing ratio is increased.

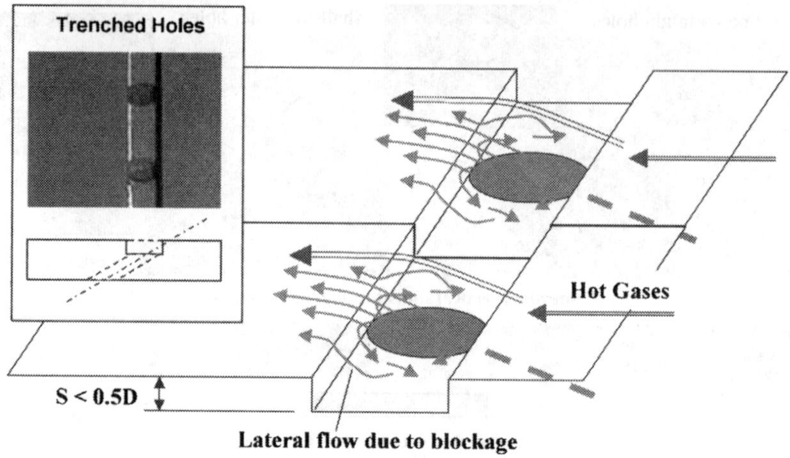


Figure 23: Flow interaction effect using trench film row (reproduced with permission from ASME).

Bunker [69] extended the work of [68] to a two-dimensional version, or shallow trench film geometry. The intention of this geometry is the same, to cause the film cooling flow to spread into the trench prior to issuing on the surface, as depicted in Fig. 23. The particular advantage of this geometry is that it can be formed using the protective coatings applied to the surface, without machining the trench into the substrate. Film effectiveness improvements of 50–75% over round holes were measured in flat plate tests, putting this on par with shaped holes. The studies of Ekkad *et al.* [70] and Waye and Bogard [71] both continued this research with additional parameter variations. Figure 24 is adapted from the results of [71] in which several variants of upstream and downstream trench edges were investigated at a blowing ratio of unity. The clearly superior geometries are those of the higher film effectiveness curves, all of which involve sharp trench edges immediately blocking the film hole downstream exit as shown in the inset sketch. Figure 24 also shows the full surface distributions for round holes and trench holes indicating the much greater lateral spreading of the coolant for the latter case.

The complete picture of film cooling must however also include the effect of the film injection on the augmentation of heat transfer coefficients due to the disruption of the boundary layer. Typically, any film cooling injection will increase heat transfer coefficients at least locally just downstream of the injection location, followed by a return to non-disrupted conditions further downstream (though not a return to the original boundary layer thickness). The augmentation of heat transfer coefficients can in some cases actually outweigh the benefit of film cooling effectiveness and cause a net heat flux increase to the surface. The goal of film cooling is always to obtain the highest possible net heat flux reduction, taking both factors into account. Using the same vane cascade test apparatus as [71], Harrison *et al.* [72] measured the heat transfer

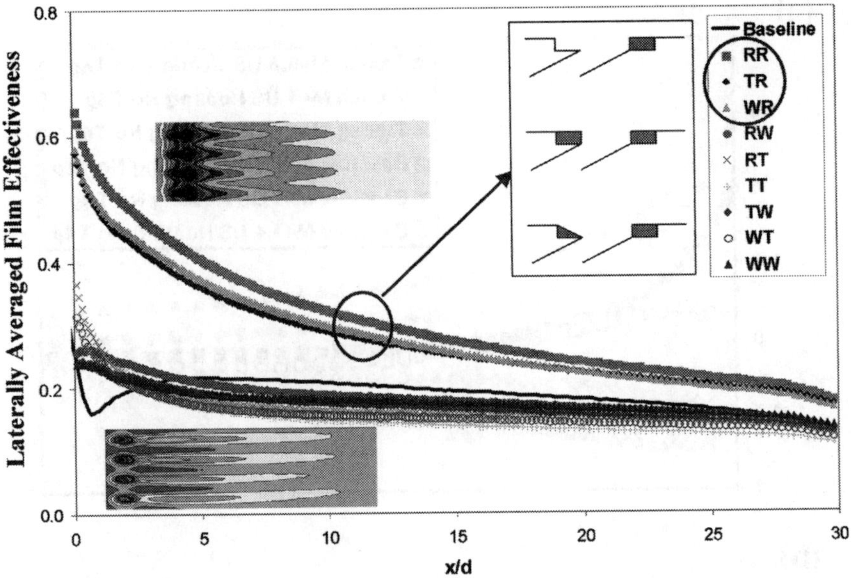


Figure 24: Laterally averaged film effectiveness for sharp edged trench (reproduced with permission from ASME).

coefficients for both round film holes and shallow trench film cooling under several imposed upstream conditions of the approaching boundary layer. With both the film effectiveness and heat transfer coefficients, and assuming a typical value for an engine wall-to-gas temperature ratio of 0.6, the laterally averaged net heat flux reductions were determined for several blowing ratios. Figure 25 shows these distributions for the representative conditions of upstream heating (a) without and (b) with a tripped boundary layer prior to the film holes. In all cases, the film effectiveness of the shallow trench holes greatly outweighs any increase in heat transfer coefficients, resulting in very good net heat flux reductions that increase with blowing ratio. In contrast, the effectiveness of the round film holes is so poor as to lead to some negative net heat flux ratios, especially at higher blowing ratios.

# 9 Conclusion

This chapter has presented several of the cutting edge, innovative cooling methods expected to further enhance the aerothermal–mechanical performance of turbine engines used for power systems. However, these methods are by no means an exhaustive or comprehensive summary. Many other variations and combinations of these techniques are anticipated as manufacturing advances become a reality. Further improvements and new techniques may become feasible as materials, systems integration and controls also advance.

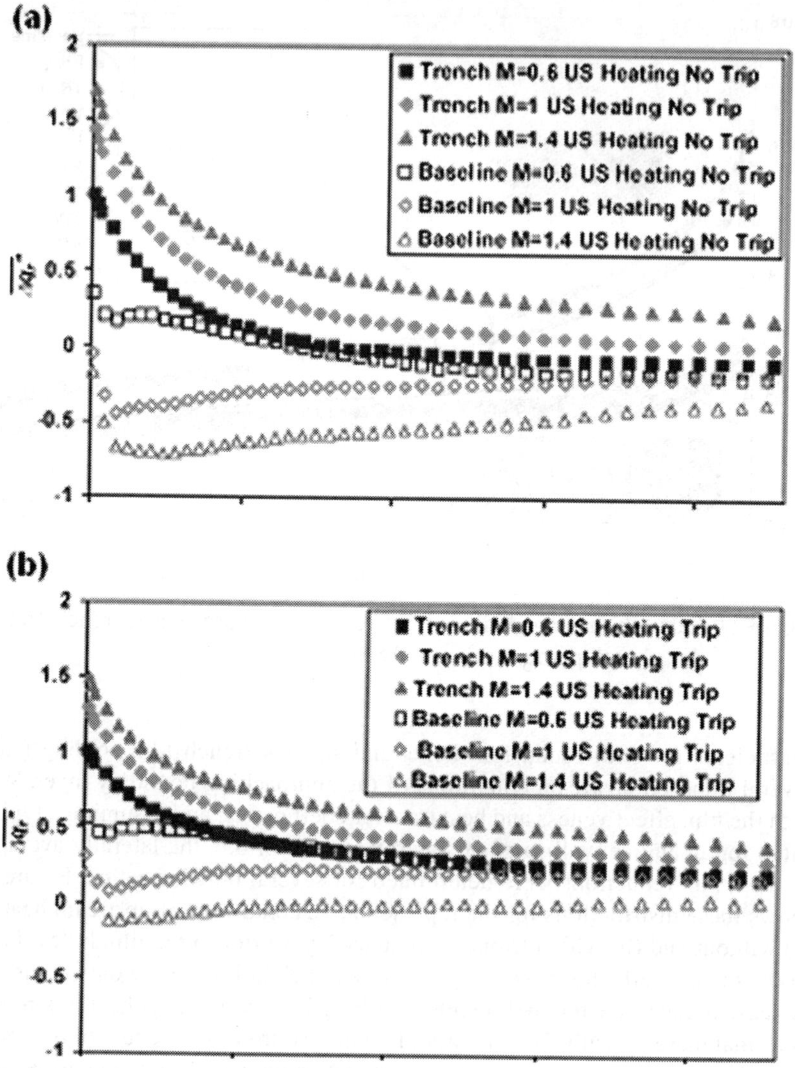


Figure 25: Net heat flux reduction comparison for round and trench film (a) without and (b) with trip (reproduced with permission from ASME).

## References

[1] Webb, R.L., Eckert, E.R.G. & Goldstein, R.J., Heat transfer and friction in tubes with repeated-rib roughness. *International Journal of Heat and Mass Transfer*, **14**, pp. 601–617, 1971.

[2] Burggraf, F., Experimental heat transfer and pressure drop with two-dimensional turbulence promoter applied to two opposite walls of a square

tube. *Augmentation of Convective Heat and Mass Transfer*, eds A.E. Bergles & R.L. Webb, ASME: New York, pp. 70–79, 1970.

[3] Han, J.C., Glicksman, L.R. & Rohsenow, W.M., An investigation of heat transfer and friction for rib-roughened surfaces. *International Journal of Heat and Mass Transfer*, **21**, pp. 1143–1156, 1978.

[4] Han, J.C., Park, J.S. & Lei, C.K., Heat transfer enhancement in channels with turbulence promoters. *Journal of Engineering for Gas Turbines and Power*, **107**, pp. 628–635, 1985.

[5] Han, J.C. & Zhang, Y.M., High performance heat transfer ducts with parallel broken and V-shaped broken ribs. *International Journal of Heat and Mass Transfer*, **35(2)**, pp. 513–523, 1992.

[6] Taslim, M.E., Li, T. & Kercher, D.M., Experimental heat transfer and friction in channels roughened with angled, V-shaped and discrete ribs on two opposite walls. *Journal of Turbomachinery*, **118**, pp. 20–28, 1996.

[7] Kiml, R., Mochizuki, S. & Murata, A., Heat transfer enhancement mechanism in a rectangular passage with V- and Λ-shaped ribs. *Journal of Flow Visualization and Image Processing*, **8**, pp. 51–68, 2001.

[8] Cho, H.H., Wu, S.J. & Kwon, H.J., Local heat/mass transfer measurements in a rectangular duct with discrete ribs. *Journal of Turbomachinery*, **122**, pp. 579–586, 2000.

[9] Cho, H.H., Lee, S.Y. & Wu, S.J., The combined effects of rib arrangements and discrete ribs on local heat/mass transfer in a square duct, Paper No. 2001-GT-175, IGTI Turbo Expo, New Orleans, USA, 2001.

[10] Chyu, M.K. & Natarajan, V., Effects of a slit in a rib on the heat transfer from a rib-mounted wall. *Journal of Heat Transfer*, **115**, pp. 792–796, 1993.

[11] Bunker, R.S., Bailey, J.C., Widener, S.K. & Johnson, T.E., Combustor cooling with angled segmented surfaces, US Patent Application US2006/0042255A1, 2006.

[12] Zhang, N., Yang, W.J. & Lee, C.P., Heat transfer and friction loss performance in flow networks with multiple intersections. *Experimental Heat Transfer*, **6**, pp. 243–257, 1993.

[13] Gillespie, D., Wang, Z., Ireland, P. & Kohler, S.T., Full surface local heat transfer coefficient measurements in a model of an integrally cast impingement cooling geometry. *Journal of Turbomachinery*, **120**, pp. 92–99, 1996.

[14] Sweeney, P.C. & Rhodes, J.P., An infrared technique for evaluating turbine airfoil cooling designs. *Journal of Turbomachinery*, **122**, pp. 170–177, 1999.

[15] Nakamata, C., Okita, Y., Matsuno, S., Mimura, F., Matsushita, M., Yamana, T. & Yoshida, T., Spatial arrangement dependence of cooling performance of an integrated impingement and pin fin cooling configuration, Paper No. GT-2005-68348, IGTI Turbo Expo, Reno-Tahoe, Nevada, USA, 2005.

[16] Hale, C.A., Plesniak, M.W. & Ramadhyani, S., Film cooling effectiveness for short film cooling holes fed by a narrow plenum. *Journal of Turbomachinery*, **122**, pp. 553–557, 1999.

[17] Battisti, L., Cerri, G. & Fedrizzi, R., Novel technology for gas turbine blade effusion cooling, Paper No. GT2006-90516, IGTI Turbo Expo, Barcelona, Spain, 2006.
[18] Novikov, A.S., Meshkov, S.A. & Sabaev, G.V., Creation of high efficiency turbine cooled blades with structural electron beam coatings, in collection of papers Electron Beam and Gas-Thermal Coatings, Paton IEW, Kiev, pp. 87–96, 1988.
[19] Bunker, R.S., Bailey, J.C., Lee, C.P. & Stevens, C.W., In-wall network (Mesh) cooling augmentation for turbine airfoils, Paper No. GT-2004-54260, IGTI Turbo Expo, Vienna, Austria, 2004.
[20] Lee, C.P. & Bunker, R.S., Thermal shield turbine airfoil, US Patent 7011502, 2006.
[21] Nagoga, G.P., Personal communication, Moscow, 2000.
[22] Nagoga, G.P., Effective methods of cooling of blades of high temperature gas turbines. Publishing House of Moscow Aerospace Institute (Russian language), p. 100, 1996.
[23] Goreloff, V., Goychengerg, M. & Malkoff, V., The investigation of heat transfer in cooled blades of gas turbines. *26th Joint Propulsion Conference*, AIAA Paper No. 90-2144, 1990.
[24] Bunker, R.S., Latticework (vortex) cooling effectiveness. Part 1: Stationary channel experiments, Paper No. GT-2004-54157, IGTI Turbo Expo, Vienna, Austria, 2004.
[25] Acharya, S., Zhou, F., Lagrone, J., Mahmood, G. & Bunker, R.S., Latticework (Vortex) cooling effectiveness: rotating channel experiments. *Journal of Turbomachinery*, **127**, pp. 471–478, 2005.
[26] Martin, H., Heat and mass transfer between impinging gas jets and solid surfaces. *Advances in Heat Transfer*, **13**, pp. 1–60, 1977.
[27] Han, B. & Goldstein, R.J., Jet impingement heat transfer in gas turbine systems. *Annals of the New York Academy of Sciences*, 934 (Heat Transfer in Gas Turbine Systems), pp. 147–161, 2001.
[28] Gardon, R. & Cobonpue, J., Heat transfer between a flat plate and jets of air impinging on it. *International Developments in Heat Transfer, Proceedings 2nd International Heat Transfer Conference*, ASME: New York, pp. 454–460, 1962.
[29] Perry, K.P., Heat transfer by convection from a hot gas jet to a plane surface. *Proceedings of the Institute of Mechanical Engineers (London)*, **168**, pp. 775–784, 1954.
[30] Schluender, E.U. & Gnielinski, V., Warme und Stoffubertragung zwischen Gut und aufprallendem Dusenstrahl (Heat and mass transfer between a surface and impinging jet). *Chem.-Ing.-Tech.*, **39**, p. 578, 1967.
[31] Petzold, K., Heat transfer on a perpendicularly impinged plate. *Wiss. Z. Tech. Univ., Dresden*, **13**, pp. 1157–1161, 1964.
[32] Chakroun, W.M., Al-Fahed, S.F. & Abdel-Rehman, A.A., Heat transfer augmentation for air impinged on rough surface, Paper 97-GT-436, IGTI Turbo Expo, Orlando, Florida, USA, 1997.

[33] El-Gabry, L.A. & Kaminski, D.A., Experimental investigation of local heat transfer distribution on smooth and roughened surfaces under an array of angled impinging jets. *Journal of Turbomachinery*, **127(3)**, pp. 532–544, 2004.

[34] Bunker, R.S. & Bailey, J.C., Turbine blade platform impingement cooling with effects of film extraction and roughness, Paper No. GT-2005-68415, IGTI Turbo Expo, Reno-Tahoe, Nevada, USA, 2005.

[35] Son, C., Gillespie, D., Ireland, P. & Dailey, G.M., Heat transfer enhancement strategy for an impingement cooling system. *Proceedings of the 8th International Symposium on Transport Phenomena and Dynamics of Rotating Machinery*, Vol. 2, Hemisphere Pub. Co., pp. 721–729, 2000.

[36] Taslim, M.E., Setayeshgar, L. & Spring, S.D., An experimental evaluation of advanced leading edge impingement cooling concepts. *Journal of Turbomachinery*, **123**, pp. 147–153, 2002.

[37] Kanokjaruvijit, K. & Martinez-Botas, R.F., Heat transfer and pressure investigation of dimple impingement, Paper No. GT-2005-68823, IGTI Turbo Expo, Reno-Tahoe, Nevada, USA, 2005.

[38] Gau, C. & Lee, C.C., Impingement cooling flow structure and heat transfer along rib-roughened walls. *International Journal of Heat and Mass Transfer*, **35(11)**, pp. 3009–3020, 1992.

[39] Bailey, J.C. & Bunker, R.S., Local heat transfer and flow distributions for impinging jet arrays of both sparse and dense extent, GT-2002-30473, IGTI Turbo Expo, Amsterdam, 2002.

[40] Gao, L., Ekkad, S.V. & Bunker, R.S., Impingement heat transfer. Part I: linearly stretched arrays of holes. *Journal of Thermophysics and Heat Transfer*, **19(1)**, pp. 57–65, 2005.

[41] Bearman, P.W. & Harvey, J.K., Golf ball aerodynamics. *Aeronautical Quarterly*, pp. 112–122, May 1976.

[42] Kiknadze, G.I., Gachechiladze, I.A. & Oleinikov, V.G., Method and apparatus for controlling the boundary or wall layer of a continuous medium, US Patent No. 6119987, 2000.

[43] Afanas'yev, V.N., Veselkin, V.Yu., Leont'ev, A.I., Skibin, A.P. & Chudnovskiy, Ya. P., Thermohydraulics of flow over isolated depressions (pits, grooves) in a smooth wall. *Heat Transfer Research*, **25(1)**, pp. 22–56, 1993.

[44] Isaev, S.A., Leont'ev, A.I. & Baranov, P.A., Identification of self-organized vortexlike structures in numerically simulated turbulent flow of a viscous incompressible liquid streaming around a well on a plane. *Technical Physics Letters*, **26(1)**, pp. 15–18, 2000.

[45] Kesarev, V.S. & Kozlov, A.P., Convective heat transfer in turbulized flow past a hemispherical cavity. *Heat Transfer Research*, **25**, pp. 156–160, 1993.

[46] Afanas'yev, V.N. & Chudnovskiy, Ya.P., Heat transfer and friction on surfaces contoured by spherical depressions. *Heat Transfer Research*, **24**, pp. 24–104, 1992.

[47] Afanas'yev, V.N., Chudnovsky, Ya.P., Leont'ev, A.I. & Roganov, P.S., Turbulent flow friction and heat transfer characteristics for spherical cavities on a flat plate. *Experimental Thermal and Fluid Science*, **7**, pp. 1–8, 1993.

[48] Bunker, R., Belen'kiy, M., Gotovskii, M., Isaev, S. & Fokin, B., Experimental and numerical investigation of heat transfer intensification in duct of non-constant area with smooth and concavity surfaces., *3rd Russian National Heat Transfer Conference*, 2002.

[49] Bunker, R., Gotovskii, M., Belen'kiy, M. & Fokin, B., Heat transfer and pressure loss for flows inside converging and diverging channels with surface concavity shape effects. *Proceedings 4th International Conference on Compact Heat Exchangers and Enhancement Technology*, September. 29–October 3, Crete Island, Greece, Begell House Inc.: New York, 2003.

[50] Belen'kiy, M.Ya., Gotovskii, M.A., Lekakh, B.M., Fokin, B.S. & Khabenskii, V.B., Experimental study of the thermal and hydraulic characteristics of heat transfer surfaces formed by spherical cavities. *Teplofizika Vysokikh Temperatur*, **29(6)**, pp. 1142–1147, 1992.

[51] Belen'kiy, M.Ya., Gotovskii, M.A., Lekakh, B.M., Fokin, B.S. & Dolgushin, K.S., Heat transfer augmentation using surfaces formed by a system of spherical cavities. *Heat Transfer Research*, **25**, pp. 196–202, 1993.

[52] Chyu, M.K., Yu, Y., Ding, H., Downs, J.P. & Soechting, F.O., Concavity enhanced heat transfer in an internal cooling passage, IGTI Turbo Expo, Paper No. 97-GT-437, Orlando, 1997.

[53] Moon, H.K., O'Connell, T. & Glezer, B., Channel height effect on heat transfer and friction in a dimpled passage, IGTI Turbo Expo, Paper No. 99-GT-163, Indianapolis, 1999.

[54] Mahmood, G.I., Hill, M.L., Nelson, D.L., Ligrani, P.M., Moon, H.K. & Glezer, B., Local heat transfer and flow structure on and above a dimpled surface in a channel, IGTI Turbo Expo, Paper No. 2000-GT-230, Munich, 2000.

[55] Bunker, R.S. & Donnellan, K., Heat transfer and friction factors for flows inside circular tubes with concavity surfaces. *Journal of Turbomachinery*, **125**, pp. 665–672, 2003.

[56] Khalatov, A.A., Vortex technologies in aerospace engineering. *Proceedings of the U.S.–Ukrainian Workshop on Innovative Combustion and Aerothermal Technologies in Energy and Power Systems*, May 20–25, Kiev, 2001.

[57] Glezer, B., Moon, H.K. & O'Connell, T., A novel technique for the internal blade cooling, IGTI Turbo Expo, Paper No. 96-GT-181, Birmingham, UK, 1996.

[58] Hedlund, C.R., Ligrani, P.M., Moon, H.K. & Glezer, B., Heat transfer and flow phenomena in a swirl chamber simulating turbine blade internal cooling, Paper No. 98-GT-466, IGTI Turbo Expo, Stockholm, Sweden, 1998.

[59] Ligrani, P.M., Hedlund, C.R., Thambu, R., Babinchak, B.T., Moon, H.K. & Glezer, B., Flow phenomena in swirl chambers, IGTI Turbo Expo, Paper No. 97-GT-530, Orlando, 1997.

[60] Glezer, B., Moon, H.K., Kerrebrock, J., Bons, J. & Guenette, G., Heat transfer in a rotating radial channel with swirling internal flow, Paper No. 98-GT-214, IGTI Turbo Expo, Stockholm, Sweden, 1998.

[61] Khalatov, A., Syred, N., Bowen, P., Al-Ajmi, R., Kozlov, A. & Schukin, A., Innovative cyclone cooling scheme for gas turbine blade: thermal-hydraulic

performance evaluation, Paper 2000-GT-237, IGTI Turbo Expo, Munich, Germany, 2000.

[62] Goldstein, R.J., Film cooling. *Advances in Heat Transfer*, **7**, pp. 321–379, 1971.

[63] Goldstein, R.J., Eckert, E.R.G. & Burggraf, F., Effects of hole geometry and density on three-dimensional film cooling. *International Journal of Heat and Mass Transfer*, **17**, pp. 595–607, 1974.

[64] Bunker, R.S., A review of turbine shaped film cooling technology. *Journal of Heat Transfer*, **127**, pp. 441–453, 2005.

[65] Moser, S., Ivanisin, M., Woisetschlaeger, J. & Jericha, H., Novel blade cooling engineering solution, Paper 2000-GT-242, IGTI Turbo Expo, Munich, Germany, 2000.

[66] Sargison, J.E., Guo, S.M., Oldfield, M.L.G., Lock, G.D. & Rawlinson, A.J., A converging slot-hole film cooling geometry. Part 1: Low-speed flat plate heat transfer and loss, Paper 2001-GT-0126, IGTI Turbo Expo, New Orleans, Louisiana, USA, 2001.

[67] Nasir, H., Acharya, S. & Ekkad, S., Film cooling from a single row of cylindrical angled holes with triangular tabs having different orientations, Paper 2001-GT-0124, IGTI Turbo Expo, New Orleans, Louisiana, USA, 2001.

[68] Fric, T.F. & Campbell, R.P., Method for improving the cooling effectiveness of a gaseous coolant stream which flows through a substrate, and related articles of manufacture, US Patent No. 6383602, 2002.

[69] Bunker, R.S., Film cooling effectiveness due to discrete holes within a transverse surface slot, Paper GT-2002-30178, IGTI Turbo Expo, Amsterdam, Netherlands, 2002.

[70] Ekkad, S., Lu, Y. & Nasir, H., Film cooling from a row of holes embedded in transverse slots, Paper No. GT-2005-68598, IGTI Turbo Expo, Reno-Tahoe, Nevada, USA, 2005.

[71] Waye, S. & Bogard, D., High resolution film cooling effectiveness measurements of axial holes embedded in a transverse trench with various trench configurations, Paper GT-2006-90226, IGTI Turbo Expo, Barcelona, Spain, 2006.

[72] Harrison, K.L., Dorrington, J.R., Dees, J.E., Bogard, D.G. & Bunker, R.S., Turbine airfoil net heat flux reduction with cylindrical holes embedded in a transverse trench, Paper No. GT 2007-27996, IGTI Turbo Expo, Montreal, Canada, 2007.

# CHAPTER 8

# Hot gas path heat transfer characteristics/ active cooling of turbine components

T. Simon[1] & J. Piggush[2]
[1]*Department of Mechanical Engineering, University of Minnesota, USA.*
[2]*Pratt and Whitney, Inc., USA.*

## Abstract

Gas turbines play an important role in the lives of all of us. They are instrumental in providing electricity or fuel to our homes and in transportation via aircraft and ships. Despite over 50 years of operation and their current ubiquity, gas turbines continue to evolve, becoming more fuel efficient, reliable, and durable. With present emphasis on further improvement in efficiency and operation with alternative fuels, significant engineering challenges remain. The gas turbine did not come about by virtue of a single, momentous breakthrough, but rather by numerous small advances on many fronts, including heat transfer, aerodynamics, materials, dynamics, lubrication, and fuels. The efforts of tens of thousands of engineers have produced the present, extensive body of knowledge that provides our basis for further development. Given the complexity of the engine, it is quite easy to become overwhelmed when trying to become acquainted with the technology on which it is based. This chapter is offered to help in one aspect of gas turbine technology, the gas path heat transfer within the turbine. It aims first to describe the physics of engine heat transfer by developing from simple ideas an introduction to the complex heat transfer phenomena within. This effort attempts to address how the engineer applies heat transfer tools available in the literature to support designs, which will advance engine life and enhance efficiency. If successful, the newcomer will establish a foothold in the technology and the more experienced engineer will be reminded of some basic concepts.

## 1 Introduction

The main objective sought by the gas turbine heat transfer engineer is to provide an airfoil and gas path design, which is both efficient and can meet the life goals. This requires a broad understanding of the system. First, an understanding of the constraints associated with the material to be used must be developed. This information (which is often proprietary and carefully protected) includes such parameters as the material maximum use temperatures associated with oxidation and corrosion life, creep, and thermal–mechanical fatigue (all functions of temperature and stress). Given the projected usage and life goals, the designer applies life equations to determine metal temperature limits throughout the airfoil and on the passage walls. The internal cooling and film cooling schemes are designed to maintain the metal temperatures below these limits.

The heat transfer engineer has several tools, which can be applied to provide a cooling scheme that meets the temperature constraints with minimal negative impact on engine aerodynamic and structural performance. These tools generally fall into two broad categories: those that cool the airfoil and endwalls inside and those that reduce the external heat load. Note that with regard to heat transfer from the passage flow to the airfoil, the two broad categories have opposite effects on heat transfer. Internal cooling generally increases airfoil surface heat transfer rates, while the external cooling aims to reduce them. They both tend to reduce airfoil and endwall temperatures. The complexity of choices to apply the various techniques to be discussed requires that the gas turbine heat transfer engineer has a thorough understanding of the entire heat transfer problem.

Internal cooling schemes generally employ one or more of the following techniques: impingement of cooling flow upon the wall to be cooled, convective cooling by channeling coolant flow through passages within the airfoil or endwall, and conduction through pin fin arrays to augment heat transfer within the solid components of the airfoil and enhance convection by increasing both flow turbulence and fluid-to-metal contact area. The choices for flow conditions and cooling path configurations associated with the cooling scheme are influenced by the pressure of the cooling flow supply, the temperature of the coolant fluid, the heat load distribution, and the stress distribution within the airfoil.

The external heat load may be controlled by one or more of the following methods. First, there is the geometry. Although the shape of the airfoil is generally defined by aerodynamic considerations, there are regions, such as the leading and trailing edges, where some improvements may be effected, while maintaining acceptable aerodynamic performance. As necessary to accommodate cooling, the airfoil is thicker than the aerodynamic design engineer would have it. Also, there are thermal barrier coatings (TBCs). These are generally ceramic-based coatings that have low thermal conductivity. They are used to reduce the heat flux through the airfoil wall. Finally, film cooling can be applied effectively on the external surface of the airfoil and on the endwall to reduce the temperature of the gas flowing immediately adjacent to the airfoil and endwall surfaces. It should be noted that film cooling injection into the gas path is usually done with an aerodynamic penalty. Effective cooling schemes must consider all impacts on the engine design.

Designers quickly realize that airfoil design is truly complex. The internal configuration has a direct influence on the external heat load distribution, while the external heat load defines what the internal thermal design must accomplish.

Finding the final configuration of the airfoil requires a systematic approach to the analysis; either beginning with the external heat load and then moving into the coolant fluid (the direction of energy flow) or beginning with the internal coolant and then documenting the external heat load (essentially moving in the general direction of the coolant). Although the choice is arbitrary, the discussion herein will move from the inside of the airfoil to the outside. As with any design process, multiple iterations through the analysis are required.

The thermal capacity of the coolant used for internal cooling is finite, defined by the mass flow rate, the coolant's specific heat, and the coolant temperature. The heat transfer engineer must carefully balance coolant flow rates and pressure drops while augmenting heat transfer coefficients with various devices to effect proper cooling. Figure 1 shows the arrangement of various coolant flows.

The external heat transfer situation is different. Energy flows past and into the airfoil from, effectively, an infinite source, the passage stream. The influence on passage energy flow of the airfoil's cool surface is minor. The main concerns are: (1) reducing the external effective heat loads between the airfoil wall or endwall and the bulk passage flow, and (2) reducing the external area. The heat transfer engineer has little opportunity to effect significant changes in the second concern. Film cooling offers the possibility of introducing a film of cooler fluid (spent from the internal cooling functions) to cover the external surface of the airfoil to help effect the first concern listed above. This film offers a cooler source temperature for convective heat transfer through the thermal boundary layer from the passage

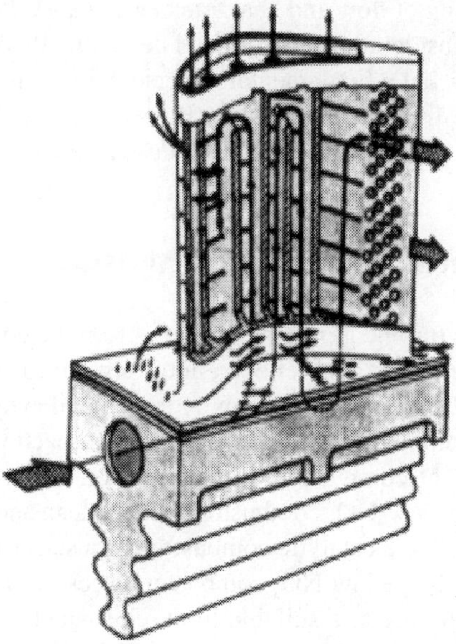

Figure 1: Coolant flows: internal channels in airfoil and endwall, impingement jets, trailing edge and tip region flows, film cooling flows (from NASA Glenn).

flow to the surface and, as such, reduces heat flux across the convective layer. Film cooling has, over the years, evolved to become a very effective method for reducing the airfoil and endwall surface temperatures. A second opportunity for reducing the surface heat flux is to introduce a solid-material coating of high thermal resistivity on the surface. This, in conjunction with the reduced heat flux due to film cooling and with effective internal cooling, gives lower temperatures on the airfoil external surfaces and load-bearing regions of the airfoil and endwall.

The internal geometries of high-performance engine airfoils tend to be very complex, incorporating many features designed to strategically cool the internals so that the entire external surface of the airfoil remains generally at a uniform temperature at a level, which is suitable for sustained operation. Complicating the problem is the matter of getting sufficient coolant supply pressure and in dealing with the thin trailing edge, the exposed tip, and the high heat transfer stagnation region at the leading edge of the airfoil. A rise in convective heat transfer coefficient in a channel flow by some devised method or another is generally accompanied by a rise in pressure drop, giving a trade-off between enhancement and flow rate. Cooling flow circuit designs optimize the distribution of cooling effectiveness for a given pressure drop. Manufacturing considerations, such as accommodating component thermal growth and maintaining clearances, as well as controlling cost (usually reflected in avoiding very fine feature sizes) are also taken into consideration. Designing a cooling scheme for a high performance engine gas path becomes a delicate balancing act between designs for internal and external heat transfer performance among many other design considerations. Tools include a strong fundamental understanding of flow and heat transfer and a working knowledge of the many other design considerations. Codes and design methods are available to the heat transfer engineer, but a fundamental understanding is paramount.

References, such as the excellent books by Han *et al.* [1] and Lakshminarayana [2] provide more detail to gas turbine heat transfer design and to turbomachinery design, respectively.

## 2 Hot gas path heat transfer characteristics

The flow entering the turbine gas path has come from the combustor, as shown in Fig. 2. Within the combustor, fuel (either a liquid similar to kerosene or diesel fuel or a gas, such as methane, syngas or high hydrogen and syngas (HSG) processed from coal) and air combine in a highly exothermic reaction that leads to high-temperature products of combustion at the high combustor pressure. A modern combustor called a dry, low-$NO_x$ combustor operates lean and without water injection. A second type called a catalytic combustor has a selective catalytic reduction in the exhaust stream. In the low-$NO_x$ combustor, excess air is used to maintain the combustor exit temperature at a suitable limit, consistent with the engine design and turbine durability limits. Combustor designs often have a strong swirl in the combustion zone to enhance mixing and shorten the reaction zone. About 30% of the airflow enters the primary combustion zone either through a swirl section, or through holes in primary zone of the combustion chamber. The remainder of

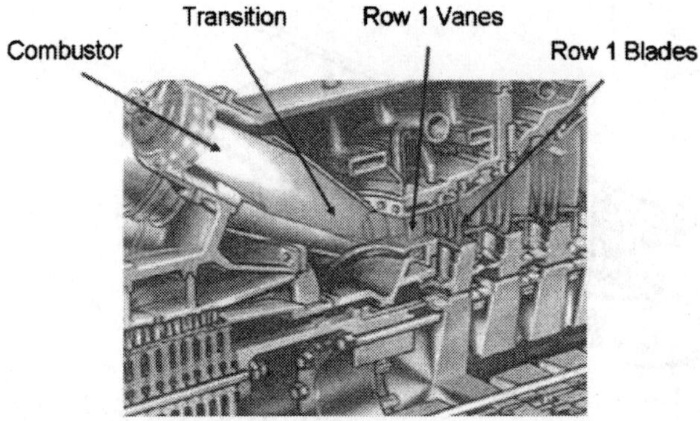


Figure 2: Hot gas path of a Siemens SGT6-5000F Gas Turbine (from Brummel *et al.* [3]).

the air, about 70% of the total airflow, is supplied through holes in the combustor liner as dilution air to mix and cool the flow and to protect the liner from reaching excessive temperature levels. In the case of a catalytic combustor, a catalytic reactor section is put downstream of the fuel/air mixing section of a pre-burner. This reactor consists of rather fine channels coated with a catalyst, such as palladium oxide. The catalytic reaction takes place at temperature levels that are below the turbine inlet temperature so a downstream combustor is needed to raise the gas temperature to the turbine inlet temperature. This combustor is similar to the dry low-$NO_x$ combustor.

The flow passes from the combustor exit to the turbine through a transition section. This section generally has a decreasing area in the streamwise direction and some wall curvature, depending on the combustor and turbine designs. The flow departs the transition section to enter the vane section of the first stage turbine. To accommodate thermal growth of components, the junction between the transition section and the turbine is fabricated with gaps in the wall, see Fig. 3. These gaps are sealed against ingression of hot gases with compressor discharge air that has been routed through a cavity in the centre of the engine.

The flow then is turned through the vane, or nozzle, row of the first stage of the turbine. In this row, the flow is given a swirl (a circumferential velocity component) and is accelerated to a higher velocity. With the rise in velocity, there is a drop in static temperature of the passage flow. This temperature change in the passage core flow can be estimated by analyzing it as a one-dimensional isentropic flow. Local velocities within the vane row can be at sonic velocity or mildly supersonic and shock waves can stand within the passage. Generally, a shock from the trailing edge of one vane crosses the passage to impinge on the suction surface of the neighboring vane. This shock will affect wall heat transfer rates on the suction surface at the location of impingement, and downstream. The fluid temperature upon entry to the vane row exceeds the maximum allowable materials temperature; thus, this row must be aggressively cooled. The gas path heat transfer engineer spends most of their time on the design of the first stage vane and rotor sections. The techniques for

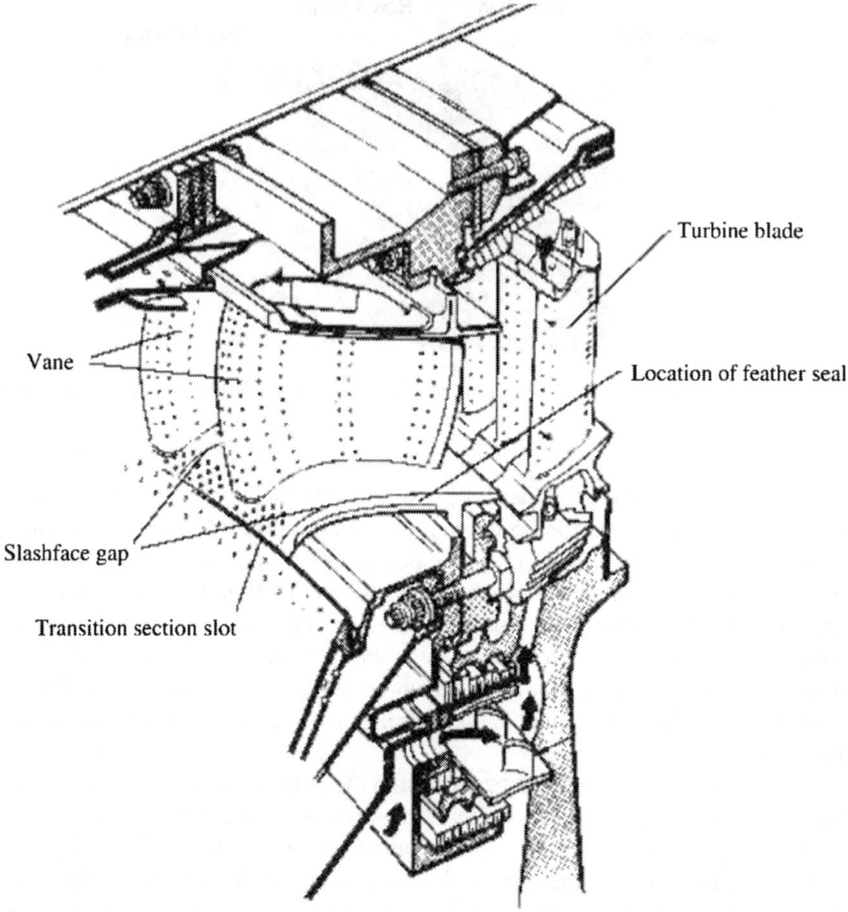


Figure 3: First stage vane and rotor section showing the transition section to vane slot as well as the slashface or gutter gap (from Cumpsty [4]).

thermal control mentioned above are applied to this section: film cooling, impingement cooling, internal convection, pin fins augmentation, and trailing edge and tip cooling. The airfoil surfaces and the endwalls are actively cooled by an appropriate choice of these techniques. Because of thermal expansion of components, the airfoils/endwall assemblies are segmented, having one or two (doublet) airfoils per assembly. The segments are separated by a gap to allow for thermal expansion, the interface gap (also known as the slashface or gutter, see Fig. 3) is sealed by air taken from the engine cavity. This sealant or leakage air is cool and mixes with the passage flow near the endwall surface and so it has important effects on the gas path cooling. Because of mixing with passage flow, which is of high velocity in the downstream half of the passage, there may be strong aerodynamic losses associated with this leakage.

The passage flow next enters the first stage rotor. In this stage, swirl is removed from the flow, while the velocity decreases and work is extracted. This work extraction results in a drop in stagnation temperature. Though the fluid velocity drops, the net effect of velocity change and work interaction is a drop in static temperature across the rotor passage (within the passage core flow). This stage is rotating; thus, the endwall, or platform, is moving relative to the endwall of the upstream vane row. The flow in the near-endwall region thus is skewed as it passes onto a moving platform from a stationary endwall. To accommodate the relative motion of the two rows and the thermal growth of the components, there is a gap between the vane and rotor sections (see Fig. 3). This gap is sealed with airflow from the cavity, again with effects on gas path heat transfer and aerodynamic losses. The degree of skewing mentioned above can be modified by introducing the sealant flow with pre-swirl. Because of rotation, the flow and heat transfer are affected by buoyancy forces within the centripetal force field and by Coriolis forces wherever there is a radial flow component, such as with the internal cooling channels.

As the flow proceeds through the remainder of the turbine, it passes through a series of vane and rotor components of subsequent stages. In each stage, work is extracted and the total temperature is reduced, while the loss penalty for the system due to cooling air usage continues to accumulate (the coolant air used in a downstream stage is not available to do work within an upstream stage). The reduction in stage temperature has a significant impact on the choice of materials and cooling schemes. Parts with lower external heat loads (such as the second blade) tend to require less aggressive internal cooling, resulting in reduced thermal gradients within the part material (compared to the gradients in the first stage blade). Also, there are fewer internal ribs used to divide internal cooling passages and, since they tend to remain much cooler than the external walls, they carry much of the centrifugal stress load. This reduces the significance of local surface failure modes, such as oxidation, and increases the relative importance of creep, which may be understood as a section average temperature and stress-related failure mode. The change in the life limiting failure mode and the reduced external heat loads may allow the designer to choose less exotic materials or apply less aggressive thermal control, thereby reducing the cost of the part. In the large castings found in industrial gas turbine engines, the greater cost benefit may be realized by choosing less exotic materials. In the case of the smaller castings found in aircraft engines, machining costs may dominate the part price, driving the designer to choose a less complex cooling scheme. In both cases, the thermal design techniques are those employed in the first stage, but are applied more sparingly in the downstream stages.

Other components demanding thermal design attention in the gas path are support struts and the engine propulsion nozzle (in the case of the aircraft gas turbine). Thermal loads on these components are more severe if the engine is equipped with an afterburner, a combustor in the gas path upstream of the nozzle. Our more detailed discussions to follow will be restricted to the vane and rotor stages of the turbine.

## 2.1 The first stage high pressure turbine vane

The combustor exit turbulence level for a low-$NO_x$ style of combustor can be as high as 20–25% [1]. The engine manufacturer generally obtains data on the flow field for a particular combustor, as it will depend on the combustor geometry.

However, several studies existing in the open literature give an indication of the turbulence levels that may be expected at the exit of the combustor. Measurements that are frequently referred on combustor exit flow conditions are by Goldstein *et al.* [5], Moss and Oldfield [6], Goebel *et al.* [7], van Fossen and Bunker [8], and Ames *et al.* [9]. A conclusion of the Moss and Oldfield study was that turbulence level and spectra are not strongly affected by whether there is combustion or not, or by the combustion rate when there is combustion. The data of Goebel *et al.*, with and without combustion, include measurements of profiles of turbulence levels and exit temperatures. They found values of around 10–12% with a small but measurable drop in turbulence intensity between the unfired case with cold inlet flow and a fired case. Their combustor was designed to create significant swirl of the mixing region flow (see Fig. 4). The drop in turbulence was more in the swirl component of turbulence than in the axial component. The turbulence components were normalized on the axial inlet velocity (no swirl component in the normalizing velocity).

The radial distribution of temperature at the combustor exit is important for the design of the turbine passage endwalls. Over recent years, redesign of the combustors to improve $NO_x$ emission and increase average turbine inlet temperatures has led to flatter profiles and, thus, hotter fluid and higher thermal loading in the near-endwall regions of the turbine. Figure 5 shows representative temperature distributions taken from Goebel *et al.* Figure 6 taken from Povey *et al.* [10] shows the radial and circumferential temperature distributions as measured in a military engine.

Next, the flow approaching the vane section is discussed. Ames *et al.* [9] noted that vane inlet turbulence depends on the nature of the flow in the core of the combustor, including residence time (as discussed above), but it is modified by the inlet contraction ratio of the transition piece between the combustor and the turbine. The cross-sectional area of the duct decreases in the streamwise direction and, thus, the flow is accelerated. As a result of this acceleration, the turbulence of the passage flow is stretched and partially dissipated and the passage wall boundary layers, which will become the vane passage endwall boundary layers, are thinned. To accommodate thermal growth of components, the passages are assembled with gaps in the wall. These gaps are sealed against ingression of hot gases with flow from the cavity of the engine (see Fig. 3). This flow will emerge into the passage and thicken the passage wall boundary layers, as it mixes with the passage flow. In addition to the sealant flow, there may be intentional film cooling flows in the approach duct, usually just upstream of the airfoil leading edge plane, which modify the approach flow boundary layer. Factors that influence the approach flow boundary layer growth are, combustor design, leakage (or sealant) flow configurations, film cooling flows, roughness, streamwise acceleration, free-stream turbulence levels and scales, and surface streamwise curvature. Thus, each design will have different wall boundary layer characteristics. Piggush and Simon [11] simulated a low-$NO_x$ combustor and a transition section, in their study of heat transfer in the first stage vane, then documented the flow characteristics 1.5 and 0.1 axial chord lengths upstream of the airfoil leading edge plane (before any leakage flow is introduced). They noted that the aerodynamic losses associated with the boundary layers on the walls that were to become the passage walls were less than one percent of the total aerodynamic losses encountered

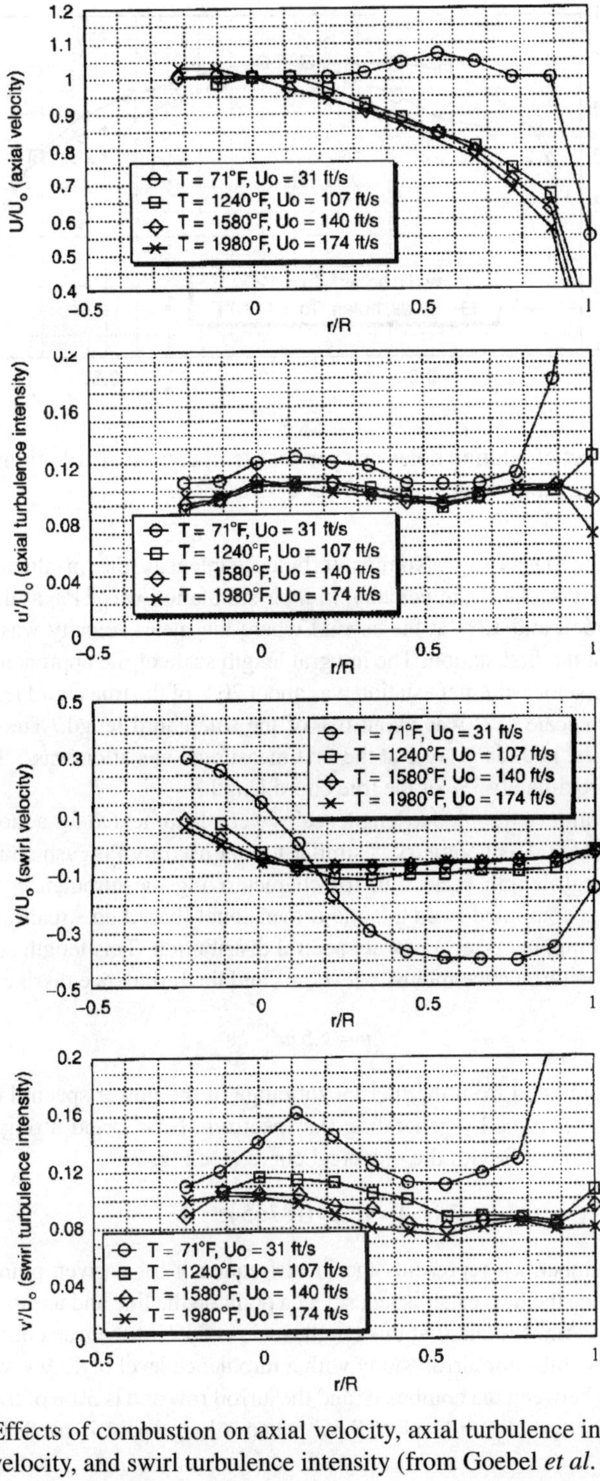


Figure 4: Effects of combustion on axial velocity, axial turbulence intensity, swirl velocity, and swirl turbulence intensity (from Goebel *et al.* [7]).

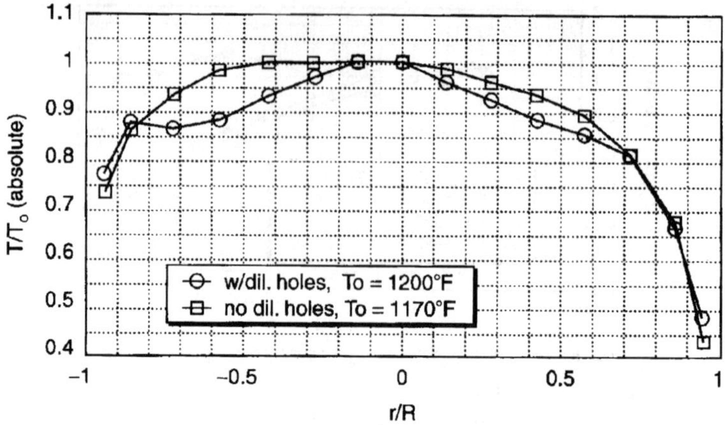


Figure 5: Effect of dilution holes on spanwise temperature distributions for similar firing temperatures (from Goebel *et al.* [7]).

within the vane row passage. Nominal turbulence intensity, the rms fluctuation of the velocity scaled on the mean velocity, in their simulation was 21% at the first measurement station and 12% at the second where the mean velocity was about 75% higher than at the first station. The integral length scale of the approach flow turbulence at the first measurement station was about 20% of the true chord length and the energy length scale [12] was about 6% of the true chord length. The momentum thickness of the boundary layer at the first measurement station was 3.24 mm [13], which is approximately 1% of the true chord length.

Cascade entrance turbulence length scales were documented for a mock combustor by Ames [14]. His geometry is similar to that used by Piggush and Simon. He used the energy length scale, Lu, for characterizing the turbulence. The energy length scale is computed from the one-dimensional (based on streamwise velocity fluctuation component only) power spectral distribution. This length scale is computed from the turbulence intensity level, $u'$, and the turbulence dissipation, $\varepsilon$, as:

$$Lu = 1.5\left|u'\right|^3 / \varepsilon \tag{1}$$

where $\varepsilon$ is computed from the inertial subrange of the power spectral density distribution. This is done by generating the spectrum, $E_1(k_1)$, and fitting the inertial subrange portion of it with the universal curve (see Fig. 7):

$$E_1\left(k_1\right) = 1.62\ \left(18/55\right)\varepsilon^{2/3}k_1^{-5/3} \tag{2}$$

The wave number, $k_1$, represents the local frequency for a given point in the 1D spectrum. Once the fit is established, select a point on the line and use its coordinates to compute $\varepsilon$. Ames found that this length scale is 23% of the true chord length for his mock-up combustor/airfoil study with a turbulence level of 12.9% when a short spool is used between the combustor and the airfoil row and is 30% of the true chord length and the turbulence level of 8.3% when a long spool is used. In the paper

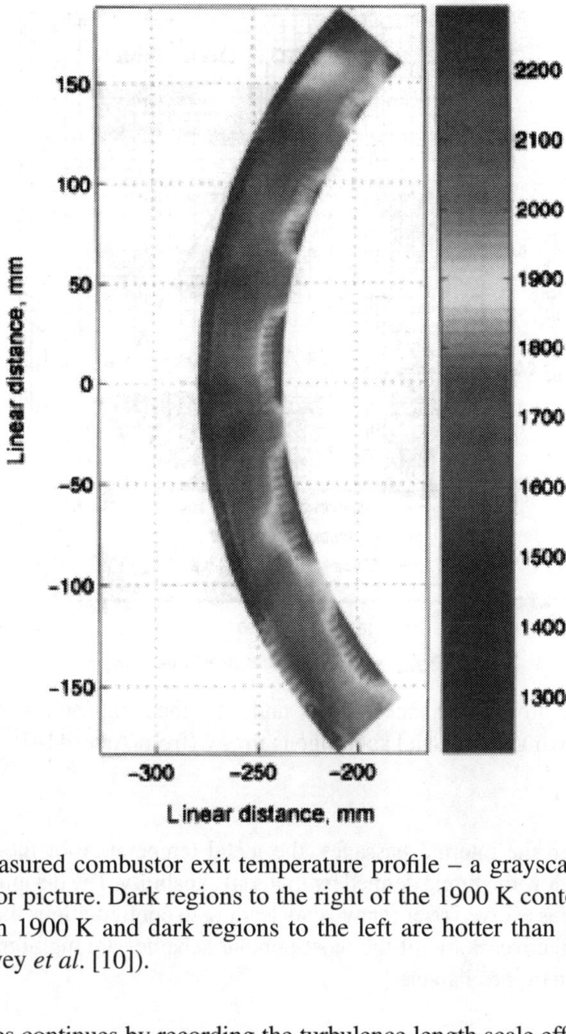


Figure 6: Measured combustor exit temperature profile – a grayscale image of a color picture. Dark regions to the right of the 1900 K contour are cooler than 1900 K and dark regions to the left are hotter than 1900 K (from Povey *et al.* [10]).

referred, Ames continues by recording the turbulence length scale effect on turbine airfoil surface heat transfer, to be discussed later in this chapter.

An example of the first stage turbine incoming flow temperature profile is provided by Povey *et al.* [10] from the data taken by Chana *et al.* [15]. Figure 8 shows a 15% variation from wall to center-span on the temperature ratio, $T/\overline{T}$ ith most of the change within 15% of the span from the wall (see "Typical engine").

## 3 Active cooling of the gas turbine components in the gas path

As discussed in the Section 1, stage temperature profiles of gas turbine engines often exceed the temperature capability of the turbine materials. To correct the problem, designers incorporate passages and heat transfer augmentation features on the internals of airfoil castings. When relatively cool fluid drawn from the compressor is

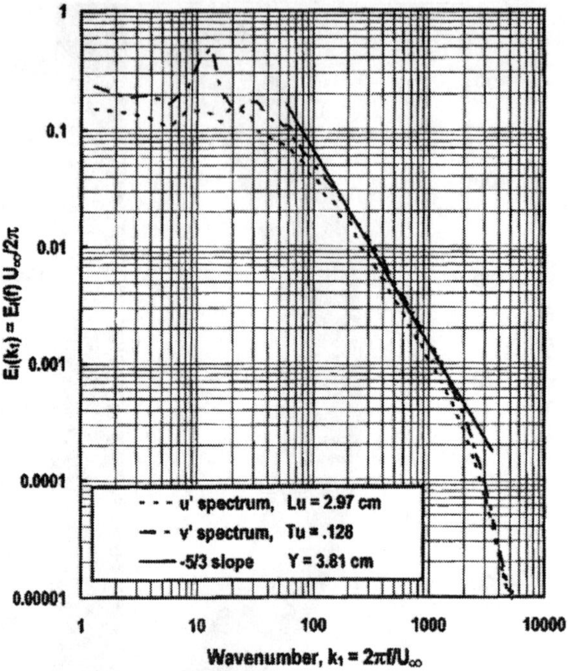


Figure 7: One-dimensional spectra of $u'$ and $v'$ for the combustor with a short spool showing the inertial subrange isotropy (from Ames [14]).

passed through the internal passages, the metal temperature of the airfoil can be maintained below the limits defined by material capability. The design of these internal passages has evolved over many years leading to configurations that are complex and highly effective. Some of the most popular schemes are highlighted in the following portion of the chapter.

### 3.1 Impingement cooling

Impingement cooling can be a very effective way to cool internal surfaces of a turbine airfoil. Properly designed impingement cooling schemes tend to have high heat transfer coefficients but also large pressure drops. As impingement cooling systems require some rather unique geometries; a feed plenum, hole plate, and an impingement target and plenum, the use of impingement cooling systems is typically confined to the first stage vane (where the high heat loads and relatively low cooling supply pressure to passage dump pressure ratio precludes the use of long-channel cooling circuits) and the leading edge and trailing edge regions of other airfoils. There are several mechanisms associated with the impinging jet that improve heat transfer. These will be discussed in the following sections.

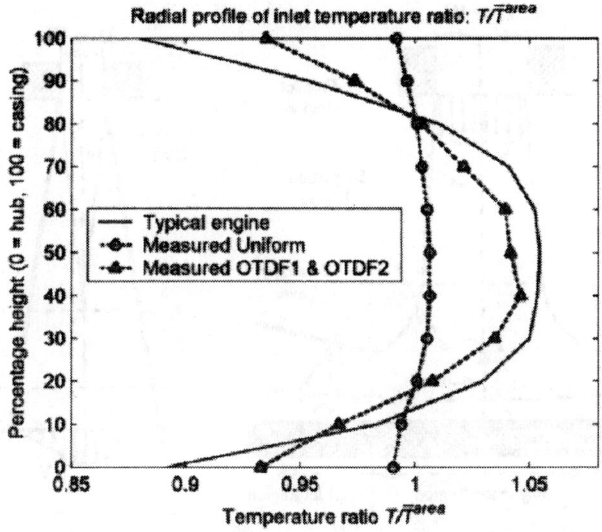


Figure 8: Circumferentially averaged temperature distributions at the turbine vane inlet. Shown is a measured profile in a real engine (typical engine), a measured profile in a simulator in which distortions due to discrete fuel and dilution air injection and endwall cooling flows are active (OTDF1 and OTDF2), and a profile measured in the simulator when the inlet temperature distortion mechanisms are not active (measured uniform).

### 3.1.1 Single jets

The structure of the impinging jet is of interest. A schematic of a representative geometry is shown in Fig. 9. In the simplest case, the jet flow originates in a plenum where the coolant fluid has little or no momentum. The fluid passes through a hole or slot into a lower pressure zone, driven by the pressure difference between the two plenums. As the fluid passes through the hole, it may be thought of as a slug flow with a negligibly thin boundary layer around the edge of the jet and a very flat velocity distribution across the jet core (as shown in Fig. 9). The thin boundary layers can be attributed to the rapid spatial acceleration of the flow approaching the hole exit plane. The jet has relatively high momentum as it passes into the low-pressure zone. The flow on the periphery of the jet shears on the low momentum fluid external to it, eventually leading to a breakdown of the shear layer into a train of large turbulent eddies. This causes the jet to widen and entrain external flow. The velocity profile of the jet becomes non-uniform, as shown in Fig. 9. If the jet length to diameter ratio is sufficient, the shear layer between the jet and the surrounding flow consumes the core of the jet, which leads to high levels of turbulence in the vicinity of the center of the jet. The large eddies in the shear layer also establish circulation in the low-pressure plenum, mixing that fluid. Finally, the jet impinges upon the target surface. Due to the high momentum of the jet perpendicular to the wall, the boundary layer at the impingement point is

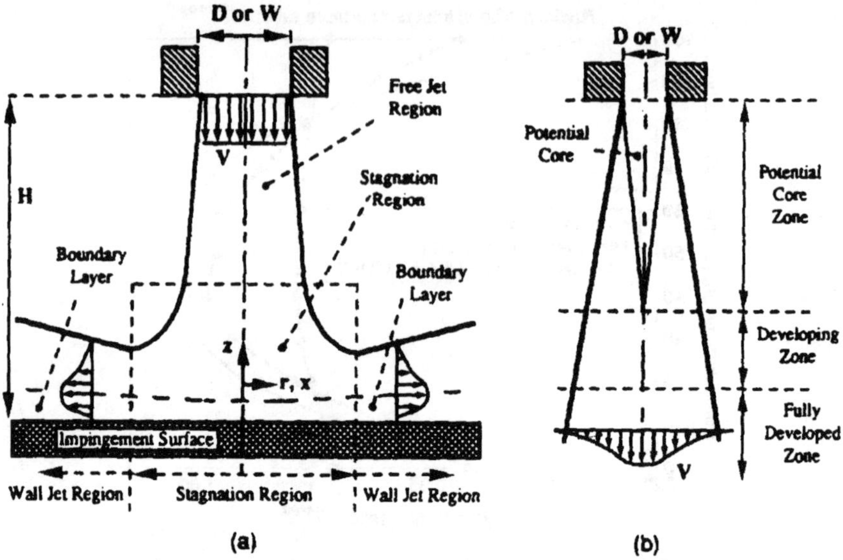


Figure 9: Different regions of an impingement jet (from Viskanta [16]).

extremely thin, while the location of this point is typically unstable. The instability is associated with the axial compressive straining of the flow as it approaches the stagnation point. As the fluid of the jet moves from the point of impact, its spatial acceleration is high, creating a thin boundary layer with a highly turbulent external flow. The net result of these processes is a high heat transfer coefficient at, and in the vicinity of, the jet impact center.

Correlations are available to describe the average heat transfer achieved with a single jet. One study is by Goldstein *et al.* [17]. In their correlations, below, the values of A, B, and C are 24, 533, and 44, respectively. The parameter $R$ is the distance from the geometrical center of the jet, $L$ is the distance from the orifice plate to the impingement wall, and $D$ is the jet diameter. The correlations indicate that a maximum average heat transfer coefficient can be achieved with a distance of $L/D = 7.75$. This is sufficiently long for the non-turbulent core of the jet to be consumed, but short enough not to lose the momentum of the jet to excessive spreading.

$$\frac{\overline{Nu}}{Re^{0.76}} = \frac{A - |L/D - 7.75|}{B + C(R/D)^{1.285}} \quad \text{for constant surface heat flux} \qquad (3)$$

$$\frac{\overline{Nu}}{Re^{0.76}} = \frac{A - |L/D - 7.75|}{B + C(R/D)^{1.394}} \quad \text{for constant surface temperature} \qquad (4)$$

Note that there is sensitivity to the thermal boundary condition of the target plate, isothermal or of uniform heat flux. The Nusselt number of a laminar boundary layer is known to be dependent on the thermal boundary conditions whereas a turbulent boundary layer is nearly independent of the thermal boundary conditions.

### 3.1.2 Impingement arrays

The single jet is of academic importance and helps introduce the reader to the general physics of the flow field. In turbine applications, impingement jets are typically arranged in two- or three-dimensional arrays. Both arrangements tend to be strongly influenced by the effect of cross flow, which develops as the spent jet air exits the stagnation zone on the target surface. Cross flow interaction with the jet causes the development of several important flow structures in addition to the jet shear layer vortex structure discussed above. The interaction of the jet fluid and the endwall boundary layer creates a counter-rotating vortex pair. The flow structure and mechanism causing the vortex pair is similar to a cylinder in cross flow and its interaction with the endwall boundary layer. The cross flow perceives the jet as an obstacle and passes around it, and a horseshoe vortex is formed. Finally, behind the jet, there exists a wake zone characterized by the presence of what Kelso *et al.* [18] term the upright vortex and also the presence of wall vortices. The vortices are shown in Fig. 10. For a more extensive treatment of the interaction of the jet and cross flow, the reader is referred to the thorough work of Kelso *et al.* [18] and also Lee *et al.* [19]

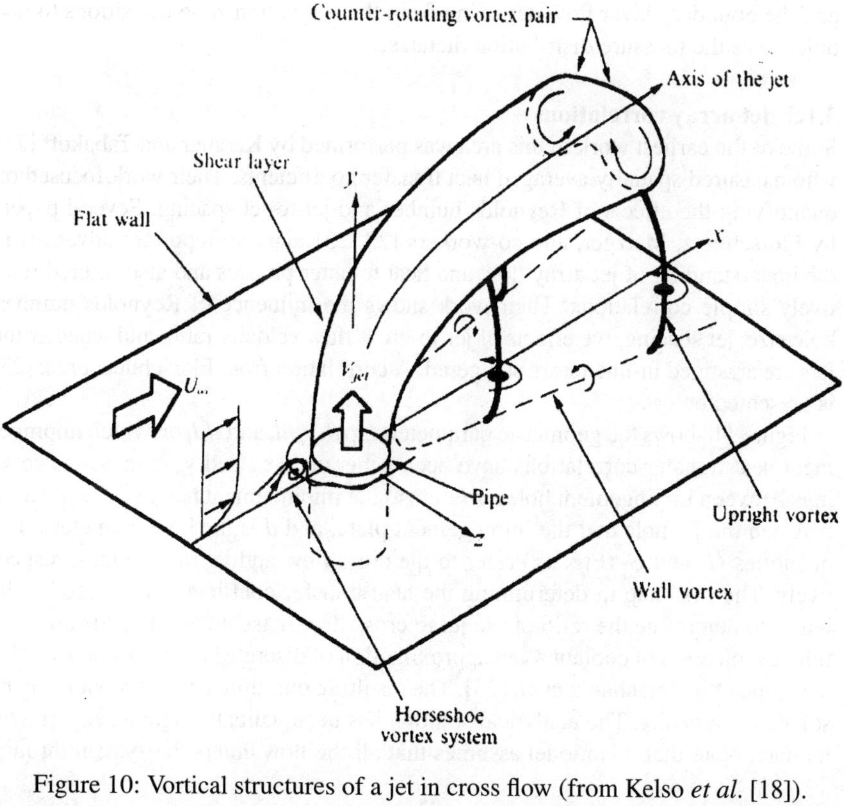


Figure 10: Vortical structures of a jet in cross flow (from Kelso *et al.* [18]).

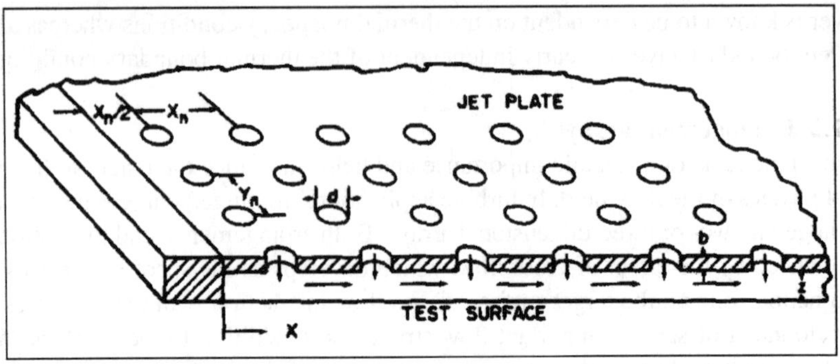


Figure 11: Sample of an impingement array showing spacing parameters ($x_n$, $y_n$, and $z$) (from Florschuetz *et al.* [23]).

It should be noted that in the case of impingement cooling, the jet may be interrupted by the target wall before all the flow structures outlined above have had a chance to develop fully. Such is the case when $L/D$ is small and the laminar core of the jet extends to the impingement point. This short $L/D$ case for a single jet was discussed by Adolfson *et al.* [20]. In this case, the impingement zone is laminar and the boundary layer flow departing from the stagnation zone transitions to turbulence as the pressure distribution dictates.

### 3.1.3 Jet array correlations

Some of the earliest work in this area was performed by Kercher and Tabakoff [21] who measured spatially averaged heat transfer coefficients. Their work focused on quantifying the effects of Reynolds number and jet-to-jet spacing. Several papers by Florschuetz, Metzger, and co-workers [22, 23] marked important advances in the understanding of jet array flow and heat transfer physics and also offered relatively simple correlations. Their work shows the influence of Reynolds number, hole size, jet spacing, the effects of jet-to-cross flow velocity ratio, and whether the jets are arranged in-line or are staggered. A correlation from Florschuetz *et al.* [23] is presented below.

Figure 11 shows the geometric parameters, $x_n/d$, $y_n/d$, and $z/d$, on which impingement heat transfer correlations have been generated ($x_n$, and $y_n$ refer to the spacings between impingement holes, $z$ refers to the impingement height, or separation between the jet hole and the impingement plate, and $d$ is the hole diameter). The quantities $G_c$ and $G_j$ (Fig. 12) refer to the cross flow and jet mass fluxes, respectively. The first step in determining the heat transfer coefficient produced by the jets is to determine the ratio of the jet to cross flow mass fluxes. The idea of continuous injection of coolant as an approximation of discrete hole injection was also developed by Florschuetz *et al.* [23]. The resulting one-dimensional model may be solved analytically. The analytical solution has been validated against experimental data. Note that this model assumes that all the flow enters the system through

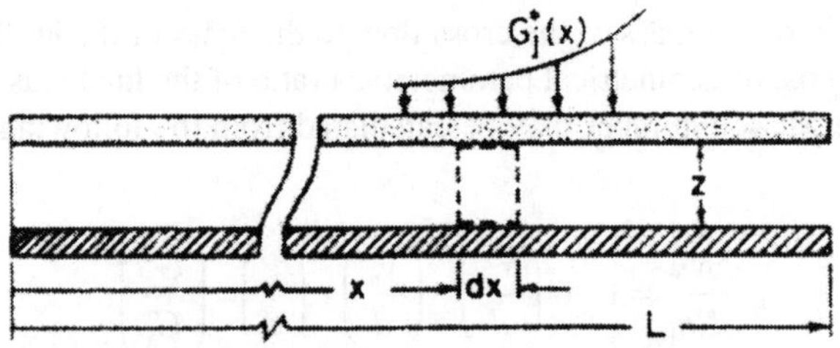


Figure 12: The continuous injection model. Here the jet flow is simulated as a continuous injection of mass along the upper plate ($G_j^*$) while the fluid element with dimensions of $dx$ and $z$ has fluid moving across its left and right faces and also into it from the jet flow injection. The discrete hole injection model shown below is an analogous configuration, which may be described by similar, discrete quantities (from Florschuetz *et al.* [23]).

the coolant jets and exits the system by moving in the direction of $L$. The model is given in Fig. 12.

The solution of the model is given by the following equation. This equation also requires that the discharge coefficient of the hole is known. Correlations for determining the discharge coefficient are available.

$$\frac{G_c}{G_j} = \frac{1}{\sqrt{2C_D}} \frac{\sinh\beta\left((x/x_n)-(1/2)\right)}{\cosh\beta\left(x/x_n\right)} \tag{5}$$

where

$$\beta = \frac{C_D\sqrt{2}(\pi/4)}{(y_n/d)(z/d)} \text{ and } x = x_n(i-1/2),\ i = 1, 2, 3, \ldots, N_c \tag{6}$$

The mass flux ratio will form the basis of the heat transfer coefficient calculation. The next step is to calculate the jet Reynolds number, which will be required in the calculation of a reference Nusselt number. The jet Reynolds number is defined as: $Re_j = G_j d/V_j$. This requires the calculation of the mass flux of a given jet. Based on the open area of the plate, pressure ratio across the jet, and discharge coefficient, the mass flux may be determined.

$$G_j = A_o C_D\left[2\rho\left(P_o - P\right)\right]^{1/2} \tag{7}$$

The parameter $A_o$, the open area ratio of the plate, is defined as

$$A_o = (\pi/4)/\left[(x_n/d)(y_n/d)\right] \tag{8}$$

The reference Nusselt number may be determined by using the following equation:

$$Nu_1 = 0.363\left(\frac{x_n}{d}\right)^{-0.554}\left(\frac{y_n}{d}\right)^{-0.422}\left(\frac{z}{d}\right)^{0.068} Re_j^{0.727} Pr^{1/3} \tag{9}$$

When the ratio of the mass of the cross flow to the mass of the jet flow is known, along with a host of geometrical parameters, a ratio of the final Nusselt number to the reference Nusselt number may be calculated from the following equation and constants.

$$\frac{Nu}{Nu_1} = 1 - C\left(\frac{x_n}{d}\right)^{n_x}\left(\frac{y_n}{d}\right)^{n_y}\left(\frac{z}{d}\right)^{n_z}\left(\frac{G_c}{G_j}\right)^{n} \tag{10}$$

| | $C$ | $n_x$ | $n_y$ | $n_z$ | $n$ |
|---|---|---|---|---|---|
| Inline | 0.596 | –0.103 | –0.380 | 0.803 | 0.561 |
| Staggered | 1.07 | –0.198 | –0.406 | 0.788 | 0.660 |

It should also be noted that the same paper includes a more complex (and slightly more accurate) version of the correlation. For the purpose of discussion and initial design work, the equation presented here is sufficient.

### 3.1.4 Jet impingement arrays: areas of study

Much work on characterizing the performance of jet arrays in less general configurations is underway. A full study of this literature could not be offered in this chapter. The following is offered as a brief introduction.

Early work on characterizing heat transfer performance of impinging jets typically considered only low speed jets, which precluded the investigation of Mach number influence on heat transfer performance. Recent efforts by Park *et al.* [24] include studies with high-speed jets that show a significant dependence of heat transfer rates on Mach number (Fig. 13). The study shows increases in both local Nusselt number and line-average Nusselt number values, as shown in Fig. 14. Correlation of the Mach number data leads to a simple correction, which may be applied to the correlation of Florschuetz *et al.* [23]:

$$\overline{\overline{Nu}} / \overline{\overline{Nu_F}} = 0.58Ma^{1.35} + 1$$

A second study of interest investigates how the heat transfer coefficient can be increased by adding dimples to the target surface. This study by Kanokjaruvijit and Martinez-Botas [25] produced significant increases in the heat transfer coefficient (as much as 64%) for some jet and dimple configurations and jet-to-plate spacings. The increase in the heat transfer rate was attributed to additional vorticity created on the target surface by the dimples. Fig. 15 shows the distribution for a configuration, which produces a maximum augmentation of 51%.

### 3.1.5 Jet impingement on curved surfaces

Although the previous set of equations are accurate for surfaces of the airfoil that are well approximated as flat plates, they do not apply to surfaces having a significant concave curvature, as in an airfoil leading edge. A separate body of research was developed to describe this flow system and to provide correlations.

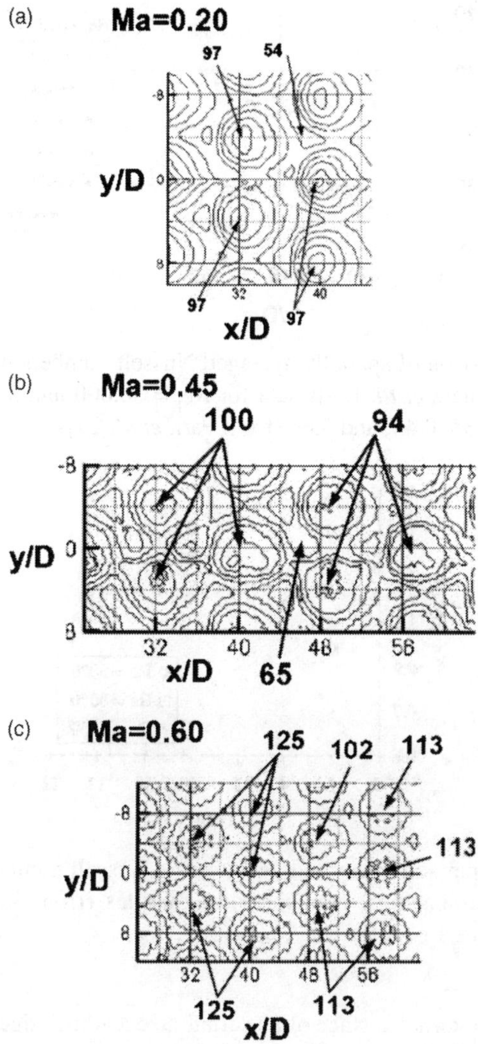


Figure 13: Spatially resolved distributions of surface Nusselt number for $Re_j = 30{,}000$ and *Ma* values of: (a) 0.20, (b) 0.45, and (c) 0.60 (from Park *et al.*[24]).

The jet flow field is similar to that already described for the flat plate jet systems. In situations which include cross flow, similar vortical structures exist, as described previously. The main geometrical differences between this system and the flat plate arrays are that the arrays typically include just a single row of jets, the target surface is curved, and the walls, which affect the spent flow departure, are positioned differently.

Chupp *et al.* [26] performed early work on the subject of leading edge jet impingement. This includes measurement and correlation of the heat transfer

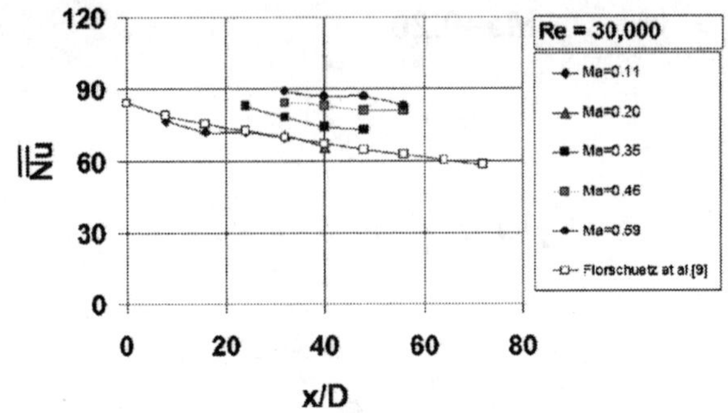


Figure 14: Comparison of spatially averaged Nusselt numbers with correlation of Florschuetz *et al.* [23]: data for $Re_j$ = 30,000 and *Ma* values of 0.11, 0.20, 0.35, 0.45, and 0.60 (from Park *et al.*[24]).

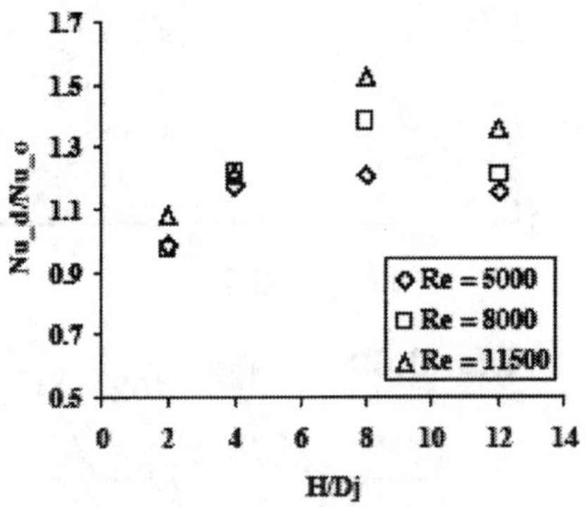


Figure 15: A comparison of normalized average Nusselt numbers at various jet-to-plate spacings, hemispherical dimples (from Kanokjaruvijit and Martinez-Botas [25]).

coefficient on the internal surface of an actual-size leading edge. The test section was configured so that the spent air exited axially (there was no cross flow). This allowed the data to be correlated on the Reynolds number and several key geometrical parameters; *d*, the diameter of the jet hole, *D*, the diameter of the target surface, *l*, the distance from the jet hole to the target surface, and *p*, the pitch of the holes. In this study, the effect of Prandtl number was considered to not be significant. The correlation for Nusselt number at the stagnation point is given by the equation below. The data also include measurements of the Nusselt numbers at locations away from the jet stagnation point allowing the average Nusselt number to be correlated as given by the second equation. The definition of the Reynolds number is given in the third equation where the diameter and mass flow rate refer to the jet.

$$Nu_{stag} = 0.44 Re^{0.7} (d/s)^{0.8} \exp\left[-0.85 (l/d)(d/p)(d/D)^{0.4}\right] \tag{11}$$

$$Nu_{avg} = 0.63Re^{0.7}(d/s)^{0.5}(d/D)^{0.6}\exp\left[-1.27(l/d)(d/p)^{0.5}(d/D)^{1.2}\right] \quad (12)$$

$$Re = 4\dot{m} / \pi d \mu \quad (13)$$

Later work by Bunker and Metzger [27] and Metzger and Bunker [28] investigated similar curved leading edge geometries. The first study indicates that heat transfer rates vary with the jet Reynolds number to the approximate power of 0.6–0.7 (a result which is similar to the rates reported by Chupp *et al.* [26]). The second study reports on the impact of coolant extraction on the leading edge heat transfer coefficients. The conclusion is that jet heat transfer is primarily dependant on the total amount of jet flow with a secondary but still significant contribution due to the extraction of coolant flow through holes near the impingement site. The spent flow from the jets produces a recirculation region in the jet exhaust plenum. Drawing off spent jet air as film cooling tends to affect this recirculation. An optimum configuration of jet holes and film coolant holes can be found. The impact of the film extraction varies with the jet Reynolds number, the largest impact being produced at low jet Reynolds numbers.

## 3.2 Pin fin cooling

Pin fins are useful tools for enhancing heat transfer coefficients. They are often found in the trailing edge region of turbine airfoils where the high heat transfer coefficients they produce are advantageous and the relatively large pressure drops they create are acceptable. Pins can also be designed to span from one side of the channel to the other, providing a conduction path, which helps maintain the uniformity of airfoil metal temperatures. Pins also provide additional surface area, which further enhances cooling performance.

### 3.2.1 Pin fin flow field

The flow field around pin fins has been the subject of several studies. The results of a select group of the studies [29–31] have been used to create Fig. 16. The pin fin produces several important flow structures, including horseshoe vortex pairs and a wake region.

The horseshoe vortex pair is a flow structure which is typically formed due to a protrusion of a body from an endwall with flow across it. It is a result of a non-uniform velocity profile of incoming flow, which causes flow in some regions to stagnate at a higher pressure than flow in other regions. In Fig. 16, two horseshoe vortices are shown, the primary and secondary vortices. The wake region of the pin fin is also readily evident. This region begins after separation of the flow moving around the pin. The shear layer between the slow moving flow behind the pin fin and the fast moving flow which has moved around the pin interact causing effective mixing of the fluid behind the pin. The interface between the fast and slow moving fluid is characterized by the presence of the arch shaped vortex (similar to the upright vortex in the impingement cooling section). Additionally, the legs of the horseshoe vortices assist in the mixing the flow near the endwall. These features are visible in Fig. 17.

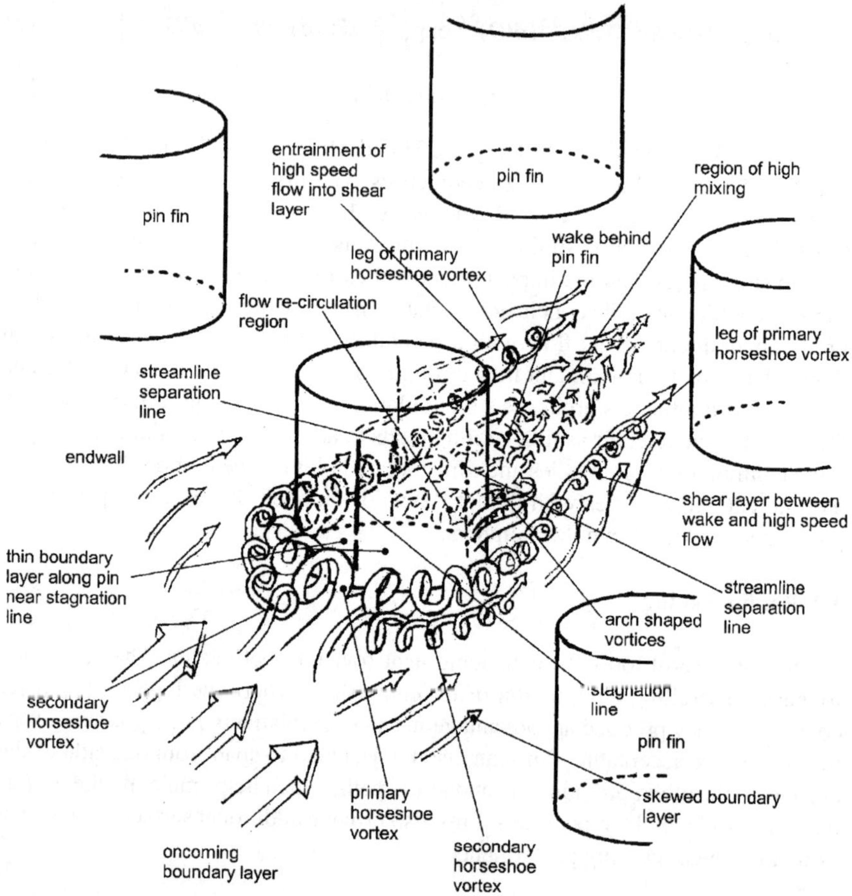


Figure 16: Flow field around pin fins within an array. The oncoming boundary layer gives rise to the primary and secondary horseshoe vortices. The pin creates a wake region with strong mixing (from Ligrani *et al.* [32])

Near the surfaces forming the pin and the endwall, the boundary layers are strongly affected by the presence of the pin. As the incoming flow stagnates on the face of the pin, the boundary layer is thin and the point of stagnation is unstable. This produces the high heat transfer coefficients well documented in other articles that show the heat transfer coefficients on cylinders in cross flow. The boundary layer builds, somewhat, as it moves around the surface of the pin until the boundary layer detaches. On the endwalls, the boundary layers skew as the flow moves around the obstruction. Behind the pin, in the wake region, the boundary layer is highly disturbed.

### 3.2.2 Pin fin heat transfer

As one may guess, the highest local heat transfer rates on the endwall are found near the leading edge of each pin in region influenced by the roll-up of

Figure 17: Oil and graphite visualization of a pin in cross flow clearly showing separation lines along the pin endwall (from Chyu and Natarajan [33]).

the horseshoe vortex. Downstream, the legs of the horseshoe vortices and the arch vortices help maintain high heat transfer rates. The work by Ligrani and Mahmood [34] clearly shows the heat transfer rates around an array of staggered pins (Fig. 18). Note that the heat transfer rates in the wake region are generally low but that the effective mixing in this region helps promote high heat transfer rates on the downstream pin. Arrays with staggered pins tend to exhibit slightly higher average heat transfer rates and a more uniform heat transfer rate distribution.

Most data on pin fin arrays are presented as row average rather than local heat transfer coefficients. This is because pin fins typically are small features and the conduction of the metal tends to smooth any metal temperature variations produced by the uneven heat transfer rates. Also, data giving the average heat transfer rate tend to be easier to collect. The average heat transfer rate at each row is a function of the Reynolds number, the configuration of the pin array (in-line or staggered), the spacing, and the size of the pins.

Row average heat transfer rates within pin arrays tend to exhibit increasing heat transfer from the first to the third or fourth row and then a slow decline in heat transfer rates (see Fig. 19). The trend is shown in the data of Metzger *et al.* [35].

The data were also correlated with the Reynolds number (based on pin diameter and the fluid velocity through the minimum flow area). For the first row of the array, the relationship is

$$Nu = 0.140Re^{0.611} \text{ for } 10^3 < Re < 10^4 \tag{14}$$

and

$$Nu = 0.022Re^{0.813} \text{ for } 10^4 < Re < 10^5 \tag{15}$$

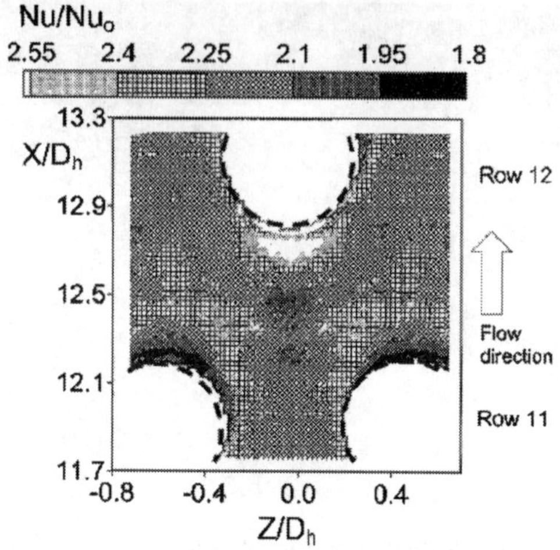


Figure 18: Local heat transfer rates around a staggered pin array, $Re = 18,300$ (from Ligrani and Mahmood [34]).

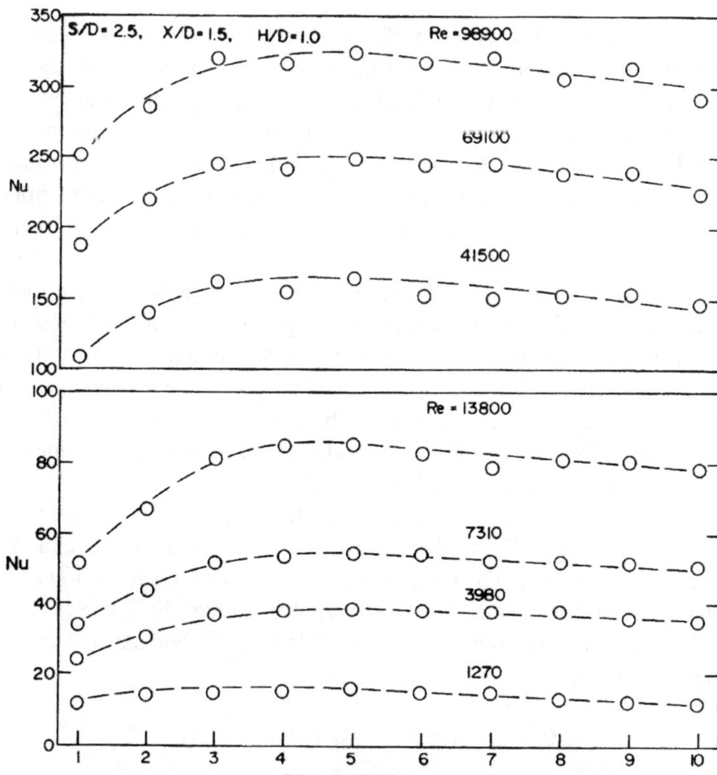


Figure 19: Nusselt number development showing the rise in heat transfer rate over the first four rows of the pin array and subsequent reduction (from Metzger *et al.* [35]).

Note that $H/D$ of the study was unity and that the spacing between the pins across the span was $S/D = 2.5$. The average Nusselt numbers for the entire array (10 rows) were also correlated. The authors give the following relationships:

$$\overline{Nu} = 0.069\overline{Re}^{0.728} \quad \text{for } X/D = 2.5 \tag{16}$$

and

$$\overline{Nu} = 0.092\overline{Re}^{0.707} \quad \text{for } X/D = 1.5 \tag{17}$$

## 3.3 Channel cooling techniques

In the mid-region of turbine airfoils, some form of channel flow cooling is often employed. This cooling may be directed through a simple channel that spans the airfoil from the root to the tip (a radial flow configuration) or it may be configured in a more complex arrangement that directs the flow outward towards the tip, then inward towards the root, then repeating, through a series of channels and turns (a serpentine configuration). The number of passes that a serpentine scheme may incorporate will vary with the specific application and will depend on several factors including the external heat load, available supply pressure, internal heat transfer coefficient that can be attained, cross sectional area of the airfoil, and manufacturing considerations. Actual turbine applications typically employ three or five passes, although patents for schemes with additional passes have been filed. The channels can be either smooth or include some form of roughening elements which typically increase both the heat transfer coefficient and the pressure loss. The turns between the passages tend to exhibit complex flow fields making heat transfer coefficient and pressure loss predictions difficult.

### 3.3.1 Smooth channel cooling

Perhaps the simplest form of channel cooling is the smooth channel. The ability to predict heat transfer and pressure drop within the smooth channel makes it an ideal starting place when considering channel flows. Also, roughened channel heat transfer and pressure drop correlations are often presented as multipliers to smooth channel correlations.

### 3.3.2 Smooth channel pressure loss and heat transfer correlations

Basic correlations for smooth channel pressure drop and heat transfer have been provided by Blasius and Dittus–Boelter. The Blasius equation may be derived from the Navier–Stokes equation and solved analytically. Discussion of the solution method may be found in many heat transfer textbooks including Kays *et al.* [36]. Solution of the equation provides the following dependence for local friction factor:

$$c_\mathrm{f} = \frac{0.664}{Re_x^{1/2}} \tag{18}$$

and mean friction factor:

$$c_{f_m} = \frac{1.328}{Re_x^{1/2}} \tag{19}$$

Used with Bernoulli's equation (the modified Bernoulli equation) the pressure drop, along a smooth channel may be determined. Many pressure loss correlations for rough channels supply factors for multiplication, which make use of smooth channel correlations as the basis.

The Dittus–Boelter equation (first proposed in 1930) has been developed to describe smooth channel heat transfer. The equation assumes fully developed flow (no entrance effects) and is often shown as $Nu = 0.023Re^{0.8}\ Pr^n$ where $n = 0.4$ for heating applications and $n = 0.3$ for cooling applications. The equation is often used with $n = 0.4$ as a basis for rough channel heat transfer correlations. The equation can be found in many heat transfer textbooks including Gebhart [37] and Incropera and DeWitt [38].

### 3.3.3 Effect of rotation on the flow field and heat transfer coefficients

Correlations for smooth channels may be used directly in the design of cooling channels on non-rotating parts. However, when the cooling channel is rotating, as in the case of a turbine blade, the smooth channel correlations must be adjusted to take into account the effects of rotation. Rotation produces a Coriolis force and, with variations in density, a buoyancy force.

The first step in understanding the heat transfer rates that will develop inside the passage is to consider the flow field. As a channel moves around an axis of rotation, fluid flowing with a radial component within the channel experiences an angular acceleration giving rise to a Coriolis force. The Coriolis force is described by the equation $\vec{F}_{Cor} = -2m\vec{\Omega} \times \vec{v}$, where $m$ is the mass of a fluid element experiencing the force. In Fig. 20, the channels are shown rotating into the page such that the axis of rotation is to the left. The direction of the axis of the negative rotation vector($-\Omega$) is therefore to the right. By taking the cross product between the direction of negative rotation and the direction of the fluid flow, $\vec{v}$, one may find the direction of the Coriolis force. In the channel flowing toward the tip, the force is out of the page (this is represented by the arrows near the center of the channel in the cross stream depiction of the flow field, Fig. 20), while in the channel which is flowing away from the tip, the force is in the opposite direction, into the page. When viewing a cross section of the channel normal to the mean flow direction, one would find that a secondary flow circulates. Fast moving flow in the centre of the channel (unaffected by the momentum loss produced in the boundary layers near the walls of the channel) feels a large Coriolis force while a small Coriolis force is felt by the slow moving fluid in the side wall boundary layers. The pressure gradient established within the core flow pushes the low-momentum fluid in the side wall boundary layer region, causing the secondary flow field within the channel.

In channels with fluid moving away from the axis of rotation, the fluid near the leading side of the channel is stabilized by rotation, reducing turbulence, while fluid near the high pressure, trailing side is unstable (see Fig. 21). The velocity gradient

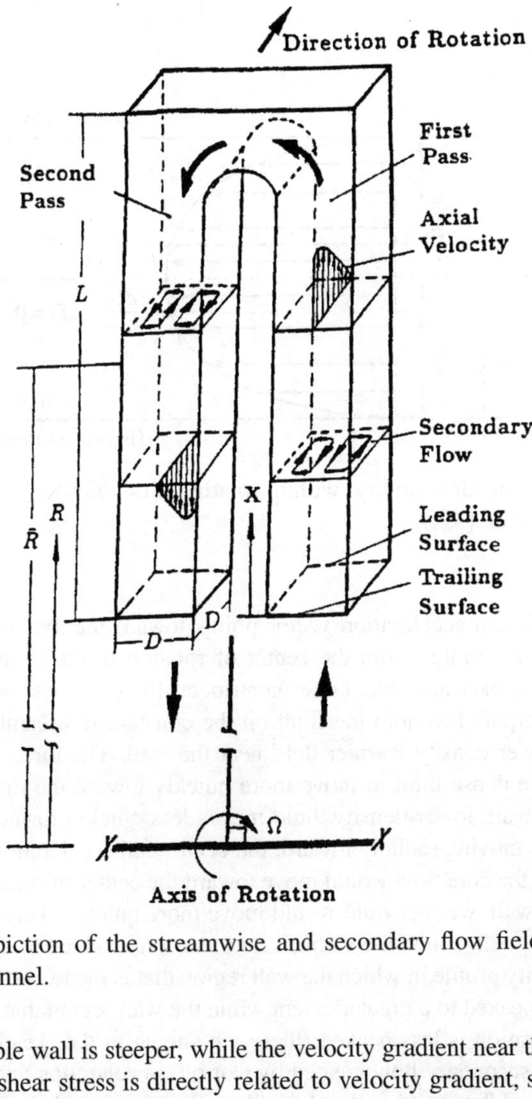


Figure 20: Depiction of the streamwise and secondary flow fields in a rotating channel.

near the unstable wall is steeper, while the velocity gradient near the stable wall is shallower. As shear stress is directly related to velocity gradient, shear stress near the unstable wall is higher, while that near the stable wall is smaller compared with those in the non-rotating channel. With these changes in turbulent mixing, we expect that heat transfer rates on the stable side will be lower than those on the unstable side.

Rotation may also induce buoyancy forces. This will affect flow if there is an imbalance of buoyancy forces, as would be the case when there is a density gradient normal to the centripetal acceleration direction. To understand the effects of buoyancy, imagine a single rotating channel with flow moving toward the tip of the channel. Let us also assume that the walls are held to the same, uniform temperature, i. e. the channel walls are isothermal. Imagine that the fluid in the core of the channel is cooler. Rotating channels impose a centripetal acceleration, $\vec{a} = -|\vec{\Omega}||\vec{\Omega}|\vec{R}$, on the

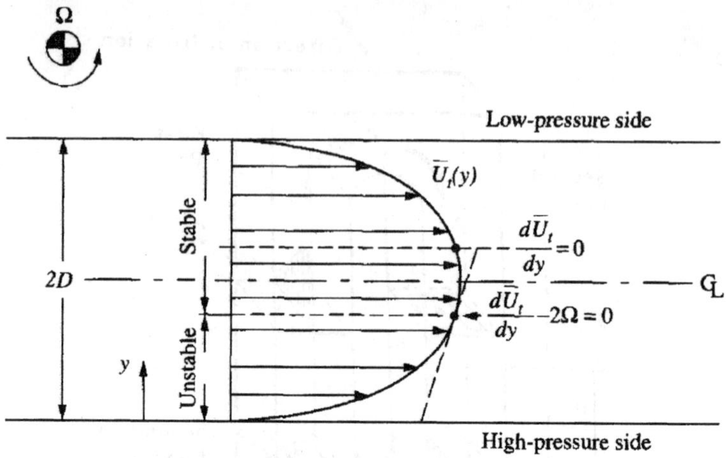


Figure 21: Regions of stability within rotating flow fields (from Lezius and Johnston [39]).

fluid (the centripetal acceleration vector points toward the axis of rotation). The parameter $R$ is the radius from the center of rotation to the location of interest. Cooler fluid is denser and thus larger centripetal force per unit volume of fluid, $\vec{F}_{cen} = \rho\vec{a}$, is required to hold the fluid on the constant-$R$ streamline when compared to the lower density warmer fluid near the wall. The force field causes the cooler and more dense fluid to move more quickly toward the tip of the channel while the near-wall, lower-density, fluid moves less quickly toward the tip. If the flow were to be moving radially inward, the centripetal acceleration would remain unchanged and the core flow would move toward the center of rotation less rapidly while the near-wall, warmer fluid would move more quickly. These two scenarios are shown in Fig. 22. When the Coriolis acceleration is added to the situation, the result is a velocity profile in which the wall region that is made unstable by Coriolis acceleration is peaked to a greater extent while the wall region that is stabilized by Coriolis acceleration is less peaked (these are shown in the sketches in Fig. 22). Because of the secondary flow created by Coriolis acceleration and its convection of heated and cool flows within the channel, the buoyancy and Coriolis effects cannot be treated separately.

One pioneering study on rotation effects was by Wagner *et al.* [40] This study investigates several important parameters, including the effect of density ratio ($DR = (\rho_{in} - \rho_{wall})/\rho_{in}$) and rotation number, *Rot*, (as presented below). The results indicate that the flow behavior described above does indeed influence the heat transfer rates.

Effects of rotation can be expressed with the following parameters:

$$\text{The Reynolds number: } Re_d = \frac{\rho d u_{\text{in}}}{\mu} \tag{21}$$

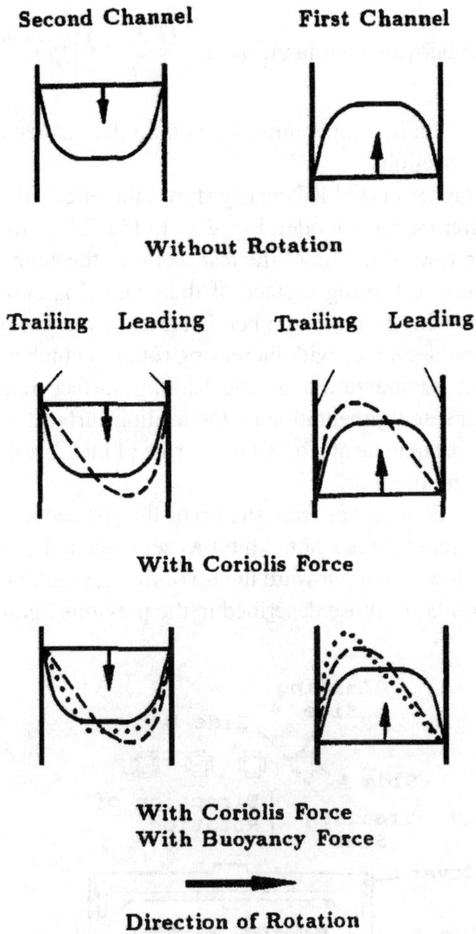


Figure 22: Effects of inertia, Coriolis, and rotational buoyancy. Note that, given the clockwise direction of rotation, the direction of the vector is out of the page (taken from Han *et al.* [1]). Solid lines, without rotation; dashed lines, with Coriolis force only; dotted line, with buoyancy and Coriolis forces.

where $d$ is the hydraulic diameter $\rho$ is the gas density $u_{\text{in}}$ is the inlet gas velocity $\mu$ is the gas viscosity.

$$\text{The rotation number: } Rot = \frac{\Omega d}{u_{\text{in}}} \tag{22}$$

where $\Omega$ is the rotational frequency, and

$$\text{The buoyancy number: } Bo = \left(\frac{(T \pm T_{\text{in}})}{T}\right) Rot^2 \frac{R_{\text{m}}}{d} \tag{23}$$

where $T$ is the local static temperature, $T_{\text{in}}$ is the inlet gas temperature, $R_{\text{m}}$ is the mean radius of test section.

The study by Wagner *et al.* [40] clearly shows the effect of rotation on smooth channel heat transfer (see test model, Fig. 23). In Fig. 24 (with flow moving from the axis of rotation toward the tip of the test section), the augmented heat transfer on the high pressure or trailing surface of the channel is clearly shown as augmented by the increasing rotation number. On the low pressure or leading surface, the heat transfer first decreases with increasing rotation number and then begins to increase. However, augmentation on the leading surface remains substantially lower than heat transfer augmentation on the trailing surface. As discussed above, heat transfer behavior may be attributed to the role of the Coriolis force on stabilizing and de-stabilizing the flow.

In Fig. 25 (where flow moves from the tip of the test section toward the axis of rotation), the heat transfer does not exhibit as sensitive a dependence on rotation as was seen in the flow moving toward the tip of the test section. The trends exhibited are roughly similar to those described in the previous figure.

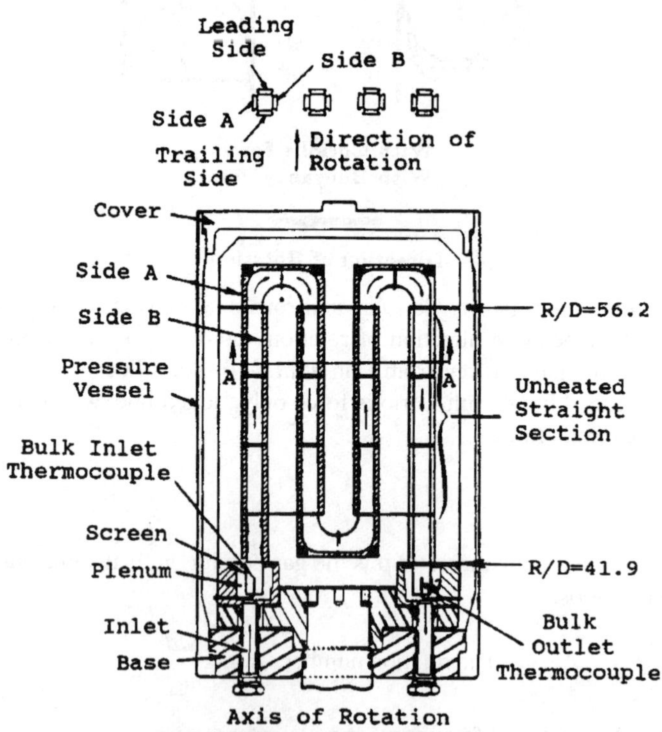


Figure 23: Serpentine coolant passage model used by Wagner *et al.* [40].

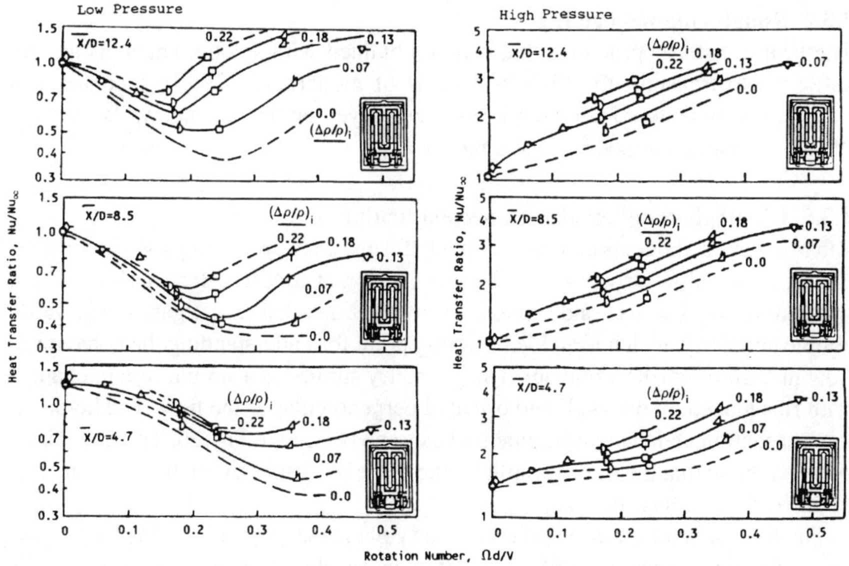


Figure 24: Effect of rotation number and density ratio on heat transfer ratio on the leading surface (low pressure) and trailing surface (high pressure) of a channel with flow moving away from the axis of rotation, $Re$ = 25,000 (from Wagner *et al.* [40]).

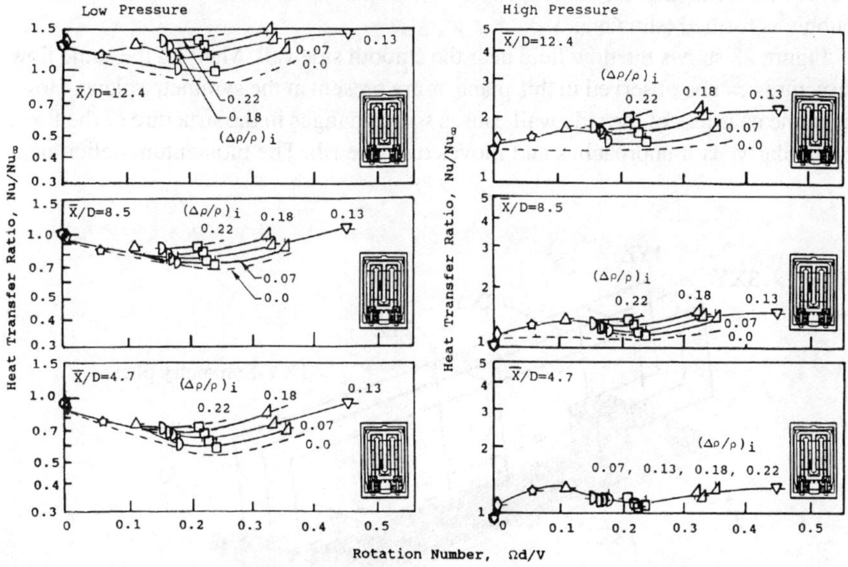


Figure 25: Effect of rotation number and density ratio on heat transfer ratio on the leading surface (low pressure) and trailing surface (high pressure) of a channel with flow moving toward the axis of rotation, $Re$ = 25,000 (from Wagner *et al.* [40]).

### 3.3.4 Rough channel cooling

The heat transfer coefficients that can be obtained with smooth channels are not sufficient to transport the high heat loads of modern airfoils. To increase heat transfer coefficients, roughening is often employed on the channel walls. Perhaps the most popular method is by adding ribs.

### 3.3.5 Channels roughened with perpendicular ribs

Rib roughened elements can take on many forms, each producing a slightly different secondary flow structure and therefore different heat transfer coefficients and pressure drops. Casarsa and Arts [41] provide a detailed investigation of a fairly simple geometry, which is an ideal starting point for understanding the behavior of flow in a rib roughened channel. The geometry studied is a simple square channel with ribs placed on one wall and oriented perpendicular to the flow. The layout of the test section and the measurement planes can be seen in Fig. 26. The flow direction is $X$. Note that the figure is split by the plane of symmetry so that only one half of each of the ribs is shown.

The first plane of interest is the symmetry plane (see Fig. 27). This plane shows the streamwise effect of the rib. As the flow leaves the downstream edge of the rib, it detaches and a recirculation bubble is formed. At some distance downstream of the rib (in this case approximately 4.3 rib heights) the mainstream flow reattaches and the boundary layer along the wall redevelops until the presence of the next rib is felt. Here, the flow begins to rise away from the wall. A small vortex resides between the wall and the rib (shown as $V_3$). At the top of the rib, a small separation bubble is formed (shown as $V_2$).

Figure 28 shows the flow field near the smooth sidewall. Many of the same flow structures can be observed in this plane as are present at the symmetry plane. However, the nearness of the side wall causes some changes in the structure of the flow; particularly, as it approaches and moves over the rib. The momentum deficit near

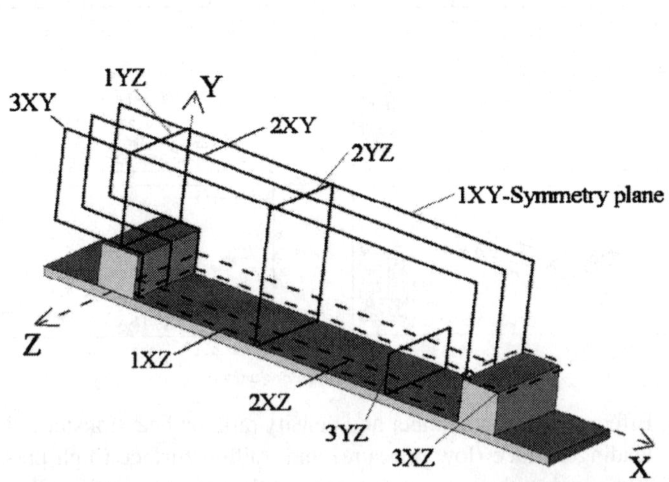


Figure 26: Layout and PIV measurements planes in Casarsa and Arts [41]).

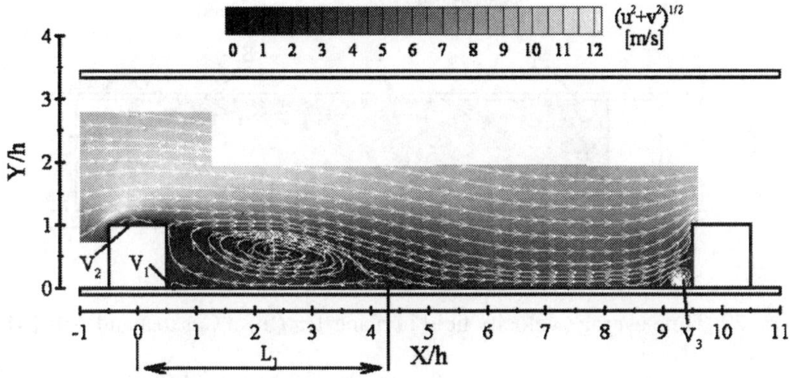


Figure 27: Time averaged velocity field in plane $1xy$ (from Casarsa and Arts [41]).

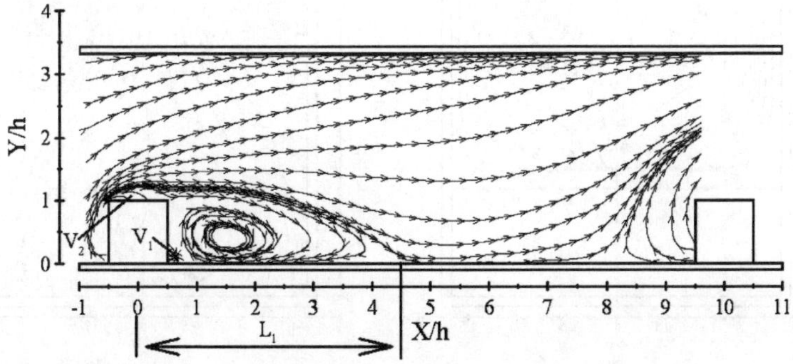


Figure 28: Stream tracer visualization of the time averaged velocity field in plane $3xy$ (from Casarsa and Arts [41]).

the side allows the vortex structure $V_3$ to expand, moving the separation point upstream toward the previous rib.

Figure 29 shows a plane near the rib-roughened wall. A second separation zone, $S_1$, is identified, located between the smooth wall and the rib. It is a result of the interaction between the low-momentum fluid near the smooth side wall, the rib boundary layers, and the recirculating flow in the separation bubble.

The final planes of interest are shown in Fig. 30. They cut across the channel and clearly show two recirculating structures, which reside at the top of the channel. They are initially created by the rib, which forces the flow upward. They intensify as the flow develops further through the passage. Away from the rib (as shown in Fig. 30, plane $2yz$) the structures move toward the smooth walls as the channel expands downstream of a rib.

### 3.3.6 Effect of rib spacing and height

Typically, correlations for heat transfer in rib-roughened channels are dependent on rib spacing and rib height. Closer rib spacing tends to increase heat transfer

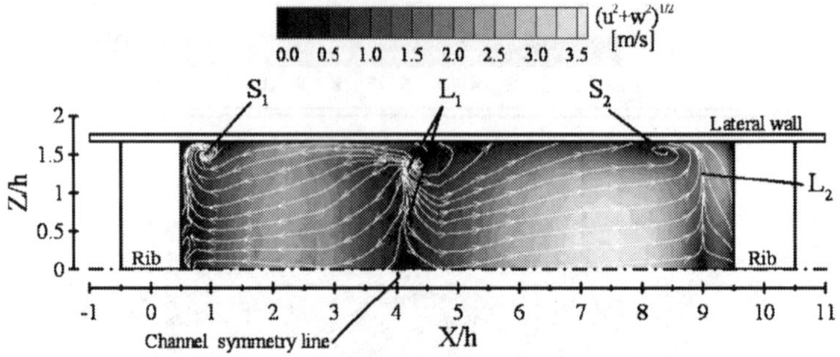


Figure 29: Time average velocity field in plane $1xz$ (from Casarsa and Arts [41]).

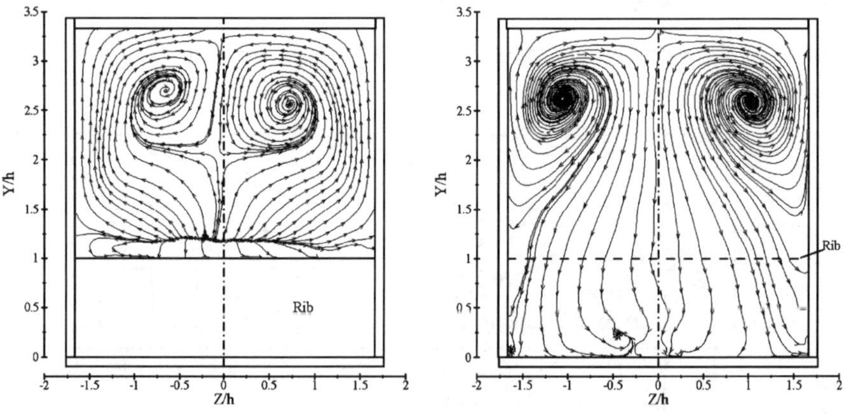


Figure 30: Time averaged flow field of plane $1yz$ (left) and $2yz$ (right) (from Casarsa and Arts [41]).

coefficients until the spacing of the ribs becomes so close that the separation zone behind a rib is not able to reattach before the next rib. At this point, correlations tend to break down and the rib roughening is no longer effective. This is often expressed as the ratio of rib height to pitch, $\varepsilon/P$. The height of the rib is also important and is often expressed as the ratio of height over channel hydraulic diameter, $\varepsilon/D_h$. Higher ribs typically yield increased average heat transfer rates. Ribs that are too low to reach above the boundary layer do less to augment heat transfer. On the other hand, ribs that are quite high may overly disturb the flow field, behaving more like baffles than roughening elements.

### 3.3.7 Channels roughened with skewed ribs

Channels roughened with skewed ribs exhibit some of the same flow field characteristics as the channels with perpendicular ribs. Flow tends to accelerate over the ribs, detach, and reattach at some point downstream of the rib. However, skewed ribs tend to be more effective at inducing secondary flows, which act to mix the coolant fluid, thereby increasing the effective heat transfer rates.

Kiml *et al.* [42] performed flow visualization with a water tunnel to explore the secondary flows created by skewed ribs. A sketch of the flow field with skewed ribs is shown in Fig. 31. The separation and reattachment zones are readily visible in the figure. Several additional flow structures highlighted in the schematic require explanation. The skew of the ribs produces a secondary flow near the wall which has the roughness elements (in this figure, the roughness elements are located on the left and right walls, though only the right wall is shown). This secondary flow causes the fluid near the roughened wall to flow from the bottom wall toward the top wall. The flow near the center of the passage moves in the opposite direction, from the top wall toward the bottom wall. This secondary flow pattern is shown projected toward the left hand side of the schematic. With this flow, a particle within the fluid may be expected to follow the course shown by the path lines. Note that the secondary flows in this configuration very effectively mix the coolant, circulating it from the center to the walls of the channel and also from the bottom wall to the top wall.

### 3.3.8 Channels roughened with ribs, heat transfer correlations

Many studies explore the heat transfer coefficients that may be obtained with rib augmentation. A review of all these studies is well beyond the scope of this chapter; therefore, only a select few are included. A more complete treatment may be found in Han *et al.* [1].

The amount of heat transfer augmentation a given configuration will produce depends on several parameters including the Reynolds number, the height and spacing of the ribs, the aspect ratio of the channel, and the angle of the ribs. The study of Fu *et al.* [43] clearly shows the effect of Reynolds number on heat transfer augmentation for both 90° ribs and 45° ribs on both the leading and trailing walls of the channel. Note the higher heat transfer coefficients observed on the trailing side of the channel. This is due to the destabilization of the trailing side boundary layer as discussed above. It can also be seen that augmentation rates on the trailing

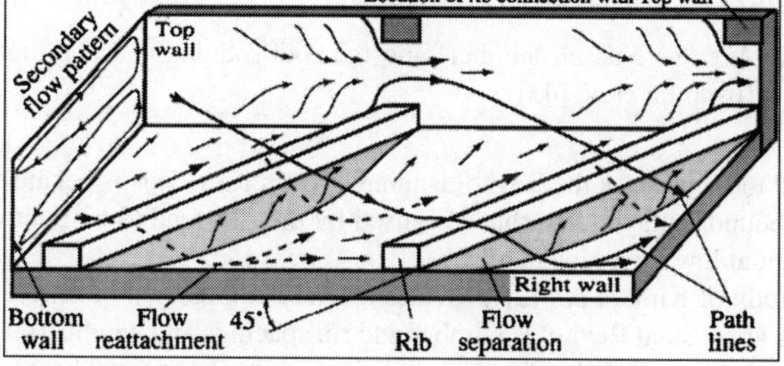


Figure 31: Flow field with skewed ribs, according to Kiml *et al.*[42]. Flow within the wall boundary moves from the upstream portion of the rib toward the downstream portion of the rib. Flow within the center of the channel moves in the other direction.

Table 1: Passage-mean Nusselt numbers and friction coefficients showing several rib configurations and a smooth channel values (from Kiml *et al.*[42]).

| Rib pattern | $Nu_{mp}/Nu_{mp0}$ | $C_f/C_{f0}$ |
|---|---|---|
| Smooth | 1.00 | 1.00 |
| 90° | 2.34 | 15.45 |
| 75° | 2.49 | 15.97 |
| 60° | 2.67 | 16.24 |
| 45° | 2.60 | 16.72 |

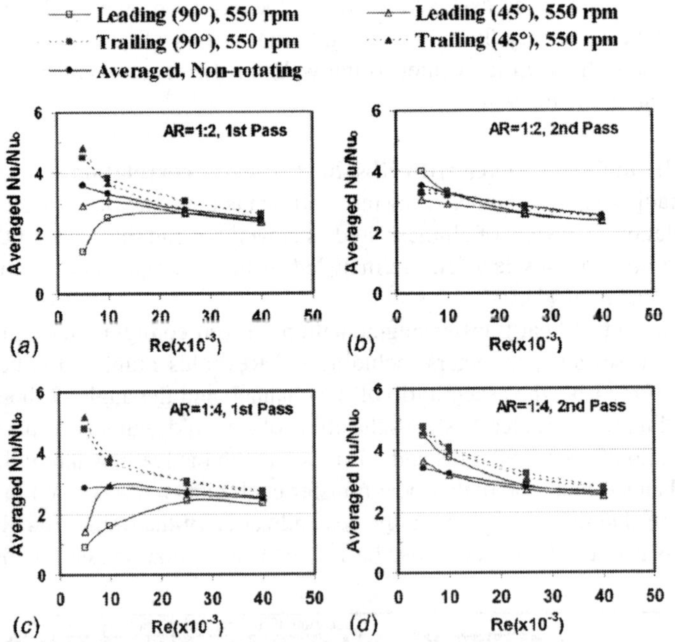


Figure 32: Average Nusselt number ratio for both rotating and nonrotating cases (from Fu *et al.* [43]).

wall tend to increase as the Reynolds number decreases. This is because the mixing and boundary layer disturbance caused by the ribs tend to be relatively more significant at low Reynolds number.

The study of Kiml *et al.*[42] provides a handy comparison of different rib orientations at constant Reynolds number and rib spacings and heights (see Table 1). It is clear that augmentation features do increase average channel heat transfer but also have a very significant pressure loss penalty. The 90° ribs are also clearly inferior to the 60° ribs, which produce the most significant heat transfer augmentation with a reasonable pressure loss.

Heat transfer over the surface of a channel is highly non-uniform, as can be seen in Fig. 33. Heat transfer rates on the channel surface near the upstream portion of

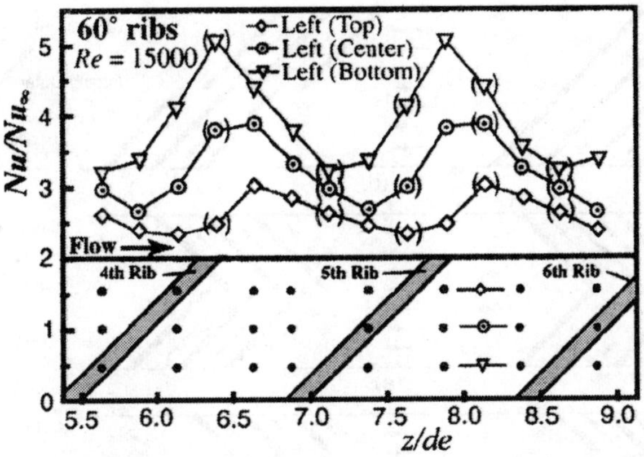


Figure 33: Local Nusselt number distribution (from Kiml *et al.* [42]).

the rib are more uniform, but are also less than the heat transfer rates on the channel surface near the downstream portion of the rib. The peaks are associated with the secondary flow along the channel wall, created in the channel by the ribs.

### 3.3.9 Channels roughened with V- and W-shaped ribs

As discussed above, one of the mechanisms that cause angled ribs to enhance heat transfer is the secondary flow system that they induce. The flow tends to move diagonally with the rib and then circulates back though the center of the channel. This causes one side of the channel to exhibit different heat transfer rates than the other. To reduce this difference, the ribs can be configured in a V shape. Rather than generate just two vortices, this configuration generates four vortices (see Fig. 34 which indicates the secondary flows developed by each rib configuration). The secondary flow then impacts the sides of the channel in a more uniform manner, providing a more uniform distribution of heat transfer coefficients. In the case of the W-shaped rib, the same concept is taken one step further. This configuration has a secondary flow dominated by eight individual vortices. Additionally ribs may be broken, forming discrete rather than continuous ribs.

With the V- or W-shaped ribs, the uniformity of the local heat transfer coefficients are improved over other rib configurations and average heat transfer augmentation of the V- and W-shaped ribs is slightly improved over that which can be obtained with angled ribs (see Fig. 35); however, this comes with an increased pressure penalty (Fig. 36) Similarly, the discrete configurations also tend to slightly outperform continuous ribs with an additional pressure loss. Due to the similarity in the performance (Fig. 37) of the rib configurations, the choice is not clear-cut and typically varies with the application. For instance, in some cases, the non-uniformity of the heat transfer coefficients obtained with angled ribs may be exploited to address a complementary non-uniformity in the external heat transfer rate while in other cases, the designer may find that the V-shaped configuration best matches external heat loads. The reader is referred to Ligrani *et al.* [32] for a handy comparison of the performances of different rib cooling configurations.

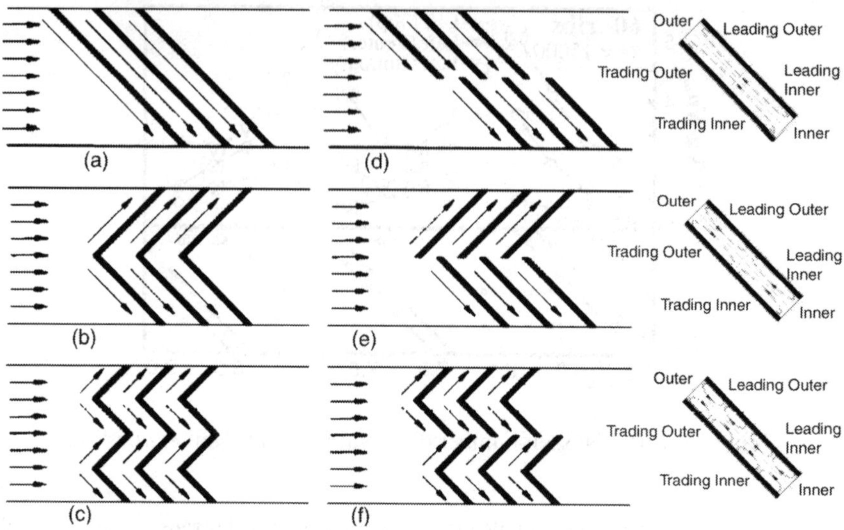


Figure 34: Secondary flows induced by different rib configurations (from Wright *et al.* [44]): (a) angled ribs; (b) V-shaped ribs; (c) W-shaped ribs; (d) discrete angled ribs; (e) discrete V-shaped ribs; (f) discrete W-shaped ribs.

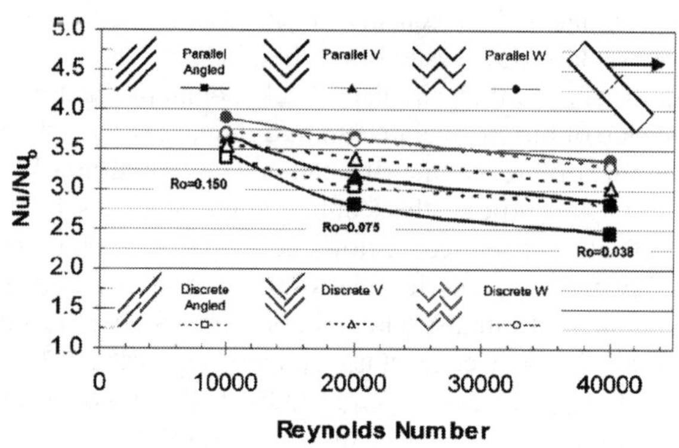


Figure 35: Channel average Nusselt number for several rib configurations in rotating channels (from Wright *et al.*[44]).

### 3.3.10 Dimpling

Dimpling is also an effective method of augmenting heat transfer rates, particularly in situations where only moderate increases in heat transfer rates are needed and very little pressure loss can be tolerated.

The flow field produced by the dimple is of particular interest as it is quite different from flows induced by other methods of augmentation. It is highly non-steady and is characterized by periodic shedding of vortices and fluid reattachment. As shown in Fig. 38, flow separates from the edge of the dimple forming a recirculation

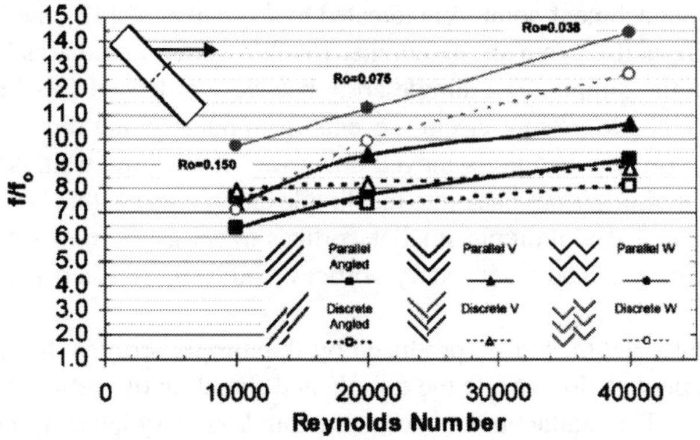


Figure 36: Average friction factor ratios for several rib configurations in rotating channels (from Wright *et al.*[44]).

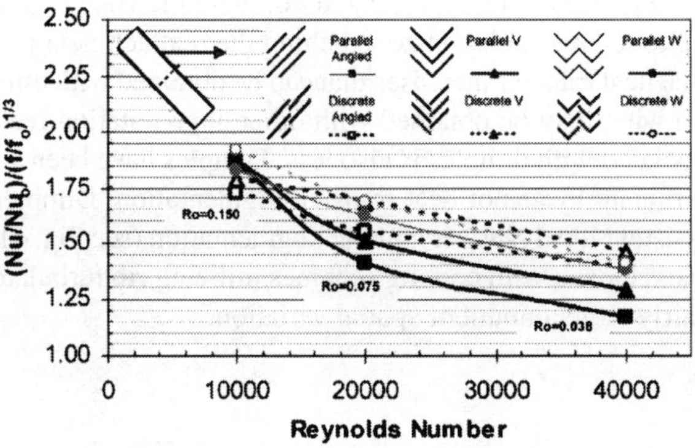


Figure 37: Thermal performance of several rib configurations in a rotating channel (from Wright *et al.*[44]).

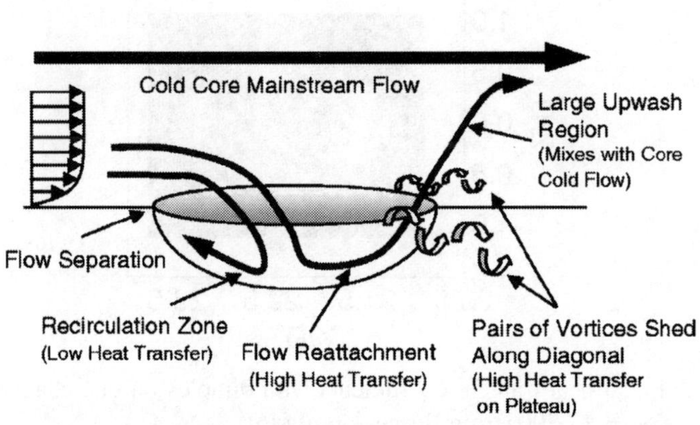


Figure 38: Secondary flow structure near a dimple (from Griffith *et al.*[45]).

zone and a reattachment point. As indicated by Ligrani *et al.* [32], the large upwash region manifests itself in a strong counter-rotating vortex pair that is intermittently ejected from the dimple. The dimple array behaves co-dependently; ejection from one dimple tends to prompt ejection events from neighboring dimples. Along the sides of the vortex pair fluid is moved downward toward the wall. Along with an ejection event, fluid rushes back into the dimple cavity causing thin boundary layers. Also, along the sides, multiple smaller vortices travel away from the downstream edge of the dimple. These secondary vortices tend to be steadier then the primary vortices.

Heat transfer augmentation produced by the dimple array is the result of both the reattachment of flow inside the dimple and shedding of vortices over the edge of the dimple. The reattachment causes the thin boundary layers typically associated with high heat transfer rates. The other dominate flow structure, namely the large counter rotating vortex which is periodically ejected from the dimple, effectively mixes boundary layer flow with the core and drives smaller vortices which enhance heat transfer along the area between neighboring dimples. Local heat transfer data are presented in Figs. 39 and 40. Note the zone of low heat transfer corresponding to the recirculation zone behind the reattachment point.

Although the heat transfer increases that can be obtained with dimples are somewhat less than what may be obtained with other augmentation features, the pressure drop associated with dimples is also less. Dimples have been shown to be one of the more efficient means of heat transfer augmentation. Dimples also produce relatively consistent local Nusselt number augmentation (see Fig. 41); an improvement over that achieved with cooling schemes utilizing rib turbulators which tend to exhibit a fairly large amount of spatial variation.

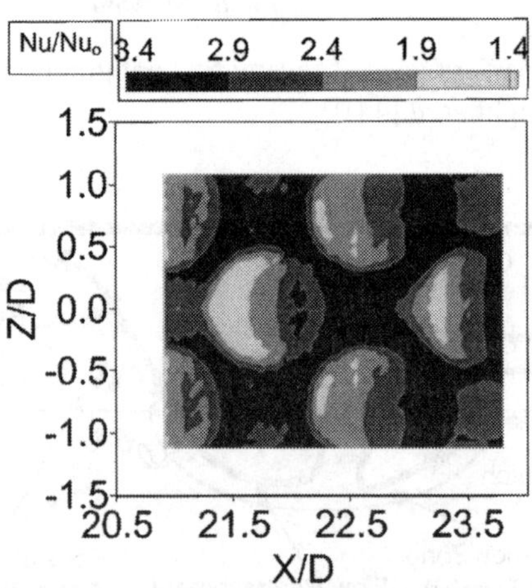


Figure 39: Local heat transfer coefficients with dimples on one channel surface, $Re_H = 17{,}200$ (from Burgess *et al.*[46]).

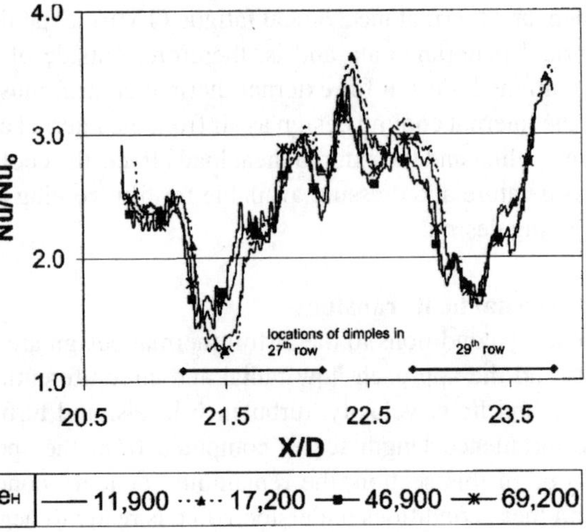


Figure 40: Local channel Nusselt numbers on a dimpled surface (from Burgess *et al.* [46]).

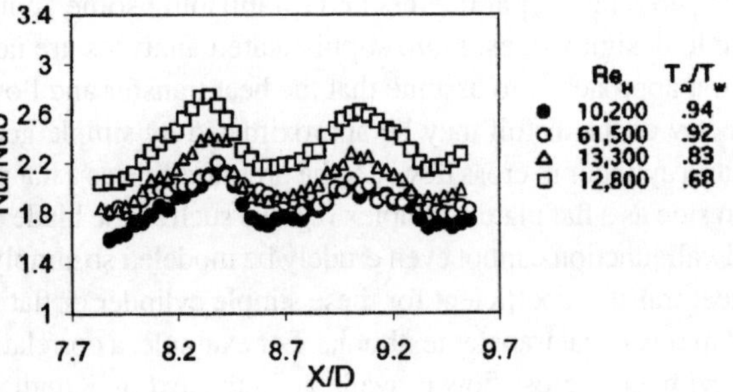


Figure 41: Spanwise average Nusselt number augmentation (from Mahmood *et al.* [47]).

### 3.4 External heat transfer

External heat transfer applies to surfaces adjacent to the passage flow, whether they are the airfoil external surfaces or the passage walls (endwalls or platforms). As noted, the airfoil resides in a flow of which the temperature exceeds the maximum materials temperature for a suitable design; thus, the surfaces adjacent to the passage flow must be aggressively cooled. This can be by conduction to the internal surfaces, which are cooled by the means discussed above, or by film cooling, which will be discussed in later sections. The first step in the design process is to describe the boundary conditions. The entry flow conditions to the first stage of the turbine have been discussed. The maximum surface temperature limits that the design must meet are determined by materials limitations, such as yield, oxidation, creep, and thermal mechanical fatigue (TMF). Establishing these limits requires empirical materials data and is, therefore, outside of the scope of this chapter. As mentioned, design for external thermal control must be done in conjunction with the internal cooling design as air from the internal cooling operations is used for film cooling and the external heat load affects the coolant heat up. Thus, the coolant temperature and pressure available for film cooling are determined by the internal cooling design.

### 3.4.1 Airfoil external heat transfer

Important boundary conditions to the airfoil thermal design are the material temperature limits and the approach flow radial and circumferential distributions of gas temperature and flow: velocity, turbulence levels, and turbulence spectra or representative turbulence length scales computed from the spectra. The first is beyond the scope of this section; the remaining boundary conditions have been discussed. With these conditions understood, one is ready to begin the analysis.

**3.4.1.1 Basic analysis** An understanding of external heat transfer coefficients may be gained by beginning with a highly simplified approach. This allows becoming oriented to the relevant parameters and establishing some approximate values. To have suitable design values, more sophisticated analyses are needed.

A simple first approach is to assume that the heat transfer and flow are steady and that the geometry of the airfoil may be approximated by simple geometries: (1) the leading edge as a cylinder in cross flow, (2) the pressure side as a flat plate surface, and (3) the suction side as a flat plate. Complex regions such as the blade tip, trailing edge, or airfoil–endwall junction cannot even crudely be modeled so simply. Equations correlating the heat transfer coefficient for these simple cylinder or flat plate geometries may be found in most heat transfer textbooks. For example, a correlation of heat transfer data for a cylinder in cross flow is available in the text by Kreith and Bohn [48]:

$$\frac{h_{c\theta} D_o}{k_f} = 1.14\left(\frac{V_\infty D_o}{\nu_f}\right)^{1/2} Pr_f^{0.4}\left[1-\left(\frac{\theta}{90}\right)^3\right] \tag{24}$$

for $0 < \theta < 80^\circ$, assuming laminar flow and a low level of approach flow turbulence intensity. The parameters are: $h_c\theta$, the convective heat transfer coefficient, variable with; $D_0$, the cylinder diameter; $k$, $Pr$, and $v$, the fluid thermal conductivity, Prandtl number, and kinematic viscosity based upon the fluid film temperature; $V_\infty$, the approach flow velocity; and $\theta$, the position around the cylinder, where the $0^\circ$ point is the stagnation line. Other than missing the effects of approach flow turbulence intensity, this will give a rather good estimate of the turbine leading edge heat transfer coefficient. Now look at computing the downstream flow. If the leading edge diameter Reynolds number is below approximately 100,000, one might expect the boundary layer to be laminar at $\theta = 80^\circ$, if above this value, a turbulent boundary layer may be assumed. This transition Reynolds number is based upon a low disturbance level external flow.

Over the region were laminar flow may be assumed, the flat plate heat transfer coefficients may be roughly calculated from a correlation of results of an analysis, as reported by, for instance, Incropera and DeWitt [49]:

$$Nu_x = \frac{h_{cx} x}{k_f} = 0.332 Re_x^{1/2} Pr^{1/3} \tag{25}$$

assuming $0.6 < Pr < 50$ and a low level of external disturbance. Choosing the origin of $x$ as the location $\theta = 80°$ on the leading edge would give conservatively high values of heat transfer coefficient. Since this correlation gives infinite values of heat transfer coefficient at the origin, there must be some fairing with the upstream leading edge heat transfer coefficient distribution at that point.

Where the flow may be assumed to be turbulent, the Chilton–Colburn analogy can be applied to compute the heat transfer coefficient. The results of this application of the analogy can be found in Incropera and DeWitt [49]:

$$St = \frac{h_c}{\rho c_p V_\infty} = 0.0296 Re_{x,\mathrm{turb}}^{-1/5} Pr^{-2/3} \tag{26}$$

assuming a low level of freestream turbulence intensity. The origin for $x$ may be the $\theta = 80°$ location on the leading edge if the diameter Reynolds number exceeds 100,000. Otherwise, the location of transition to turbulence of the above laminar boundary layer would be a suitable choice, if it were known. A value of $Re_{x,\mathrm{trans}} = 5 \times 10^5$ is often used as the transition Reynolds number, but the transition location is highly variable with streamwise pressure gradient, surface curvature, surface roughness, and external flow disturbance level, all of which are present and strong in turbine external surface flow analyses. Prediction of flow transition on turbine surfaces remains a major unsolved problem of turbine flow and heat transfer analysis. Nevertheless, models exist and one will be discussed below. Often, the first row of film cooling holes beyond the leading edge region can be counted upon to trip transition to turbulence on the suction surface. Thus, the origin for $x$,turb may be the streamwise location of that row of holes. On the pressure surface, film cooling injection may trip transition. However, if the injection is in a region of strongly accelerating flow, the boundary layer may remain laminar. Assuming that a trip to turbulence is at the film cooling hole would lead to conservatively high heat transfer coefficients. Again, since this correlation gives infinite values of heat transfer coefficient at the origin of $x$, turb, there must be some fairing of the heat transfer coefficient distributions at that point. However, it is known that with an abrupt transition, the heat transfer coefficient can rise rather steeply, so the fairing could be applied over a short streamwise distance. The above equations may be used together to see the relevant parameters and give a very crude approximation to the local heat transfer coefficient distribution on an airfoil.

When tripping to turbulence by film cooling injection cannot be assumed, a transition model must be applied. One rather simple model that has been shown to apply well to airfoils [50], even when applied in a quasi-steady fashion during the passing of wakes from upstream airfoils, is the correlation of Mayle [51]. In this model, the Reynolds number based upon the momentum thickness, $\delta_2$, and the freestream velocity, $V_\infty$, at the point of transition is given as a simple function of the freestream turbulence intensity, $Tu$, (give as a percent) as:

$$Re_{\delta_2,\text{transition}} = 400 \cdot Tu^{-5/8} \tag{27}$$

The effects of acceleration are captured in the value of the momentum thickness.

Now we introduce some modifications to the simple heat transfer models that lead to more accurate computed values of heat transfer rates.

Additional accuracy can be gained on the calculation of the heat transfer coefficient in the leading edge region of an airfoil by including the effect of approach flow turbulence. Ames *et al.* [9] noted that elevated levels of turbulence augment heat transfer in the laminar region of airfoil boundary layers, especially at the stagnation region. Turbulence levels affect suction surface heat transfer mostly by way of the influence on the transition location. Increased free-stream turbulence will move the transition onset location upstream, as the Mayle correlation shows. Increases in turbulence length scales increase pressure surface heat transfer. One correlation for the effect of turbulence on the stagnation region, given by Van Fossen *et al.* [52], is based on measurements in a flow with isotropic, grid-generated turbulence.Application of it begins with a modification of the stagnation line heat transfer coefficient:

$$Fr(0)_{\text{turbulence-enhanced}} = Fr(0)_{\text{low-turbulence}}\left(0.00851\sqrt{Tu \cdot Re_D^{0.8}\left(\frac{\Lambda_x}{D}\right)^{-0.574}} + 1.0\right) \tag{28}$$

where the Frossling number at the stagnation line,

$$Fr(0) = \frac{Nu_D(\theta = 0^\circ)}{Re_D^{1/2}} = \frac{(h_{c,\theta=0^\circ} D_o / k_f)}{(V_\infty D_o / \nu_f)^{1/2}} \tag{29}$$

of the low-turbulence flow can be taken from eqn. (25). The other parameters are: $Tu$ = the turbulence intensity and $\Lambda_x$ = the integral scale of turbulence based upon the streamwise velocity fluctuation component [53]. The paper gives a value for the coefficient of 0.00851. It was later changed to 0.00792 as explained in Giel *et al.* [54]. The integral length scale is not often known in design applications. The Van Fossen et al. correlation is based upon data in the range: $0.05 < \Lambda_x/D < 0.3$. Van Fossen and Bunker [55] measured stagnation region heat transfer on a circular leading edge residing in the exit flow of a dry, low-$NO_x$ can combustor. The flow turbulence intensity level was 28.5% and integral length scale to leading edge diameter ratio was 0.5. These can be taken as representative of that type of combustor–turbine arrangement. Their measured stagnation line heat transfer coefficient was 14% higher than that given by the Van Fossen et al. correlation. They could not say why, but noted the high degree of swirl in this combustor as a possibility. The turbulence intensity range of the correlation is: 1.1% < Tu < 15.9%. One author notes [56] better performance of the correlation when applied with turbulence levels that are closer to the correlation range.

A correlation that includes the effects of turbulence level and scale on laminar boundary layers with a constant free-stream strain rate is given by Dullenkopf and Mayle [57]. An effective Nusselt number for the stagnation region, given in terms of effective turbulence intensity, is:

$$Nu_{\mathrm{a}} Pr^{-0.37} = 0.571 + 0.01 Tu_{\lambda} \tag{30}$$

where the effective Nusselt number is

$$Nu_{\mathrm{a}} = \frac{Nu_{\mathrm{D}}}{\sqrt{a_1 Re_D}} \tag{31}$$

the parameter $a_1$, the dimensionless strain rate, is 4.0 for a cylinder in cross flow. The effective turbulence intensity, $Tu_{\lambda}$, is given by

$$Tu_{\lambda} = \frac{Tu_{\mathrm{a}} \sqrt{L_{\mathrm{a}}}}{(1 + 0.004 L_{\mathrm{a}}^2)^{5/12}} \tag{32}$$

the dimensionless length scale, $L_{\mathrm{a}}$, is given by

$$L_{\mathrm{a}} = \frac{\Lambda_x}{D} \sqrt{a_1 Re_D} \tag{33}$$

where $\Lambda_x$ is the integral length scale based upon the streamwise velocity fluctuation component. The turbulence term is given as:

$$Tu_{\mathrm{a}} = \frac{Tu \sqrt{Re_D}}{\sqrt{a_1}} \tag{34}$$

where $Tu$ is based upon the streamwise component of approach flow velocity fluctuation.

Additional accuracy on the pressure and suction surfaces where the flow is turbulent can be gained by applying the simple flat plate turbulent boundary layer equation discussed above but with multipliers that account for pressure gradients. The inputs to the development of these multipliers are taken from analysis, correlation of data, or engine experience. For instance, the effects of pressure gradient and temperature ratio for the turbulent boundary layer flow on an airfoil were correlated by Rued and Wittig [58] as a modifier to a turbulent flow flat plate correlation, which is similar to the one above:

$$St = \frac{h_{\mathrm{c}}}{\rho c_{\mathrm{p}} V_{\infty}} = 0.0295 Re_{x,\mathrm{turb}}^{-1/5} Pr^{-0.4} \left(1 - F_{\mathrm{K}} \cdot \frac{K(x)}{St}\right) \cdot F_{\mathrm{T}} \tag{35}$$

where $F_{\mathrm{K}}$ is 165 and $F_{\mathrm{T}}$ is 1.08 for a wall to freestream temperature ratio of 0.64. The acceleration parameter, $K(x)$, is defined as $(V_{\infty} / V_{\infty}^2)/dV_{\infty}/dx$. The Reynolds number captures some flow history with the definition:

$$Re_{x,\text{turb}} = \int_{x,\text{turb}}^{x} \frac{V_\infty(x)dx}{\nu_\infty(x)} \tag{36}$$

When the acceleration parameter, $K$, exceeds ~$0.5 \times 10^{-6}$, heat transfer is reduced perceptibly, as eqn. (35) shows. When it exceeds $3.5 \times 10^{-6}$, the boundary layer appears to be entirely laminar [36].

**3.4.1.2 More advanced analysis** Numerical prediction of heat transfer coefficients offers the gas turbine designer the opportunity to include more effects, such as those due to acceleration, curvature, roughness, elevated turbulence intensity, and, possibly, turbulence scale. The parabolic finite volume code TEXSTAN [36], based on the SIMPLE velocity and pressure solution method [59] provides additional accuracy over the method of defining the heat transfer coefficients described above. Other computational avenues are available for modeling the heat transfer coefficient. If the flow is complex, perhaps three-dimensional, unsteady, or has regions of separation, an elliptic solver is needed. Some examples of codes frequently used for gas turbine analysis are: Fluent [60], Star-CD [61], and CFX [62]. These require the construction of an appropriate two- or three-dimensional mesh and the application of a turbulence model. Several methods of turbulence modeling are available, as outlined by Han *et al.* [1] and Lakshiminarayana [2]. They discussed the Prandtl mixing length, $k$–$\varepsilon$, $k$–$\omega$, algebraic stress, Reynolds stress, and large eddy simulation (LES) turbulence models and direct Navier–Stokes simulation. The LES simulation needs a turbulence closure model for only the scales of turbulence that are smaller than the computational grid scale. Subgrid scale models are expected to be applicable over a wider array of situations than the other turbulence models listed. More recently, the $v^2f$ [63] model has been applied successfully to turbomachinery flows. Discussions on the application of these models for the simulation of turbulent flows are available in the open literature. An excellent review of such computations and comparisons of their results to experiments is given by Dunn [64]. Dunn notes that unsteady calculations often offer improved prediction of heat transfer coefficients on airfoils that are downstream of airfoil rows, such as the first stage rotor. He cites the work of Abhari and Epstein [65], which shows that the time-average of the unsteady heat transfer predictions is significantly greater, as much as ~230% of the steady-state predictions. Unsteady computations tend to represent a significant increase in computational time and software complexity over steady computations.

**3.4.1.3 Heat transfer coefficient distributions** The discussion of 2D airfoil external heat transfer coefficient distributions is complete. Measurements of a representative distribution, along with a computer simulation, are given by Giel *et al.*[66] in Fig. 42 as the 50% span case. One can see high levels at the stagnation point on the leading edge, decreasing levels on both suction and pressure surfaces, a rapid rise in values on the suction side where transition is observed, and a milder rise on the pressure side where velocity is increasing and acceleration is retarding transition. Acceleration, compressibility, transition, curvature effects are in play. There is no film cooling or roughness effect in this study. Shown are the results

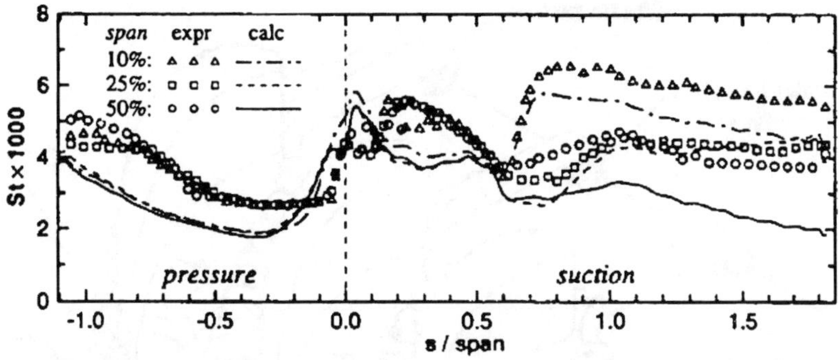


Figure 42: Predicted and measured Stanton number distributions for the blade. Reynolds number based upon inlet velocity and axial chord length = $1.0 \times 10^6$, exit Mach number = 0.98, inlet turbulence intensity = 7%. Experiments are with the liquid crystal technique; computations are with the 3D code RVC3D (from Giel *et al.* [66]).

of a complex array of effects on Stanton number. The figure also shows how the distribution varies from the mid-span region (50% span) to near the endwall (10% span). Clearly, 3D effects cannot be ignored in airfoil design.

**3.4.1.4 Additional external heat transfer considerations** Correlation-based heat transfer coefficients and heat transfer coefficients calculated using two-dimensional numeric codes may be improved upon by considering additional flow features that are not well modeled in two dimensions. They are considered under the general category of secondary flows and can have significant effects on heat transfer.

Factors that influence the near-wall flow field, such as skewing and mixing by secondary flows and with film cooling and leakage (or sealant) flows tend to influence heat transfer coefficients. Also, such effects as surface roughness or freestream turbulence (level and scale) significantly influence heat transfer.

**3.4.1.5 Secondary flows** Secondary flows within turbine passages are complex. Much research has been devoted to characterizing them. They are generally the result of pressure gradients induced by strong turning of the flow as it moves through the passage. This is expected at junctions between airfoils and endwalls or in regions of strong jetting, such as leakage flows or tip bypass flows. Several of the more important flows are highlighted below.

The endwall cross flow, as identified by Langston [67] in Fig. 43 is induced by the strong pressure gradient between the pressure side of the passage and the suction side of the passage. This is due to curvature of the streamlines within the passage. In the endwall boundary layer that resides within the passage, velocities are generally low and these pressure gradients create a movement of near-wall fluid from the pressure side of the passage to the suction side. The intensity of this cross flow is dependent on several factors including the degree of turning and the aspect ratio (the ratio of the passage height, or airfoil span, to the airfoil chord length).

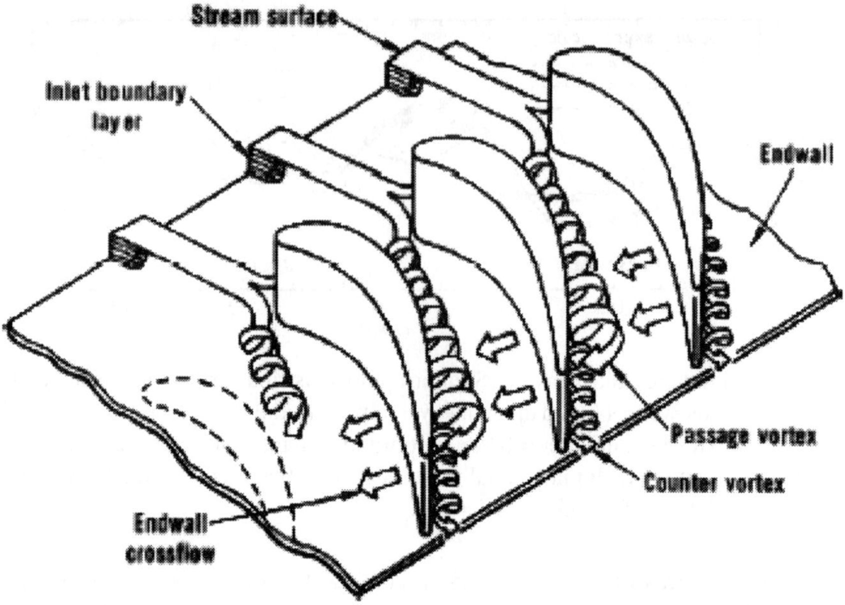


Figure 43: Secondary flows in the endwall region for a turbine airfoil row (from Langston[67]).

Another component of secondary flow in the passage is the horseshoe vortex, two legs of which are shown in Fig. 43. This structure begins at the leading edge of the airfoil at the junction between the airfoil and the endwall. The fluid stagnates against the leading edge of the airfoil. Since the approach flow near the endwall has a boundary layer, the stagnation pressure varies from a larger value at the outer edge of the boundary layer on the stagnation line and the minimum value at the endwall-airfoil leading edge junction. This pressure gradient at the leading edge drives the flow down the leading edge region toward the endwall (as shown in Fig. 43). The net result is a vortex that initiates at the leading edge of the airfoil and continues through the passage. If it were not for curvature of the passage, this vortex would assume a horseshoe shape. On the suction side of the airfoil, the vortex is quickly stretched by flow acceleration and begins to dissipate. On the pressure side, the vortex is influenced by the passage cross flow, which carries it to the suction side of the passage. There, it impinges against the suction side of the airfoil.

The passage flow augments the strength of the vortex. At the upper reaches of the pressure side leg of the vortex, the core passage flow velocity is high. The shear resulting from the interaction of the boundary layer flow (with the vortex embedded within it) with the core passage flow causes the vortex to grow in size and strength. The result is a much larger and stronger structure associated with this leg as it reaches the suction side of the passage, compared to the size and strength of the suction side of the horseshoe vortex. This pressure-side leg of the horsehoe vortex, which has been carried to the suction side, combines with the cross flow vortex (which is of the same rotational direction) to be called the passage vortex.

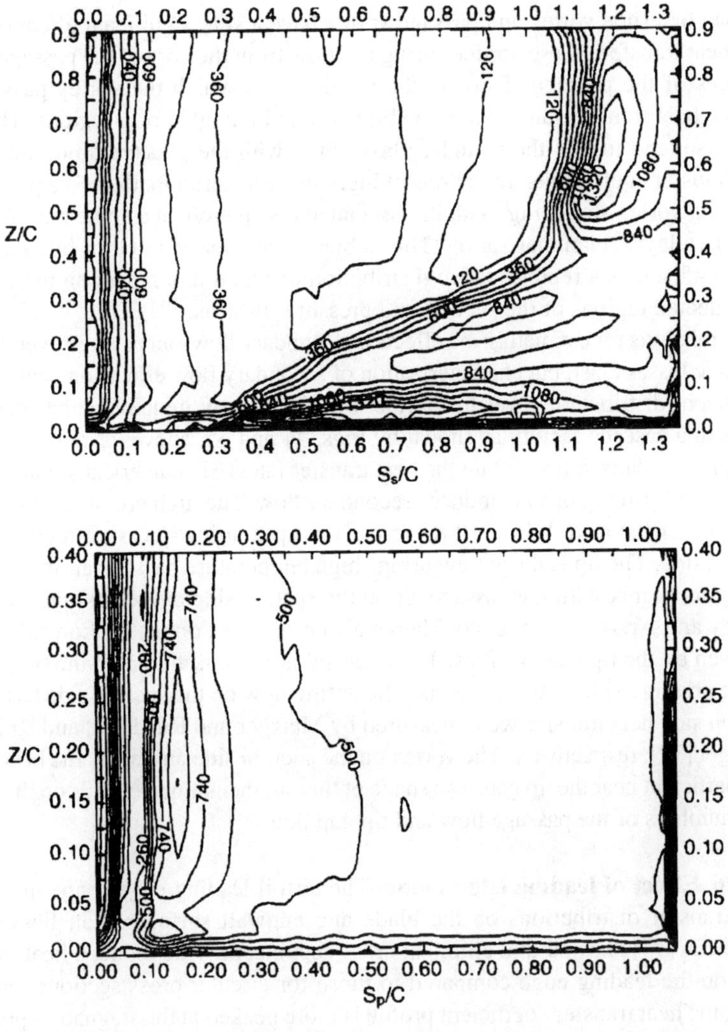


Figure 44: Detailed Sherwood numbers on blade suction and pressure surfaces near the endwall (from Goldstein *et al.* [68]).

Downstream, the passage vortex moves off the passage endwall and up the suction sidewall. Flow on the pressure side of the airfoil is drawn toward the endwall of the airfoil. Figure 44 shows mass transfer measurements in a cascade. With the analogy between mass and heat transfer, one can convert the mass transfer Sherwood number values, shown, to heat transfer Nusselt numbers. One can see on the suction side the effect on mass transfer of the vortex impinging on and traveling up the suction wall in the downstream portion of the passage. On the pressure wall, the Sherwood number distribution is rather two-dimensional, indicating that the passage vortex is carrying major components of the near-endwall region secondary flow away from the airfoil pressure surface.

The horseshoe vortex and passage vortex have a very significant effect on endwall heat transfer. These vortices bring hot flow from the core of the passage to the surfaces of the passage. Thus, in the regions over which the vortex passes, the surface heat transfer rates tend to be higher than in neighboring regions. The vortices also tend to mix the boundary layer fluid with the passage fluid, increasing heat transfer coefficients. In regions of high flow acceleration, there is a possibility of strong vortex stretching, with its associated dissipation, and some possibility of boundary layer relaminarization. This subject is addressed in Graziani *et al.* [69]. Figure 45 shows a representative distribution of Stanton numbers on the endwall and selected regions of the suction and pressure surfaces.

Correlations for estimating the effects of secondary flows on passage overall aerodynamic losses exist, but characterization of secondary flow effects on heat transfer coefficient distributions over the surfaces of the passage requires either measurements in a cascade facility as shown by Figs. 44 and 45, above, or rather complex and finely nodalized (to capture the heat transfer rates) 3D numerical simulations.

The blade tip region also induces secondary flow. The high pressure on the pressure side of the airfoil drives flow across the tip to a low-pressure region on the suction side. The tip leakage flow brings high temperature passage air through this tip region to mix with the passage air on the suction side of the passage as shown by Key and Arts [70] in Fig. 46. Shown also is a vortex on the suction side which is driven by the tip leakage flow. The effects that the sink flow into the tip gap has on the pressure side heat transfer and the jetting flow on the suction side has on the suction side heat transfer were measured by Metzger and Rued [71] and Rued and Metzger [72], respectively. The vortex on the suction side augments the heat transfer coefficient near the tip gap. The shape of this augmentation depends on the Reynolds numbers of the passage flow and tip gap flow.

**3.4.1.6 Effect of leading edge shape** The airfoil leading edge shape influences heat transfer distributions on the blade and endwall surfaces. Van Fossen and Simoneau [73] showed that elliptical cross-sections provide reduced heat transfer rates on the leading edge compared to those for circular cross sections. Furthermore, the heat transfer coefficient profile is more peaked at the stagnation point for the elliptical leading edge and the heat load downstream of the stagnation point is reduced from that of the circular cross section. The test pieces and representative data are shown in Fig. 47. The Frossling number was introduced previously. The values $Fr(0)$ were taken under low freestream turbulence conditions.

The leading edge shape at the airfoil to endwall junction can have a strong effect on the endwall heat transfer rate as shown computationally by Saha *et al.* [74]. Figure 48 shows the effect of one leading edge fillet.

**3.4.1.7 Effect of surface roughness** Turbine airfoils start out with a "smooth" (as-cast or as-coated) condition. With engine operation, the airfoil surface typically becomes rougher. Roughness may come about by dirt accumulation, surface corrosion and oxidation, erosion, spallation of TBC layers (often by thermal stress effects), and by material deposition, often by particles or condensables in the fuel.

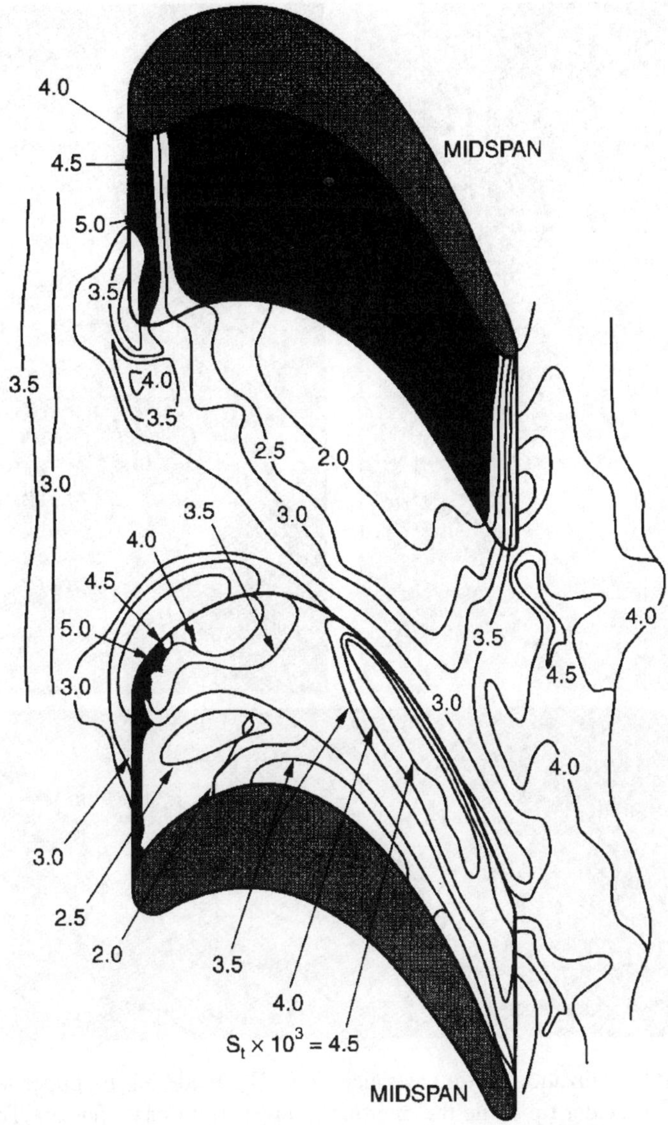


Figure 45: Endwall and blade surface Stanton number distributions for a thick boundary layer (from Graziani *et al.* [69]).

Different engines, operators, environments, materials and temperature histories affect how much roughness will develop and what shape the roughness will assume. Dirt accumulation is typically worse for a first-stage vane than for downstream airfoils, since this is the first stage in which the flow experiences strong streamline curvature and resulting separation of solids from the passage gas.

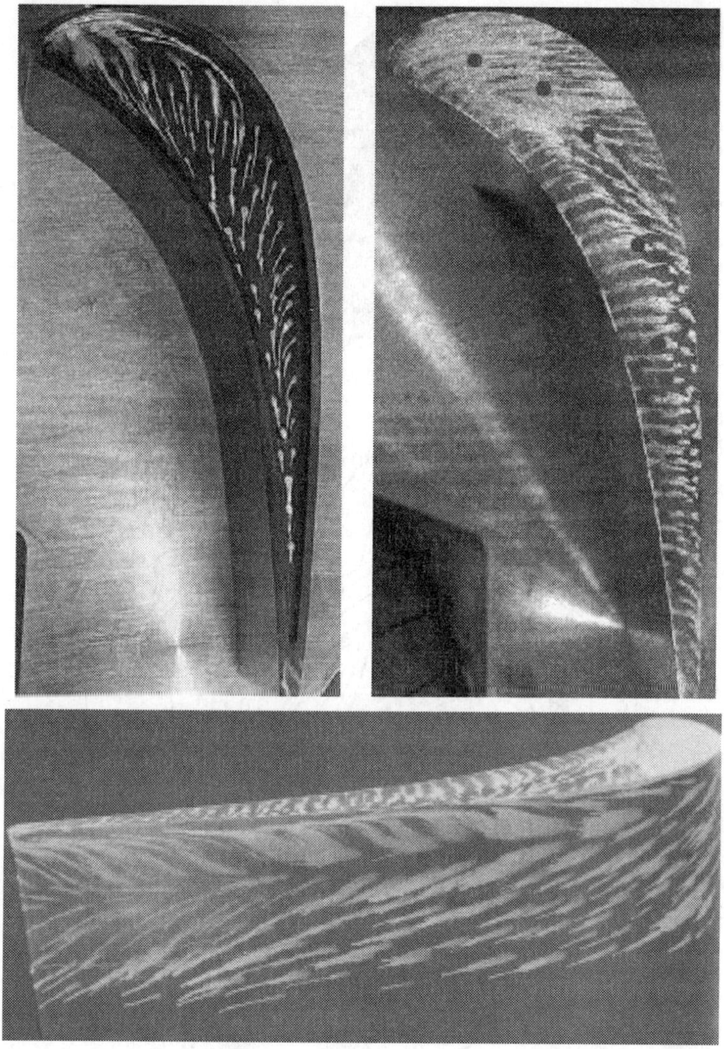

Figure 46: Visualization of tip clearance flow. The blade on the upper left has a squealer tip while the one on the upper right has a flat tip. The lower figure shows the vortex on the suction side, driven by the tip leakage flow (for the flat tip) (from Key and Arts [70]).

Effects of surface roughness on flow and heat transfer have been under study for many years, beginning with comprehensive experiments with sand-grain roughened walls and a single length scale for characterizing roughness. Roughness on the airfoil is not characterized well with uniform sand-grain roughness, however, since deposition and erosion do not occur uniformly. As found by Bons *et al.* [75], airfoils that have seen service show surface micro-geometry features caused by erosion, deposits, pitting, and TBC spallation, which are quite different from sand grain

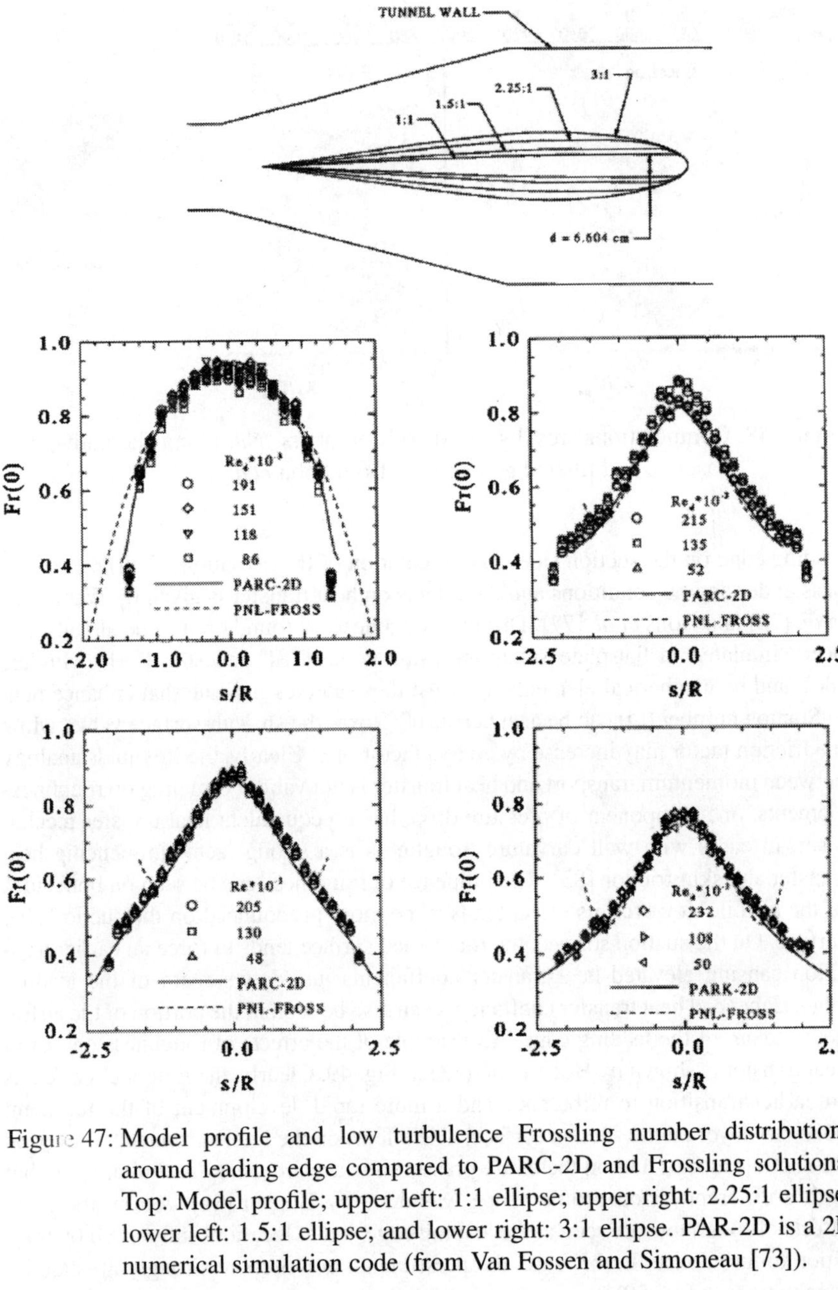


Figure 47: Model profile and low turbulence Frossling number distributions around leading edge compared to PARC-2D and Frossling solutions. Top: Model profile; upper left: 1:1 ellipse; upper right: 2.25:1 ellipse; lower left: 1.5:1 ellipse; and lower right: 3:1 ellipse. PAR-2D is a 2D numerical simulation code (from Van Fossen and Simoneau [73]).

roughness, and from one another. It would be expected that the different types cannot be modeled as an equivalent sand grain roughness. They note that for uncoated turbine airfoils, corrosive deposits and erosion are the primary causes of roughness, occurring near the pressure side tip and hub regions, while spallation is the dominant cause of roughness for coated airfoils. The spallation effect was pronounced near the

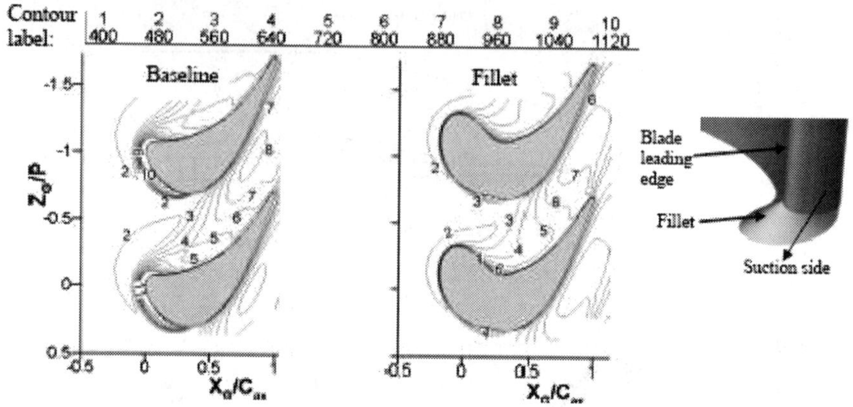


Figure 48: Computational results of Nusselt numbers, *Nu*, along the endwall of baseline and filleted geometries (from Saha *et al.* [74]).

leading edge on the suction side. Documentation of the evolution of surface roughness under engine conditions and the effects on heat transfer is given by Wammack *et al.* [76] and Bons *et al.* [77]. Discrete roughness, as found on turbine airfoils, has been simulated in flat-plate situations using cones [78], pedestals [79], cylinders [80], and hemispherical elements [81]. All these studies indicate that enhancement in Stanton number (*St*) can be as much as 60% over the smooth-surface value, while the friction factor may increase by up to a factor of 3. Clearly, the Reynolds analogy between momentum transport and heat transfer is not valid. Form drag on roughness elements, one component of pressure drop, has no equivalent heat transfer mechanism. In cases with wall curvature, roughness is a strong factor influencing heat transfer and skin friction [83]. The influence of roughness can be seen on both sides of the airfoil. However, its effect tends to be more pronounced on the suction side surface. On the suction surface, the roughened surface tends to force an earlier transition causing elevated heat transfer coefficients just downstream of the leading edge. Enhanced heat transfer coefficients can also be seen in the portion of the airfoil downstream of the leading edge. An example of the effects of roughness on airfoil heat transfer is shown by Hoffs *et al.* (83) in Fig. 49. Clearly, the roughness effect is an earlier transition to turbulence and a more rapid development of the turbulent boundary layer on the suction surface. The effects on the laminar boundary layer on the suction surface and on the entire pressure surface are weak. A similar study but with a different airfoil profile shape was done by Bunker [84]. Again, the main effects of roughness are seen at the leading edge and at locations influenced by transition to turbulence. In this case, the pressure surface is more strongly affected by roughness (see Fig. 50).

**3.4.1.8 Effect of freestream turbulence** The combustor exit flow typically shows a high level of freestream turbulence with large length scales, as discussed

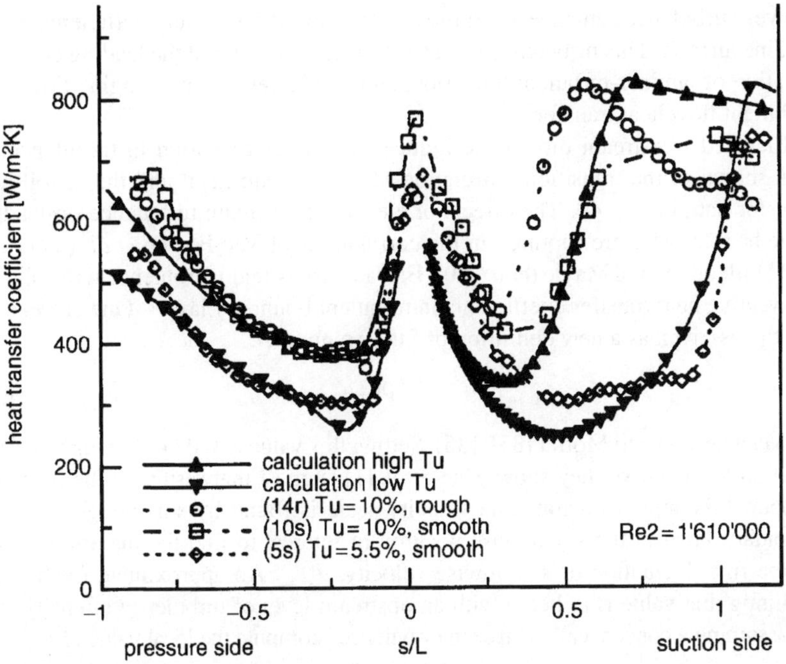


Figure 49: Effect of surface roughness and free-stream turbulence on airfoil heat transfer coefficient distribution (from Hoffs *et al.* [83]).

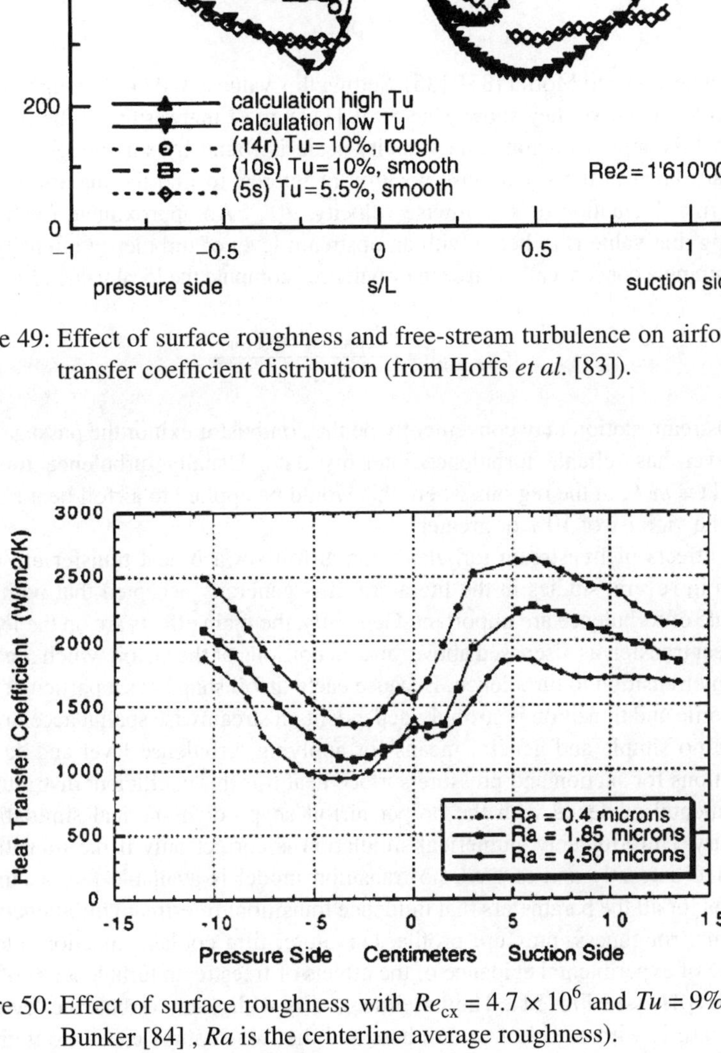


Figure 50: Effect of surface roughness with $Re_{cx} = 4.7 \times 10^6$ and $Tu = 9\%$ (from Bunker [84] , *Ra* is the centerline average roughness).

above. Turbulence can have a significant effect on heat transfer coefficients on the turbine airfoils. This turbulence can affect the heat transfer at the leading edge, the location of laminar-turbulent transition, and to a lesser extent in engine flows, the turbulent flow heat transfer.

Elevated freestream turbulence induces an earlier transition to turbulence, as was shown in the transition correlation of Mayle (eqn (27)) and the airfoil heat transfer data in Fig. 49. The effects of elevated freestream turbulence on leading edge heat transfer are captured in the correlations of Van Fossen *et al.* (eqn (28)) and Dullendopf and Mayle (eqn (30)). Elevated freestream turbulence will enhance convective heat transfer coefficients in turbulent boundary layers. One correlation for this is given as a new definition of Stanton number:

$$St' = h / \rho C_p u'_{max} \tag{37}$$

by Maciejewski and Moffat [85], [86]. Setting this value to 0.024 is a rough approximation to the curve they show. Han *et al.* [1] suggest that results in the literature support this approximation. The velocity scale is found by surveying across the boundary layer at the streamwise position of interest to find the maximum value of the rms fluctuation of streamwise velocity, $u'_{max}$. An approximate method for attaining that value is to begin with an upstream level of turbulence intensity and, by assuming a concept called "frozen turbulence" compute the local value of $u'_{max}$ as:

$$u'_{max,\ local} = \frac{TI_{upstream}\bar{U}_{upstream}}{\bar{U}_{local}} \tag{38}$$

The upstream station may conveniently be the combustor exit or the passage inlet, whichever has reliable turbulence intensity data. Usually turbulence intensity levels, $TI \equiv u/\bar{U}$, in the regions where this would be applied to airfoil heat transfer are in the vicinity of 10% or greater.

The effects of freestream turbulence on airfoil surface heat transfer are documented in several studies in the literature. It is generally accepted that both level and scale of turbulence are important. Generally, the main effects are on the leading edge heat transfer, as discussed above, and on portions of the airfoil which are influenced by transition to turbulence. Because each airfoil shape has a particular pressure profile and transition is strongly dependent on streamwise spatial acceleration, there is no simple and general means for applying turbulence level and scale to correlations for suction and pressure surface heat transfer coefficient distributions. Experimental evidence with the proper airfoil shape or numerical simulation is necessary. Unfortunately, numerical simulation is correct only if the transition is computed correctly and to date, no transition model is available that accurately accounts for all the parameters that influence transition: freestream turbulence level and scale, roughness, pressure profile, curvature, film coolant injection, etc. An example of experimental evidence of the effects of freestream turbulence is offered by Ames [87] (see Fig. 51). The baseline case is with a low turbulence level, the "grid" case is with a grid of square bars and the two "comb" cases are with two mock-up arrangements of different combustor and combustor-to-turbine transition

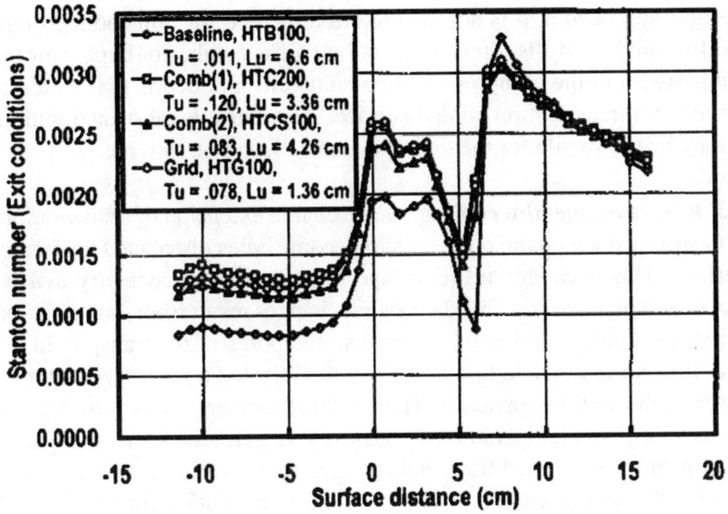


Figure 51: Effects of turbulence on Stanton number distribution, $Re_{ex} = 790{,}000$ (from Ames [87]).

duct geometries. Each case has a specific combination of turbulence level and scale. The turbulence scale is the energy length scale discussed previously.

### 3.4.2 Film cooling

**3.4.2.1 Objectives of film cooling** Film cooling by injection of coolant through porous walls, slots, and holes has been used for fifty years as a means of protecting gas turbine engine surfaces from adjacent hot gas streams. With film cooling, the surfaces are covered with a layer of relatively cool air. The effectiveness of protection is dependent upon the amount of coolant provided, the coolant supply temperature, and how the coolant interacts and mixes with the hot gas stream into which it is injected. In design, one wishes to determine where and how film cooling will be introduced so that each engine component is operating at the appropriate temperature. Of course, this is done in coordination with the internal cooling design. Film cooling design is not easy, for injection is to a gas turbine flow path environment that has a very complex flow field.

**3.4.2.2 The transpired boundary layer** Transpiration cooling is the simplest form of film cooling. It provides uniform injection of coolant flow though a porous wall. The injection temperature is the wall temperature. Injection will cause the boundary layer to grow, and the wall shear stress to drop, with distance from the beginning of injection. If the injection rate is large enough or blowing length is long enough, the shear stress approaches zero and the injected flow blows the passage flow off the wall. This would be a case of excellent film cooling coverage, but the coolant flow demand may be excessive and aerodynamic losses associated with injection may be substantial.

Though transpiration cooling appears attractive and some combustor liner cooling designs approach it, it is not the preferred method for film cooling of gas turbine airfoils and endwalls. Since film cooling air is combustor bypass air that had been ingested into the compressor, it contains dirt and debris that would fill the fine pores of a transpiration cooled surface. Also, airfoils fabricated with porous walls may be too fragile for the duty they are called upon to serve.

**3.4.2.3 Full coverage film cooling** A reasonable extension of transpiration cooling is to simply open up the pore sizes to become holes distributed uniformly over the surface. This is called "full-coverage film cooling," a geometry that is often approached in practice (see Fig. 3). Several changes result from transpiration cooling. First, the delivery holes are larger than the pores of the transpiration-cooled surface. It is not reasonable to assume that the injected flow will enter the boundary layer at the wall temperature. Thus, a third temperature is introduced to the problem, the coolant temperature, $T_c$, yielding three relevant temperatures: $T_s$, $T_\infty$ and $T_c$, for surface, external flow, and coolant, respectively. A second characteristic, which separates discrete hole cooling from transpiration cooling is the degree of mixing between the gas stream flow and the injected flow. Since the coolant flow enters the gas passage stream as discrete jets (see Fig. 52), the mixing region becomes highly three-dimensional, the emerging jet appears as a flow blockage to the approaching gas stream flow and the mainstream is forced to divert around and over the injected flow. An undesirable outcome is much stronger mixing of the injected flow and the gas stream flow. This mixing, of course, degrades the performance of film cooling. This also leads to increases in aerodynamic losses and some augmentation of surface convective heat transfer coefficients. If the injection

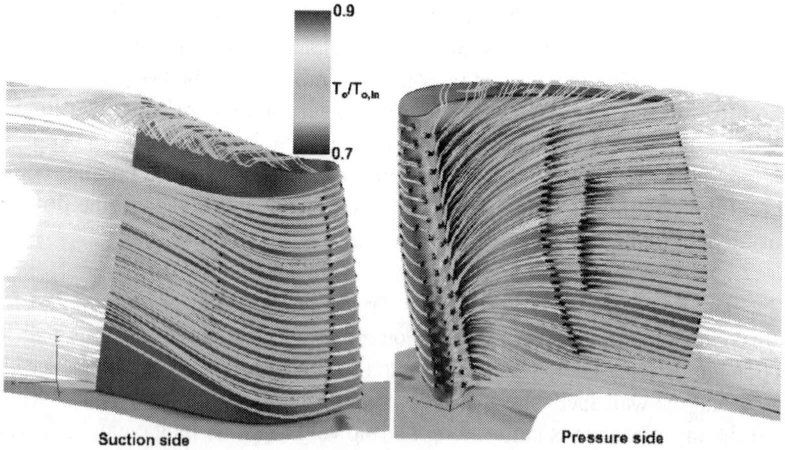


Figure 52: Streamlines emanating from holes over the cooled blade surface (from Garg [88]).

rate becomes high enough, the jets would "tunnel through" the boundary layer flow and the cooling effectiveness becomes quite poor, the aerodynamic losses are high, and the heat transfer coefficients are substantially enhanced.

**3.4.2.4 Describing film cooling in terms of the adiabatic wall temperature** When film cooling is being applied, the heat transfer analysis must be modified. Having three temperatures raises a question as to how the heat transfer coefficient might be described in terms of heat source (passage) and sink (wall) temperatures. In a traditional boundary layer flow, heat transfer rates are given by Newton's law of cooling (simply a definition of the heat transfer coefficient, $h$):

$$\dot{q}'' \equiv h\left(T_{\mathrm{w}} - T_{\infty}\right) \tag{39}$$

In the case of discrete-hole film cooling, h is modified, somewhat, by the injected flow. This is particularly true immediately downstream and in the near vicinity of the injection point. Farther downstream, one can often use, with little error, the h value one would expect without film cooling. In the case of discrete-hole film cooling, the thermal driving potential becomes the difference between the effective temperature within the zone where the coolant and passage flows mix and the wall temperature:

$$\dot{q}'' \equiv h(T_{\mathrm{film,\,effective}} - T_{\mathrm{w}}) \tag{40}$$

The effective temperature, $T_{\mathrm{film,effective}}$, is determined by a study of mixing of external flow and the injected flow. It is dependent upon the flow field. Correlations are formulated to isolate "$h$" and $T_{\mathrm{film,effective}}$ by first determining $T_{\mathrm{film,effective}}$ in the absence of wall heat transfer (set the wall to be adiabatic) but under the same flow conditions. Thus, a reasonable value for $T_{\mathrm{film,effective}}$ could be $T_{\mathrm{aw}}$, the wall temperature established when the wall is adiabatic. To continue the separation, one then determines "$h$" with a heated wall but with the "film cooling" and gas passage flow temperatures being the same. This "$h$" value is altered by injection solely via mixing due to the interaction of the two streams. This value is usually normalized as the Stanton number:

$$St = \frac{h}{\rho u_{\infty} c_{\mathrm{p}}} \tag{41}$$

or Nusselt number

$$Nu_x = \frac{hx}{k} \tag{42}$$

The Nusselt number would require an appropriate choice of length scale, perhaps $x$, the streamwise distance from the hole.

The adiabatic wall temperature is normalized as the film cooling effectiveness:

$$\eta = \frac{T_{\infty} - T_{\mathrm{aw}}}{T_{\infty} - T_{\mathrm{c}}} \tag{43}$$

The effectiveness is 1.0 when the adiabatic wall temperature is the coolant temperature and zero when it is the freestream temperature. We therefore develop expressions for $\eta$ and $St/St_0$ (the subscript "0" indicates the situation in which there is no film cooling present) in terms which capture the effects of:

- Geometric features: slot width, hole size, distance from the injection point, radius of streamwise curvature, hole pattern geometry parameters, hole delivery channel geometry parameters, etc.
- Flow parameters: mass fluxes of injection and gas passage flows, momentum fluxes of injection and gas passage flows, boundary layer thickness at the point of injection, velocity distribution of the emerging film cooling flow, turbulence intensity levels and length scales in the gas passage flow and the coolant flow, parameters which capture secondary flow features within the gas passage, etc., and
- Fluid properties: density, viscosity, specific heat, etc.

**3.4.2.5 Discrete hole injection parallel to the surface** A simple film cooling realization is through a slot oriented so that injection is tangent to the wall. This is a discrete injection with no wall-normal component of velocity component. An analysis of it from Librizzi and Cresci [89] yields the following expression:

$$\eta = \frac{1}{1+(c_{p,\infty}/c_{p,c})\left[0.329(4.01+\varsigma)^{0.8}-1\right]} \tag{44}$$

where

$$\varsigma = \frac{x}{s}\frac{\rho_\infty u_\infty}{\rho_c u_c}\left(\frac{\mu_c}{\mu_\infty}\left[\frac{\rho_c U_c s}{\mu_c}\right]\right)^{-1/4} = \frac{x}{Ms}\left(\frac{\mu_c}{\mu_\infty}\left[\frac{\rho_c U_c s}{\mu_c}\right]\right)^{-1/4} \tag{45}$$

Once we establish the origin of $x$, we can compute $\eta$ from the correlation. The most important parameters to the correlation are the mass flux ratio, $M = \rho_c u_c / \rho_\infty u_\infty$ and the streamwise distance from the hole in terms of the slot length, $x/s$. If we are sufficiently far from the injection zone, we might assume that h is unaffected by injection.

Tangential slot injection is usually not applied for airfoil surface film cooling because of the difficulty of implementing a slot into the airfoil geometry and because of the reduction of airfoil strength it would create. Some combustor cooling schemes have injection geometries that approach tangential injection, however.

**3.4.2.6 Injection with a wall-normal component of momentum** A more practical injection geometry is discrete injection with single, double, or several rows of holes (see Fig. 3). Though injection may be over a short streamwise distance, protection continues downstream of the injection zone until the coolant is mixed out. Downstream coverage by injection through rows of discrete holes is visible in

Fig. 52. Also visible is how the various coolant paths are influenced by secondary flows in the hub and tip regions.

If the coolant has a component of momentum flux in the wall-normal direction, the wall–normal momentum flux relative to the gas stream momentum flux may be a parameter of interest for it might suggest greater penetration of the coolant flow into the gas passage flow. This would influence the mixing process. One might expect that a value of wall-normal momentum flux approaching the gas stream momentum flux evaluated at the elevation where the interaction of the two streams is most intense would signal a possibility of coolant lift-off (or separation) from the surface, with a corresponding precipitous degradation of cooling performance. So long as the injected jet is not lifting off the wall, variations on the slot injection correlation given above can be applied to hole injection to give an estimate of the laterally averaged effectiveness. One could develop film cooling design correlations of the form of eqn. (44) where $x/s$ is the streamwise distance in terms of the hole diameter, $s = d$, or effective slot length, $S_e = \pi d^2/4p$ for a single row of holes ($p$ is the pitch). Of course, the actual processes are complex, so specific forms of the equation apply to specific geometries and sets of flow conditions. Such correlation forms are for the lateral-averaged effectiveness. Actual distributions are quite two-dimensional, as shown in Fig. 53.

When there are more than one row of film cooling holes, $i > 1$, and the blowing is sufficiently low that the coolant is not separating from the surface, the following superposition expression can be used for an approximation to the net film effectiveness:

$$\eta_f = \sum_{i=1}^{N} \eta_{f_i} \left[ \prod_{j=0}^{i-1} (1 - \eta_{f_j}) \right] \tag{46}$$

**3.4.2.7 Net benefit of film cooling** Film cooling reduces the source temperature for convective heat transfer from the gas to the wall, but it also augments the convective heat transfer coefficient, h. It will result in an overall thermal benefit if the net of the two is a lower heat flux $q''/q'<1.0$.

$$\frac{q''}{q''_o} = \frac{h\,(T_f - T_w)}{h_o(T_g - T_w)} = \frac{h}{h_o}\left(1 - \eta\frac{T_g - T_c}{T_g - T_w}\right) \tag{47}$$

where q" is the heat flux with film cooling $q_o$ is the heat flux without film cooling $h$ is the heat transfer coefficient with film cooling $h_o$ is the heat transfer coefficient without film cooling $T_w$ is the local wall temperature $T_c$ is the coolant temperature, and $T_g$ is the mainstream gas temperature.

The net heat flux ratio therefore depends upon the engine operational parameters, $T_c/T_g$, $T_w/T_g$, and $h_0$, as well as the film cooling parameters, $\eta$ and $h/h_0$.

**3.4.2.8 Hole shape and orientation effects** The film cooling hole shape is important as it determines the film coverage downstream of a row of cooling holes

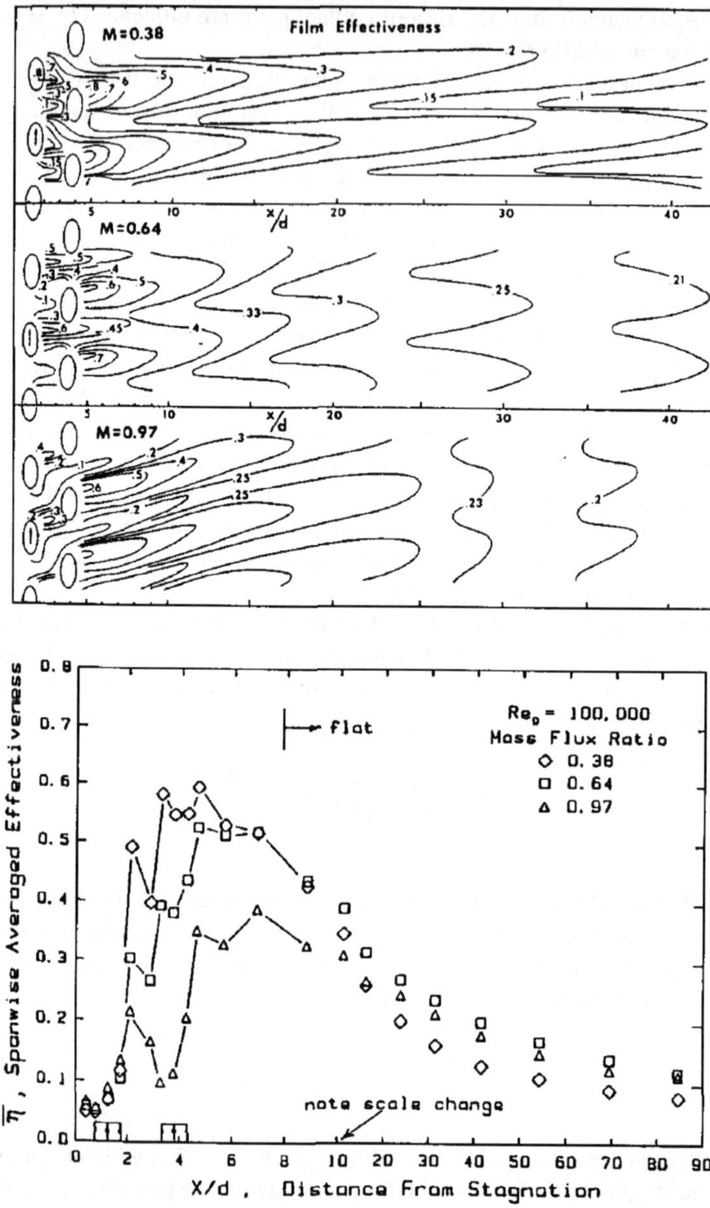


Figure 53: Detailed and spanwise-averaged film effectiveness distributions for three injection rates. The contours in the upper figure are fractions, i.e. the center contour is 0.30. This figure shows effectiveness values for injection through holes of diameter d on a half-cylinder leading edge of diameter *D*. Beginning at 90° of arc from the stagnation line are flat plates representing the suction and pressure sides. The diameter ratio, *D*/*d* is 10. The Reynolds number, $Re_D$, is based on the incidence velocity and the leading edge diameter (graphics from Mick and Mayle [90]).

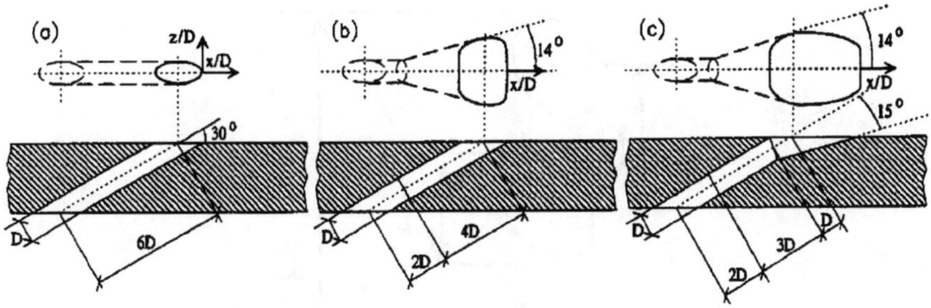


Figure 54: Some shaped hole configurations (from Gritsch *et al.* [92]).

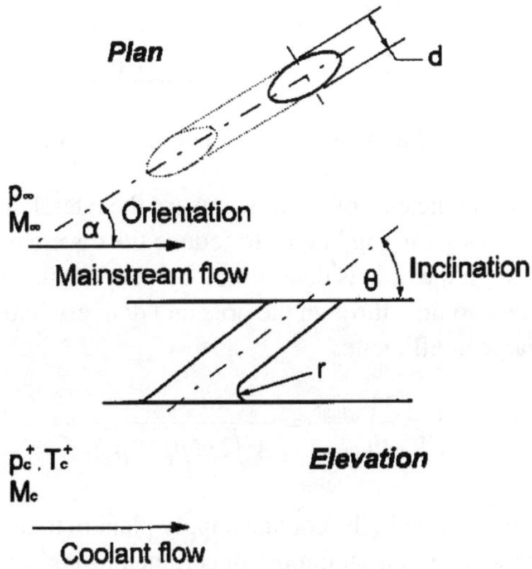


Figure 55: Film cooling geometry and flow parameters (from Hay and Lampard [93]).

and the degree of mixing between coolant and passage flow. Common shapes are round holes or shaped holes, such as expanded (or fan-shaped) holes or laid-back holes (see Fig. 54). Shaped holes are designed to allow the flow to diffuse as it exits the hole. This improves the coverage of the flow. Conical holes have also been applied. Common orientations are in-line, lateral, and compound. With lateral injection, the hole axis is perpendicular to the oncoming flow ($a = 90°$) in Fig. 55. Compound injection is usually with a $a = 45°$, half-way between in-line and lateral. The holes are inclined relative to the surface, generally with an angle of 30–35° ($\theta$ in Fig. 55). This is a compromise between the desire to introduce the flow parallel to the surface and ease of manufacturing. One parameter for discussing the film coverage is the ratio of breakout, $t$, to film hole pitch, $p$, as shown in Fig. 56. Generally, larger ratios have better film protection, though if they are

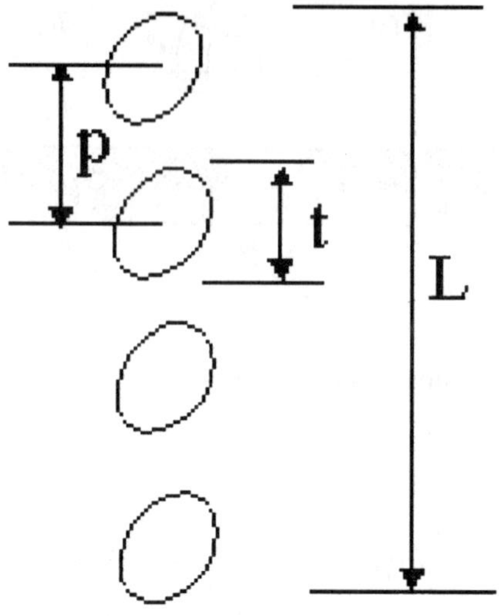


Figure 56: Parameters used to describe coverage.

made larger by swinging the axis of the hole toward the lateral direction, there may be a point at which strong mixing begins to reduce the coverage.

The flow rate through the hole is determined by the pressure gradient across the hole and the resistance to flow through the hole and delivery line. This is described in terms of a discharge coefficient:

$$c_d = \frac{\dot{m}_{\text{actual}}}{\dot{m}_{\text{ideal}}} = \frac{\dot{m}_{\text{actual}}}{A\sqrt{2p(P_{t,c} - P_s)}} \tag{48}$$

where $P_{t,c}$ is the total pressure in the coolant supply plenum to the hole and $P_s$ is the static pressure over the hole. It is an important parameter for setting the cooling flow rate from the hole. The value of the coefficient will depend on the hole geometry and also the conditions, which exist upstream and downstream of the cooling hole. An example comes from Burd and Simon [91] (see Fig. 57), which shows the effects of hole orientation and direction of the supply flow as it approaches the hole. Gritsch *et al.* [92] showed that when cross flow is applied to the channel from which a cooling hole receives its flow, the discharge coefficient of the hole tends to increase (see Fig. 58). At high pressure ratios, shaped cooling holes show less of an increase. This is due to the pressure recovery produced by the diffusing portion of the hole.

**3.4.2.9 Obstructions** During operation, there is a potential for deposits (dirt or condensed material carried in the passage gas) to impinge upon the surface of the airfoil. In time, these deposits may obstruct the film cooling holes. Dirt may accumulate where streamline curvature is strong so that particles separate from the flow and impinge upon the surface. Also, local wall shear stresses are low enough that the particles that impinge are inclined to stay. Build-up by condensation of condensable matter in the passage gas is expected where the passage gas is near a

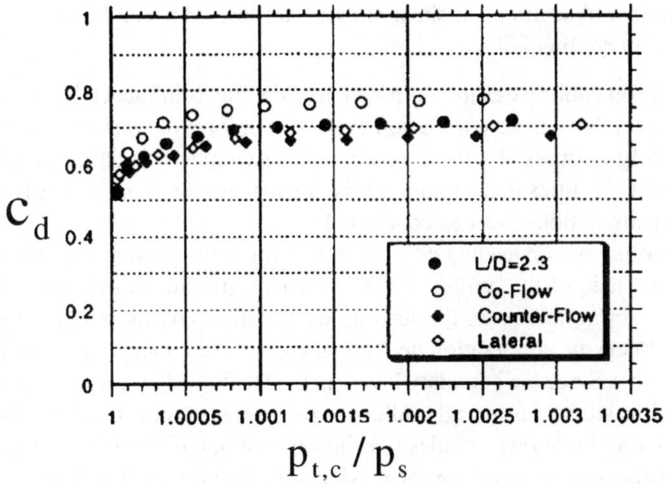


Figure 57: Film cooling discharge coefficients for cases in which there is no momentum of the fluid in the supply plenum (cases 1 and 4), momentum parallel to and in the direction of the passage flow (case 2), and in the opposite direction to the passage flow (case 3). For all, $L/D = 2.3$. Cases 1, 2, and 3 have in-line injection and case 4 has lateral injection (graphics from Burd and Simon [91]).

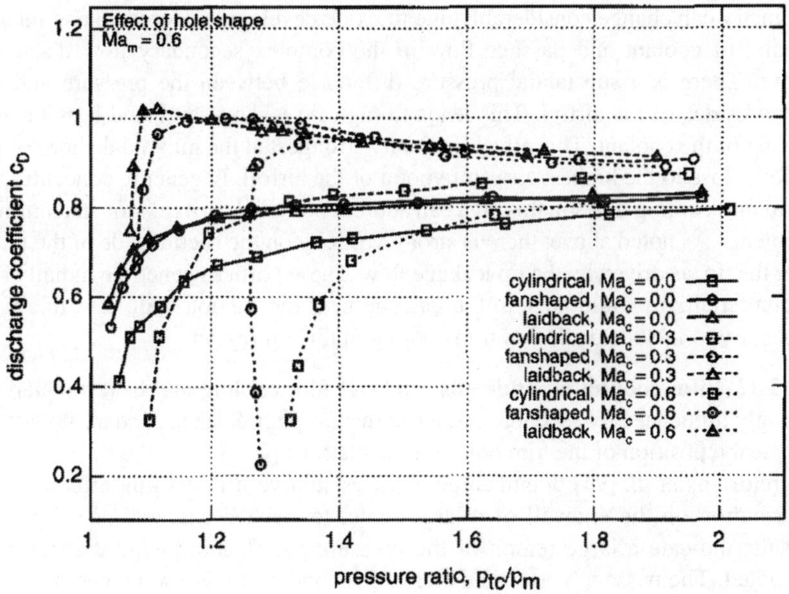


Figure 58: Effect of hole shape on discharge coefficients (graphics from Gritsch et al. [92]). Flow rates for the mainstream and coolant in the supply plenum are given in terms of Mach numbers, $Ma_m$ and $Ma_c$, respectively.

locally cool wall section and shear stresses are low. This can be near the leading edges of film cooling holes.

**3.4.2.10 Film cooling design** As noted, film cooling is influenced by many effects (listed above). No simple correlation exists of the form of eqns. (44) and (45) that appropriately captures all effects in a general. Accurate correlations require measurements in facilities that capture all the important effects, or similarly complete numerical simulations must be conducted.

Film cooling on the leading edge is important, as the leading edge often has the highest heat load of the turbine. As noted above, freestream turbulence level and scale have strong effects in the leading edge region and the radius of curvature strongly affects the acceleration field and therefore the coolant interaction with the approach flow. Since the boundary layers in the leading edge region are thin, roughness and deposition can strongly influence performance. In general, for the leading edge with a multiple row of holes ("a showerhead"), the film superposition model discussed above is not used, for the mixing with the passage flow is too intense.

Correlation of film cooling data on the pressure side must account for the rapid decay in film cooling effectiveness due to concave curvature, free stream turbulence level, and pressure side particular pressure profile.

Correlations on the suction side must account for the slower decay of film cooling performance due to the more stable, accelerated boundary layer, and convex curvature. Again, it should be done for each particular design.

**3.4.2.11 Film cooling on the tip** The tip of the airfoil is difficult to cool for its geometry can change considerably due to oxide development, erosion and intense mixing of coolant and passage flow in the complex secondary flow discussed above. There is a substantial pressure difference between the pressure and the suction sides of the airfoil. This leads to high tip leakage flow and it associated mixing of the coolant. This effect is usually strongest at the mid-axial chord of the airfoil, close to the highest curvature point of the airfoil. In general, concentration of tip film cooling holes in this area can address potential tip oxidation and burning problems. As noted above, there is strong vorticity on the suction side of the blade near the tip, as driven by the tip leakage flow. These vortices penetrate radially and circumferentially quite far into the passage near the suction wall. It is therefore necessary to include dedicated film cooling in this tip region.

**3.4.2.12 Film cooling on endwalls** Endwall film cooling and heat transfer are strongly influenced by secondary flows in the gas path. Such secondary flows will dictate the position of the film holes on the platform.

Freidrichs *et al.* [94] documented a representative film cooling effectiveness distribution on the endwall of a large-scale, low-speed cascade (see Fig. 59). Results indicate a large region of the pressure side leading edge that remains uncooled. The passage vortex and endwall secondary flows sweep coolant from the pressure side to the suction side and can disturb the mid passage coolant. Although specific results will vary significantly from one configuration to the next, the study indicates that cooling holes placed near the passage pressure side may be expected to produce little benefit. Figure 52 shows the effects of tip and

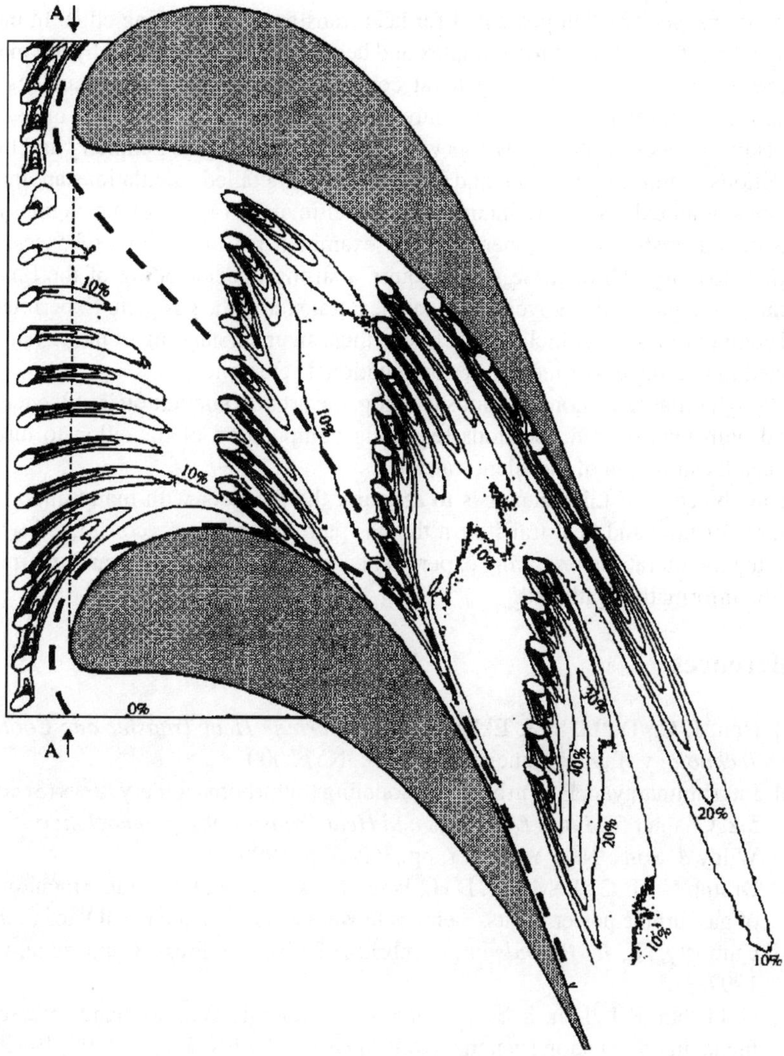


Figure 59: Endwall film cooling effectiveness distributions (graphics from Friedrichs *et al.* [94]).

endwall secondary flows on film cooling. Much of the discussion of endwall secondary flows in Section 3.4.1.5 can be applied toward making choices regarding film cooling injection hole locations on the endwall.

# 4 Concluding remarks

Important aspects of gas turbine hot gas path heat transfer technology have been reviewed. In some aspects, preliminary designs can proceed with rather simple correlations, such as that presented for heat transfer on the leading edge. In most design aspects, the flow is too complex and heat transfer is influenced by too many factors for application of any general correlations. Thus, in support of engine design, (1) information in the literature is studied and interpreted; (2) measurements in cascades or other flow rigs which simulate the engine geometry and flow conditions of interest are taken and interpreted; (3) detailed calculations are made and the computed results are interpreted and; ultimately, (4) prototype engines are constructed, instrumented, operated, and examined for indications of excessive thermal loading. All of these steps require a strong understanding of the fundamental principles which govern flow and heat transfer in the gas path. Experiments and computations never include all effects; thus, an understanding of the physics is needed in making experimental or computational setup choices.

Though reliance on computations is rising, the art of computation is still encumbered with unanswered questions regarding computation of transition to turbulence and estimation of turbulent transport.

The objective of this chapter is to acquaint the designer with many important features of flow and heat transfer in the gas path that are to be considered when entering the literature, preparing experiments and computer models, and interpreting the information gathered.

## References


[1] Han, J.-C., Dutta, S. & Ekkad, S., *Gas Turbine Heat Transfer and Cooling Technology*, Taylor Francis: New York, NY, 2000.

[2] Lakshminarayan, B., Turbulence modeling for turbomachinery flows (Section 5.2, Chapter 5). *Fluid Dynamics and Heat Transfer of Turbomachinery*, John Wiley & Sons: New York, NY, pp. 362–375, 1996.

[3] Drummel, H.-G., LeMeiux, D.H., Voigt, M. & Zombo, P.J., Online monitoring of gas turbine power plants, Siemens Power Generation Technical Paper, 2005.

[4] Cumpsty, N., *Jet Propulsion*, Cambridge University Press, Cambridge, UK, 1997.

[5] Goldstein, R.J., Lau, K.Y. & Leung, C.C., Velocity and turbulence measurements in combustion systems. *Experiments in Fluids*, **1**, pp. 93–99, 1983.

[6] Moss, R.W. & Oldfield, M.L.G., Measurement of hot combustor turbulence spectra, ASME paper 91-GT-351, 1991.

[7] Goebel, S.G., Abuaf, N., Lovett, J.A. & Lee, C.P., Measurements of combustor velocity and turbulence profiles, ASME paper 93-GT-228, 1993.

[8] van Fossen, G.J. & Bunker, R.S., Augmentation of stagnation heat transfer due to turbulence from a DLN can combustor, ASME paper GT-2000-215, 2000.

[9] Ames, F.E., Wang, C. & Barbot, P.A., Measurement and prediction of the influence of catalytic and dry low $NO_x$ combustor turbulence on vane surface heat transfer. *Journal of Turbomachinery*, **125**, pp. 210–220, 2003.

[10] Povey, T., Chana, K.S., Jones, T.V. & Hurrion, J., The effect of hot-streaks on HP vane surface and endwall heat transfer: An experimental and numerical study, ASME paper GT2005-69066, 2005.

[11] Piggush, J. & Simon, T.W., Flow measurements in a first stage nozzle cascade having endwall contouring, ASME paper GT2005-68340, 2005.

[12] Hancock, P.E. & Bradshaw, P., The effect of free-stream turbulence on turbulent boundary layer flow and heat transfer. *Journal of Fluids Engineering*, **105**, p. 284, 1983.

[13] Oke, R.A., *Measurements in a Gas Turbine First Stage Nozzle Guide Vane Cascade with Film Cooling and Endwall Contouring*, PhD Thesis, Mechanical Engineering Department, University of Minnesota, 2001.

[14] Ames, F.E., The influence of large-scale high-intensity turbulence on vane heat transfer. *Journal of Turbomachinery*, **119**, pp. 23–30, 1997.

[15] Chana, K.S., Hurrion, J.R. & Jones, T.V., The design, development and testing of a non-uniform inlet temperature generator for the QinetiQ transient turbine research facility, ASME paper GT2003-38469, 2003.

[16] Viskanta, R., Heat transfer to impinging isothermal gas and flame jets. *Experimental Thermal and Fluid Science*, **6**, pp. 111–134, 1993.

[17] Goldstein, R.J., Behbahani, A.I. & Heppelmann, K.K., Streamwise distribution of the recovery factor and the local heat transfer coefficient to an impinging circular air jet· *International Journal of Heat and Mass Transfer*, **29**, pp. 1227–1235, 1986.

[18] Kelso, R.M., Lim, T.T. & Perry, A.E., An experimental study of round jets in crossflow. *Journal of Fluid Mechanics*, **306**, pp. 111–144, 1996.

[19] Lee, S.W., Lee, J.S. & Ro, S.T., Experimental study on the flow characteristics of streamwise inclined jets in crossflow on flat plate. *Journal of Turbomachinery*, **116**, pp. 97–105, 1996.

[20] Adolfson, D.A., Simon, T.W. & Ibrahim, M.B., Measurements, computation and analysis of a semi-constrained axisymmetric impinging jet, ASME paper GT2004-54215, 2004.

[21] Kercher, D.M. & Tabakoff, W., Heat transfer by a square array of round air jets impinging perpendicular to a flat surface including the effect of spent air. *Journal of Engineering for Power*, **92**, pp. 73–82, 1970.

[22] Metzger, D.E., Florschuetz, L.W., Takeuchi, D.I., Behee, R.D. & Berry, R.A., Heat transfer characteristics for inline and staggered arrays of circular jets with cross flow of spent air. *Journal of Heat Transfer*, **101**, pp. 526–531, 1979.

[23] Florschuetz, L.W., Truman, C.R. & Metzger, D.E., Streamwise flow and heat transfer distributions for jet array impingement with cross flow. *Journal of Heat Transfer*, **103**, pp. 337–342, 1981.

[24] Park, J., Goodro, M., Ligrani, P., Fox, M. & Moon, H., Separate effects of Mach number and Reynolds number on jet array impingement heat transfer. *Journal of Turbomachinery*, **129**, pp. 269–280, 2007.

[25] Kanokjaruvijit, K. & Martinez-Botas, R.F., Parametric effects on heat transfer of impingement on dimpled surface. *Journal of Turbomachinery*, **127**, pp. 287–296, 2005.

[26] Chupp, R.E., Helms, H.E., McFadden, P.W. & Brown, T.R., Evaluation of internal heat transfer coefficients for impingement cooled turbine blades. *Journal of Aircraft*, **6(1)**, pp. 203–208, 1969.

[27] Bunker, R.S. & Metzger, D.E., Local heat transfer in internally cooled turbine airfoil leading edge regions: Part I – Impingement cooling without film coolant extraction. *Journal of Turbomachinery*, **112**, pp. 451–458, 1990.

[28] Metzger, D.E. & Bunker, R.S., Local heat transfer in internally cooled turbine airfoil leading edge regions: Part II – Impingement cooling with film coolant extraction. *Journal of Turbomachinery*, **112**, pp. 459–466, 1990.

[29] Chyu, M.K., Hsing, Y.C. & Natarajan, V., Heat transfer on the base surface of three-dimensional protruding elements. *International Journal of Heat and Mass Transfer*, **39(4)**, pp. 2925–2935, 1996.

[30] Uzol, O. & Camci, C., Elliptical pin fins for gas turbine blade cooling applications. Part 2: Wake flow field measurements and visualization using particle image velocimetry, ASME paper 2001-GT-181, 2001.

[31] Mahmood, G.I., Won, S.Y. & Ligrani, P.M., Local Nusselt numbers and flow structure in a rectangular channel with cylindrical pin-fins, Department of Mechanical Engineering, University of Utah, Salt Lake City, UT, 2002.

[32] Ligrani, P.M., Oliveira, M.M. & Blaskovich, T., Comparison of heat transfer augmentation techniques. *AIAA Journal*, **41(3)**, pp. 337–362, 2003.

[33] Chyu, M.K. & Natarajan, V., Heat transfer on the base surface of three-dimensional protruding elements. *International Journal of Heat and Mass Transfer,* **39(14)**, pp. 2925–2935, 1996.

[34] Ligrani, P.M. & Mahmood, G.I., Variable property Nusselt numbers in a channel with pin fins. *Journal of Thermophysics and Heat Transfer,* **17(1)**, pp. 103–111, 2003.

[35] Metzger, D.E., Berry, R.A. & Bronson, J.P., Developing heat transfer in rectangular ducts with staggered arrays of short pin fins. *Journal of Heat Transfer*, **104**, pp. 700–706, 1982.

[36] Kays, W.M., Crawford, M.E. & Weigand, B., *Convective Heat and Mass Transfer*, McGraw-Hill Book Company, Inc.: New York, 2005.

[37] Gebhart, B., *Heat Transfer*, McGraw-Hill Book Company, Inc.: New York, 1961.

[38] Incropera, F.P. & DeWitt, D.P., *Fundamental of Heat and Mass Transfer*, 5th edn, John Wiley and Sons: Hoboken, NJ, 2002.

[39] Lezius, D.K. & Johnston, J.P., Roll-cell instabilities in rotating laminar and turbulent channel flows. *Journal of Fluid Mechanics*, **77**(part 1), pp. 153–175, 1976.

[40] Wagner, J.H., Johnson, B.V. & Kopper, F.C., Heat transfer in rotating serpentine passages with smooth walls. *Journal of Turbomachinery*, **113**, pp. 321–330, 1991.

[41] Casarsa, L. & Arts, T., Experimental investigation of the aerothermal performance of a high blockage rib-roughened cooling channel. *Journal of Turbomachinery*, **127**, pp. 580–588, 2005.

[42] Kiml, R., Mochizuki, S. & Murata, A., Effects of rib arrangements on heat transfer and flow behavior in a rectangular rib-roughened passage: application to cooling of gas turbine blade trailing edge. *Journal of Heat Transfer*, **123**, pp. 675–681, 2001.

[43] Fu, W.L., Wright, L.M. & Han, J.C., Heat transfer in two-pass rotating rectangular channels (AR = 1:2 and AR = 1:4) with 45 degree angled rib turbulators. *Journal of Heat Transfer*, **127**, pp. 164–174, 2005.

[44] Wright, L.M., Fu, W.-L. & Han, J.-C., Thermal performance of angled, V-shaped, and W-shaped rib turbulators in rotating rectangular cooling channels (AR = 4:1). *Journal of Turbomachinery*, **126**, pp. 604–614, 2004.

[45] Griffith, T.S., Al-Hadhrami, L. & Han, J.C., Heat transfer in rotating rectangular cooling channels (AR = 4) with dimples. *Journal of Turbomachinery*, **125**, pp. 555–564, 2003.

[46] Burgess, N.K., Oliveira, M.M. & Ligrani, P.M., Nusselt number behavior on deep dimples surfaces within a channel. *Journal of Heat Transfer*, **125**, pp. 11–18, 2003.

[47] Mahmood, G.I., Hill, M.L., Nelson, D.L., Ligrani, P.M., Moon, H.-K. & Glezer, B., Local heat transfer and flow structure on and above a dimpled surface in a channel. *Journal of Turbomachinery*, **123**, pp. 115–123, 2001.

[48] Kreith, F. & Bohn, M.S., *Principles of Heat Transfer,* Brooks/Cole: Pacific Grove, CA, 2000.

[49] Incropera, F.P. & DeWitt, D.P., *Fundamentals of Heat and Mass Transfer,* John Wiley & Sons: New York, NY, 2001.

[50] Jiang, N. & Simon, T.W., Modeling laminar-to-turbulent transition in a low-pressure turbine flow which is unsteady due to passing wakes: Part I – Transition onset, ASME paper GT2003-38787, 2003.

[51] Mayle, R.E., The role of laminar-turbulent transition in gas turbine engines. *Journal of Turbomachinery,* **113**, pp. 509–537, 1991.

[52] Van Fossen, G.J., Simoneau, R.J. & Ching, C.Y., Influence of turbulence parameters, Reynolds number, and body shape on stagnation-region heat transfer. *Journal of Heat Transfer,* **117**, pp. 597–603, 1995.

[53] Tennekes, H. & Lumley, J.L., *A First Course in Turbulence,* The MIT Press: Cambridge, MA, 1972.

[54] Giel, P.W., Boyle, R.J. & Bunker, R.S., Measurements and predictions of heat transfer on rotor blades in a transonic turbine cascade. *Journal of Turbomachinery*, **126(1)**, pp. 110–121, 2004.

[55] Van Fossen, G.J. & Bunker, R.S., Augmentation of stagnation region heat transfer due to turbulence from a DLN (dry low-$NO_x$) can combustor, ASME paper 2000-GT-215, 2000.

[56] Bunker, R.S., Personal communication, 2007.

[57] Dullenkopf, K. & Mayle, R.E., An account of free-stream-turbulence length scale on laminar heat transfer. *Journal of Turbomachinery*, **117**, pp. 401–406, 1995.

[58] Rued, K. & Wittig, S., Laminar and transitional boundary layer structures in accelerating flow with heat transfer, ASME paper 86-GT-97, Dusseldorf, Germany, 1986.

[59] Patankar, S., *Numerical Heat Transfer and Fluid Flow*, Hemisphere Publishing Corp.: New York, NY, 1980.

[60] Fluent Users' Manual: http://jullio.pe.kr/fluent6.1/help/html/ug/main_pre.htm

[61] Star-CD web site: http://www.cd-adapco.com/products/STAR-CD/index.html

[62] CFX web site: http://www.ansys.com/products/cfx.asp

[63] Durbin, P., Near wall turbulence closure modeling without 'damping functions,' *Theoretical Computational Fluid Dynamics*, **3(1)**, pp. 1–13, 1991.

[64] Dunn, M.G., Convective heat transfer and aerodynamics in axial flow turbines. *Journal of Turbomachinery*, **123**, pp. 637–686, 2001.

[65] Abhari, R.S. & Epstein, A.H., An experimental study of film cooling in a rotating transonic turbine, ASME paper 92-GT-201, 1992.

[66] Giel, P.W., Van Fossen, G.J., Boyle, R.J., Thurman, D.R. & Civinskas, K.C., Blade heat transfer measurements and predictions in a transonic turbine cascade, ASME paper 99-GT-125, 1999.

[67] Langston, L.S., Crossflow in a turbine cascade passage. *Journal of Engineering for Power*, **10**, pp. 866–874, 1980.

[68] Goldstein, R.J., Wang, P.H. & Jabbari, M.Y., The influence of secondary flows near the endwall and boundary layer disturbance on convective transport from a turbine blade. *Journal of Turbomachinery*, **117**, pp. 657–663, 1995.

[69] Graziani, R.A., Blair, M.F., Taylor, J.R. & Mayle, R.E., An experimental study of endwall and airfoil surface heat transfer in a large scale turbine blade cascade. *Journal of Engineering for Power*, **102**, pp. 257–267, 1980.

[70] Key, N. & Arts, T., Comparison of turbine tip leakage flow for flat tip and squealer tip geometries at high-speed conditions. *Journal of Turbomachinery*, **128**, pp. 213–220, 2006.

[71] Metzger, D.E. & Rued, K., The influence of turbine clearance gap leakage on passage velocity and heat transfer near blade tips: Part I – Sink flow effects on blade pressure side. *Journal of Turbomachinery*, **111**, pp. 284–292, 1989.

[72] Rued, K. & Metzger, D.E., The influence of turbine clearance gap leakage on passage velocity and heat transfer near blade tips: Part II – Source flow effects on blade suction sides. *Journal of Turbomachinery*, **111**, pp. 293–300, 1989.

[73] Van Fossen, G.J. & Simoneau, R.J., Stagnation region heat transfer: the influence of turbulence parameters, Reynolds number and body shape. *General Papers in Heat and Mass Transfer, Insulation, and Turbomachinery*, eds S.H. Chan, I.S. Habib, L.E. Hochreiter, R.L. Mahajan, E.V. McAssey, J.Y. Murthy, M.S. Sohal, R.J. Couvillion, W.B. Krause, R.S. Amano & R.H. Pletcher, ASME HTD-Vol. 271, 1994.

[74] Saha, A.K., Mahmood, G.I. & Acharya, S., The role of leading-edge contouring on end-wall flow and heat transfer: computations and experiments, ASME paper GT2006-91318, 2006.

[75] Bons, J., Taylor, R., McClain, S. & Rivir, R., The many faces of turbine surface roughness. *Journal of Turbomachinery*, **123**, pp. 739–748, 2001.

[76] Wammack, J.E., Crosby, J., Fletcher, D., Bons, J.P. & Fletcher, T.H., Evolution of surface deposits on a high pressure turbine blade, Part I: Physical characteristics, ASME paper GT2006-91246, 2006.

[77] Bons, J.P., Wammack, J.E., Crosby, J., Fletcher, D. & Fletcher, T.H., Evolution of surface deposits on a high pressure turbine blade, Part II: Convective heat transfer, ASME paper GT2006-91257, 2006.

[78] Bogard, D., Schmidt, D. & Tabbita, M., Characterization and laboratory simulation of turbine airfoil surface roughness and associated heat transfer. *Journal of Turbomachinery*, **120**, pp. 337–342, 1998.

[79] Barlow, D., Kim, Y. & Florschuetz, L., Transient liquid crystal technique for convective heat transfer on rough surfaces. *Journal of Turbomachinery,* **119**, pp. 14–22, 1997.

[80] Goldstein, R.J., Eckert, E.R.G., Chiang, H.D. & Elovic, E., Effect of surface roughness on film cooling performance. *Journal of Engineering for Gas Turbines and Power*, **107(1)**, pp. 111–116, 1985.

[81] Taylor, R., Scaggs, W. & Coleman, H., Measurement and prediction of the effects of nonuniform surface roughness on turbulent flow friction coefficients. *Journal of Fluids Engineering*, **110**, pp. 380–384, 1998.

[82] Turner, A., Hubbe-Walker, S. & Bayley, F., Fluid flow and heat transfer over straight and curved rough surfaces. *International Journal of Heat and Mass Transfer*, **43**, pp. 251–262, 2000.

[83] Hoffs, A., Drost, U. & Bolcs, A., Heat transfer measurements on a turbine airfoil at various Reynolds numbers and turbulence intensities including effects of surface roughness, ASME paper 96-GT-169, 1996.

[84] Bunker, R.S., Separate and combined effects of surface roughness and turbulence intensity on vane heat transfer, ASME paper 07-GT-135, 1997.

[85] Maciejewski, P.K. & Moffat, R.J., Heat transfer with very high free-stream turbulence Part I: Experimental data. *Journal of Heat Transfer*, **114**, pp. 827–833, 1992.

[86] Maciejewski, P.K. & Moffat, R.J., Heat transfer with very high free-stream turbulence Part II: Analysis of results. *Journal of Heat Transfer*, **114**, pp. 834–839, 1992.

[87] Ames, F.E., The influence of large-scale high-intensity turbulence on vane heat transfer. *Journal of Turbomachinery*, **119**, pp. 23–30, 1997.

[88] Garg, V.K., Heat transfer on a film-cooled rotating blade. *International Journal of Heat and Fluid Flow*, **21**, pp. 134–145, 2000.

[89] Librizzi, J. & Cresci, R.J., Transpiration cooling of a turbulent boundary layer in an axisymmetric nozzle. *AIAA Journal*, **2**, p. 617, 1964.

[90] Mick, W.J. & Mayle, R.E., Stagnation film cooling and heat transfer including its effect within the hole pattern. *Journal of Turbomachinery*, **110**, pp. 66–72, 1998.

[91] Burd, S.W. & Simon, T.W., Measurements of discharge coefficients in film cooling. *Journal of Turbomachinery*, **121(2)**, pp. 243–248, 1999.

[92] Gritsch, M., Schulz, A. & Wittig, S., Adiabatic wall effectiveness measurements of film cooling holes with expanded exits. *Journal of Turbomachinery*, **120**, pp. 549–556, 1998a.

[93] Hay, N. & Lampard, D., Discharge coefficient of turbine cooling holes: a review, ASME paper 96-GT-492, 1996.

[94] Friedrichs, S., Hodson, H.P. & Dawes, W.N., Distribution of film cooling effectiveness on a turbine endwall measured using the ammonia and diazo technique. *Journal of Turbomachinery*, **118**, pp. 613–621, 1996.

# CHAPTER 9

# Design and optimization of Turbo compressors

C. Xu & R.S. Amano
*Department of Mechanical Engineering,*
*University of Wisconsin-Milwaukee, USA.*

## Abstract

A compressor has been refereed to raise static enthalpy and pressure. A successful compressor design greatly benefits the performance of the whole power system. Lean design methodologies have been used for industrial power system design. The compressor designs require benefit to both OEM and customers, i.e. lowest cost for both OEM and end users and high efficiency in all operating range of the compressor. The compressor design and optimization are critical for the new compressor development and compressor upgrade. The design experience and design considerations are also critical for a successful compressor design. The design experience can accelerate compressor lean design process. An optimization process is discussed to design compressor blades in turbo machinery. The compressor design process is not only an aerodynamic optimization, but structure analyses also need to be combined in the optimization. This chapter discusses an aerodynamic and structure integration optimization process. The design method consists of an airfoil shape optimization and a three-dimensional gradient-based optimization coupled with Navier–Stokes solvers. A model airfoil of a transonic compressor is designed by using this approach, with an efficiency improvement. Airfoil sections were stacked up to a three-dimensional rotor blade of a compressor. The efficiency is improved over a wide range of mass flow. The results indicate that the optimization process can provide improved design and can be integrated into a compressor design procedure. Initial design considerations of turbo compressor are commonly performed with experience base, although computer technology and numerical methods have made a significant progress. Three-dimensional computational fluid dynamics codes are still not major design tools for compressors. Major design systems for industrial compressor design are largely based on one-dimensional mean line and two-dimensional through flow tools. The experience of designers is one of the key factors to drive design success. The discussions and information on compressor design considerations in open literature are also reviewed. This chapter consists of three parts: Part I provides an overview of the turbo compressors and basic design procedures. Part II discusses a blade optimization process. An axial compressor blade design was used as an example to illustrate the optimal design. Part III provides some design experience of centrifugal compressors.

# 1 Part I: Compressors and their design

## 1.1 Introduction

Compressors are devices to pressurize working gases. One of the basic applications for the compressors is to increase the gas pressure and deliver gas to the downstream of the compressor. For example, compressors can provide higher-pressure gas for combustion, high-pressure gas to drive gas tools, transport process gas through pipelines, or compressor gas for heat recovery or gas conditioning circles.

Compressors can be separated into two basic types: The positive displacement and continuous flow compressors. The reciprocating and rotary compressors are two major positive displacement compressors. Ejectors and turbo compressors are the two main continuous compressors. Axial and centrifugal compressors are the two basic types of turbo compressors. The main purpose of this part is to describe an industrial turbo compressor design process.

Gas turbine and compressor manufactures are competing to produce more efficient and wide operating range machines. Correspondingly, many research works and developments were carried out to meet the manufacturers' needs [1–8]. Compressor blade design is one of the critical processes to ensure the overall compressor have good performance. For all the turbo compressors, there are two major parts: rotating blade and stationary blade. The basic functions of compressor moving blades are added the work to gas and turn the gas to the required angles to have diffusion in the flow path to increase the static pressure and discharge to next component. For the stationary blades, the basic function is to turn the gas to the required angles to have a right diffusion level for the gas. The design goal for modern compressor blades are to achieve the desired flow turning with minimum losses and can tolerance high incident within the constraint of geometric orientation of the blade row requirements defined by overall compressor design.

Both rotating and stationary blades involve the airfoil and blade designs. The basic compressor blade performance was mainly determined by five key factors: shape of the blades, airfoil or blade section stagger angle, solidity, inlet and discharge flow angle, airfoil or blade section loading. Unlike turbine blades through which flows are accelerated, the adverse pressure gradient due to the flow diffusion in compressor blades. The adverse pressure is an unfavorable force on the boundary layer development. It is very difficult to achieve a thin boundary layer. The boundary layer separation often happens inside the compressors. Therefore, the main challenge for compressor blade design is to generate a blade shape to meet the flow angle requirements and reasonable loading distributions.

Efficiency had been a major criterion in the design of compressor and turbine components for many years. The search for maximum efficiency is traditionally made by successive modifications of the geometries and verification by wind tunnel tests and flow calculations. Such a process can be time consuming and outcome strongly depended on the designs' experiences. This approach is difficult to have advanced design such as shock-free transonic blades and blades with optimized dihedral.

Many design methods were proposed in both research stage and industrial applications [1–9]. The blade design methods can be categorized into two basic methods. One is direct design and another is inverse design method. For direct design method, blade was design first based on the designer experience and then a Navier–Stokes equation code or a fast inviscid code was used to evaluate the airfoil performance. The process of design and performance calculations are completed until a satisfy airfoil was built. The designed airfoils are then stack-upped to produce a three-dimensional blade. Normally, some three dimensional optimization processes are used to continue optimizing the design [1–9]. In inverse design method [5], normally Euler equations with modified boundary conditions were used for an iterative procedure to search the desired design.

With the rapid progress in computational fluid dynamics (CFD) and computer technologies, compressor design has been developed from purely empirical methods to apply more CFD simulation into design process. The CFD method allows the aerodynamic designer to have detailed flow information and to optimize the blades. However, at the beginning stage of the design, the experience data are very helpful to accelerate the design and provide the basic information, to engineers in other discipline. In this chapter, we will start with basic introduction of the compressor, and then talk about design process and optimization, and finally the design experience the author will be discussed.

## 1.2 Types of turbo compressors

Turbo compressors can be categorized into two types: compressors and turbines. Compressors absorb power to increase the fluid pressure or head, and turbines produce power by expanding fluid to a lower pressure or head. Compressors have a wide application in order to provide the high-pressure gas for combustion in jet engines, to transfer process fluid in pipelines, to provide high-pressure gas to driving tools, etc. The compressor system is one of the important parts of a power system. Both centrifugal compressors and axial compressors are continuous flow compressors.

Compressors are categorized into three types according to the nature of the flow path through the passages of the rotors. When the fluid through flow path is wholly or mainly parallel to the axis of rotation, these types of compressors called axial compressors as shown in Fig. 1a. Centrifugal compressors having through flow paths are wholly or mainly in a plane perpendicular to the rotational axial as shown in Fig. 1b. When the through flow in the compressors has a significant amount of velocities in both radial and axial directions, we call this type of compressors as mixed flow compressors as shown in Fig. 1.c. Sometimes, mixed flow compressors are also reference as centrifugal compressors.

The axial compressor was first patented by Sir Charles Parsons in 1884 [10]. His compressor concept just simply reversing an axial turbine for use as a compressor; however, the efficiency of compressors from reversing turbines was less than 40%. The difficulties associated with the development of axial compressors stemmed from the fundamentally different nature of the flow process compared with that in axial-flow turbine. It was not until 1926 that any further improvement on axial compressor efficiency was undertake when A.A. Griffith [11] outlined the basic principles of airfoil theory of compressor and turbine design.

Axial flow compressors are the most common types of compressor for aircraft engines and big industrial gas turbine. The gas in an axial compressor flows in an axial direction through a series of rotating rotor blades and stationary stator vanes. The flow path of an axial compressor decreases in cross-section area in the direction of flow, reducing the volume of the gas as compression progresses from stage to stage of compressor blades.

In axial compressors as shown in Fig. 2, the gas being delivered to the face of compressor by the gas inlet duct, the incoming gas passes through the inlet guide vanes (IGVs). Gas entering the first set of rotating blades and flowing in axial direction, is deflected in the direction of rotation. The gas speed is slowed down and turns as it is passed onto a set of stator vanes, following which it is again picked up by another set of rotating blades, and so on, through the compressor. The pressure of the gas increases each time it passes through a set of rotors and stators. The increase in the stage pressure almost inevitably leads to some aerodynamic constraints. The main constraint is increased Mach number, possibly giving rise to shock induced boundary layer separation or increased losses due to poor diffusion of the flow. The maximum stage pressure achieved was about 2.1 based on the recent development.

The aerodynamic principles are applied to the compressor design to increase axial compressor efficiency. The axial compressor blades are treated as lifting surfaces like aircraft wings or propeller blades. The cascade effect is a primary consideration

(a)

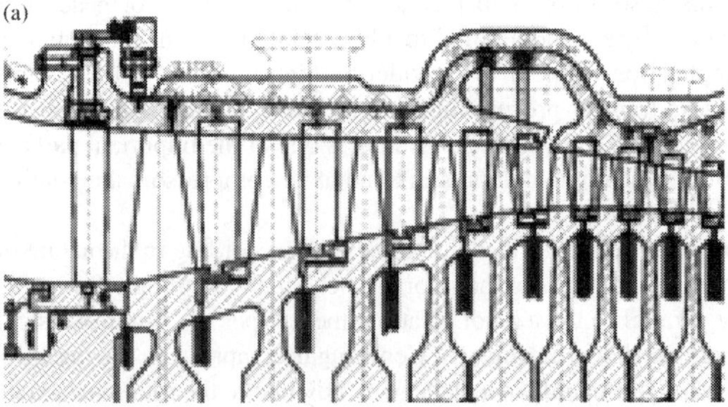

Figure 1: Types of compressors: (a) axial, (b) centrifugal, and (c) mixed flow compressors.

(b)

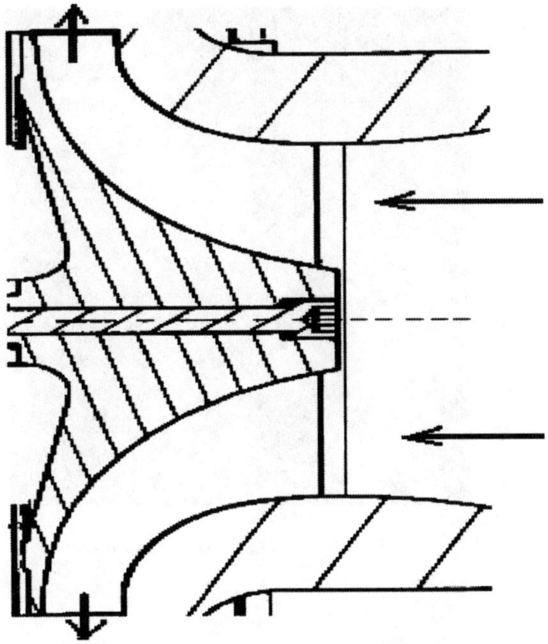

(c)

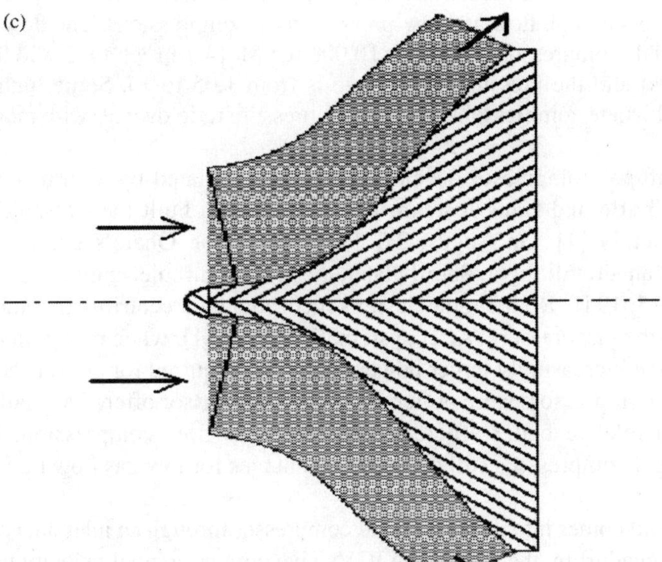

Figure 1: Continued

in determining the airfoil section, angle of attack, and the spacing between blades to be used for compressor blade design. The blade must be designed to withstand the high centrifugal forces as well as the aerodynamic loads to which they are subjected. The clearance between the rotating blades and their outer case is also very important. The rotor assembly turns at a high speed and must be rigid, well

Figure 2: Axial compressors.

aligned and well balanced. Axial compressors have more benefits than centrifugal compressors at high flow and low-pressure ratio compressions. The flow range for most axial compressors is from 10,000 CFM (4.7 $m^3$/s) to 2,000,000 CFM (943 $m^3$/s) and the pressure ratio range is from 1.05 to 15. Some high-pressure ratio multistage compressors can produce pressure ratio over 30 with more than 20 stages.

Centrifugal compressor development was stimulated by aircraft propulsion. Frank Whittle and Hans Joachim Pabst von Ohain built their turbojet engines independently [1]. On August 27, 1939, Hans von Ohain's engine first time powered an aircraft flight. Shortly later, the Whittle turbojet engine was first flown on May 15, 1941. Both engine designs incorporated a centrifugal compressor to increase the gas pressure. After systematic studies [1], when researchers noticed that for the increasingly larger powered engines required for aircraft propulsion, the axial compressors were preferred. Axial compressor offered a smaller frontal area and relative better efficiency for large gas flow compression. However, centrifugal compressors still offer the advantages for low gas flow rate compressors.

The fluid comes into the centrifugal compressor through an inlet duct and can be given a prewhirl by the IGVs. The IGVs give circumferential velocity to the fluid before it flows inside of the compressor inlet. A positive vane angle produces prewhirl in the direction of the impeller rotation, and a negative vane angle produces prewhirl in the opposite direction. The positive prewhirl decreases the relative inlet Mach number. One purpose of installing the IGVs is to decrease the relative Mach number at the induce tip inlet because the highest relative velocity at induce inlet is at shroud section. The disadvantage of this application is that a decrease in the relative Mach number also reduces the energy transfer. Another purpose of installing IGV is to increase the compressor operating range. When the compressor flow reduces, the positive prewhirl can reduce the inlet incidence to delay the flow separation to enlarge the operating range.

Centrifugal compressors operate by taking in outside gas and rotating it by means of an impeller as shown in Fig. 1. The impeller, which is usually an aluminum alloy, or steel or even titanium, guides the gas toward the outer circumference of the compressor, building up the velocity of the gas by means of high rotational speed of the impeller. The impeller consists of two basic parts: an inducer such as an axial-flow rotor part, and the blades in the radial direction where energy is imparted by centrifugal force. Impellers have two types: shrouded and unshrouded. The shrouded impeller normally used for a multistage compressor. The unshrouded impeller allows a much higher impeller rotational speed application: for a steel impeller, allowable tip speed would be as high as 450 m/s for an unshrouded impeller but only below 350 m/s for a shrouded impeller. The shrouded impellers are mainly used in multistage compressor for which it would be difficult to maintain an acceptable small clearance between the impellers and stationary shroud.

A significant impact to turbo machinery design and analysis is the introduction of the quasi-three-dimensional calculation methods by Hamrick *et al.* [12] and Wu [13]. Hamrick *et al.* proposed a streamline curvature method by handle momentum and continuity equations separately in a stream function; however, Wu developed his matrix method by coupled both equations via a stream function. Both methods introduced stream surfaces of first (S1) and second (S2). Three-dimensional flows were solved two-dimensional way in mathematics. The methods were called Quasi-Three-Dimension method. Wu [13] defined that S1 surface was in the surface from pressure side to suction side and S2 surface is from hub to shroud in impeller. It was assumed that flow only follows the stream surface and there was no flow crossing perpendicular from one surface to another. The losses were taken into account by making use of a prescribed entropy increase, which was obtained from empirical equations.

With the development of the computer technology in mid-1970s, Quasi-three-dimensional calculations and three-dimension viscous solution were developed. Bosman and El-Shaarawi [14] used a computer program to calculate the flow inside of centrifugal compressor rotors. The calculations started with a middle S2 surface at the mid pitch position. The flow quantities were calculated in this stream surface. After calculating the flow characteristics in S1 surface, the streamlines obtained from S2 were rotated until intersecting with the pressure and suction surfaces. Losses inside centrifugal compressors were considered by using a prescribed entropy increase taken from empirical equations. The three-dimensional viscous code was first developed for subsonic flows in a simple geometry [15]. The three-dimensional codes developed at this time were difficult to quantitatively predict the flow inside centrifugal compressor blade.

In 1980s, with the introduction of three-dimensional numerical method to solve the complex geometries with viscous solvers [16–18], the flows inside of the centrifugal compressors were better understood. The turbulence models used very often for centrifugal flow simulations were one-equation or two-equation turbulence models. The secondary flow patterns were simulated and reported. The flow simulation results were agreed with experimental studies in overall flow patterns.

After 1990, considerable progress has been made in understanding the complicated flows inside centrifugal compressors. The detailed flow measurements were available by using Hot-wire, [19, 20] fast response high-resolution pressure transducers [21–25], and LDV and PIV [26]. Strong passage vortices due to secondary flows were found. The impeller and diffuser flow pattern were reported. The detail measurements of the flow inside of the centrifugal compressors were used to guide compressor design.

Recently, with the developments of new method for CFD solution [27–29], many optimization methods started to combine into compressor impeller blade and diffuser airfoil designs. The advance of the modern design will continue to push compressor design toward the wide operating range high stage pressure ratio centrifugal compressor. The CFD simulation had pushed the compressor stage efficiency to a very high level. The future potential of the efficiency improvement is about 2% for state-of-art designs with current manufacturing capabilities. The potential is about 10% if we can have a perfect design. The better three-dimensional design may improve the operating range. However, the compressor efficiency may drop when we improve the operating range.

### 1.3 Aerodynamic design

The design of the turbo compressor begins with a specification and proceeds through preliminary design to blade design and analysis using design tools. The design process begins with mainline calculation, which is based on the required design flow, and stage pressure ratio or head rise to select the compressor type, number of stages and find the velocity triangles at inlet and outlet of each components of the compressor. Then the performance and basic flow parameters of each component of the compressor are analyzed based on mean line information, then the two-dimensional airfoil sections are designed and stacked up, and finally three-dimensional flow analyses are performed to do further optimization. The objective of aerodynamic design of the turbo machines is to obtain both maximum overall machine efficiency and wide operating range within the limitations imposed by stress, cost, manufacturing and other considerations. Before the 1970s, compressors were mainly designed by empirical methods, based on the accumulated experience of designers and their company. The compressors some were successful design and some time, the designed compressor could not meet efficiency and operating range target. After 1970s, CFD had played an important rule in designing the turbo machinery. Significant improvements in efficiency and operating range have been achieved. Improved experimental measurements have become available over the last three decades [30]. The combinations of the numerical predications and detail flow measurements had led to a great improvement in flow physics in the centrifugal compressors.

Recently, a competitive product demands a high level of design optimization of compressor and involves several different disciplines considerable historical design experience and a variety of design tools. Although most of the compressor design systems allow the efficient transfer of design information, how to obtain the design information especially three-dimensional flow field information is still a big challenge. Most of the compressor design involves considerable iterations between aerodynamics, structure and manufacturing cost. During the compressor design, the knowledge of the designers and experience still play an important role. Although many integrated design system and optimization codes permit some design optimizations, many design correlations still need designers' input based on designers' experience. Part II introduces an aerodynamic design and optimization procedure. Part III summarizes the authors' centrifugal compressor design experience to help compressor designers who are in the initial stages of their careers to perform aerodynamic design.

## 2 Part II: Blade design and optimization

### 2.1 Introduction

With the development of the CFD, it has been implemented in the turbo machine design process. Great efforts are devoted to improving the efficiency of the gas turbine components. The airfoil designs for turbine and compressor airfoils plays an important role in increasing the turbine efficiency. In the airfoil design, there are two types of implements the aerodynamic design engineer often considers. One is to design and employ custom-design blade profiles with minimum losses and controlled blade boundary layers. The second and even more complex part is to minimize losses resulting from secondary flows near hub and casing. Recently, three-dimensional blade design concepts were proposed to help control secondary flows [31]; however, the complex flow is difficult; to simulate even by using fully three-dimensional Navier–Stokes flow solvers. And the validation of the Navier–Stokes solvers needs lot of experimental data and it is time consuming and expensive. Therefore, almost all the aerodynamic designs are based on the two-dimensional design. The inviscid analyses of the two dimensional airfoil sections still play an important role in the design process. Authors recently developed an airfoil design process based on the Bezer curves to produce the custom airfoil sections based on the flow field requirements of the airfoil [2]. In this study, the airfoil design process was implemented with optimization based on the two-dimension viscous turbulent code.

It was known that for a blade row in an annulus, the stream surface between two annular walls is twisted for most cases. These twists are induced by either shed vorticity or by secondary from arising from inlet vorticity. Stream surface twist can arise in an irrotational flow owing to either span wise components of velocity or span wise blade forces. Many efforts had been adopted to reduce the stream surface twist and reduce the secondary flow losses, such as sweep [32], lean [33], bow [34] and twist [35, 36] the blades or design a non-axis symmetric end wall. However, there is little information in the available literature for using three-dimensional design and almost no information is available to show how to integrate the three-dimensional features into design process. Most of the study was still on the academic research and was based on the particular machines or blades. Moreover, most of the studies were based on the particular blade and flow situations. For example, Singh *et al.* [35] argued that closing the blade throat near the end walls could obtain significant efficiency improvements and Wallis and Denton [36] also obtained an efficiency increase from almost the opposite type of blade twist near the end wall. For different machines and different designs, may different techniques should be used according to the flow field nature of the designs. It is very important to design a blade design procedure and optimize the design.

The increased use of CFD tools has been driven mainly by two factors. First, from performance standard point of view, efficiency has steady increased. Second, the turbo machinery industry as a whole has been pushed toward reduced cost designs. The cost reduction is in terms of development, modification, production and operating costs. The cost reduction drives a turbo machine toward high loading in order to reduce stage count, while maintaining or exceeding past performance goals. The current design of the new stage is already outside of the standard airfoil database. Most of the airfoil needs to be designed and the development of the design tools to meet this requirement becomes critical. The application of CFD methodology to improve the turbo machinery design is becoming established within the turbo machinery community. However, only a limited number of publications suggest how to use CFD to help design and modify processes especially during the blade design. This paper serves to present a design process, which contains a novel two- and three-dimensional viscous turbulent code and optimization process.

## 2.2 Design system

Expensive manpower is invested to find configurations that are stable and efficient in the work range in the turbine and compressor design. One of the most important methods is so called stream-surface balding where two-dimensional blade profiles are to be found that insures the desired working range stability and efficiency. During the design, the constraints arising from aerodynamics, aeromechanical, mechanical, heat transfer, and manufacturing considerations have to be satisfied.

The design of turbine and compressor blade had made a great progress [31–36]. Many advanced design method and CFD tools had been incorporated into design procedure. However, most of the design procedures only focus on the flow predication and there are few papers that describe the design's overall processes and design implements. For example, Wellborn and Delaney [34] described a compressor design system used by Rolls-Royce plc, which comprise three tools, through-flow, 3D isolated blade row calculation and 3D multistage predication. The turbo machinery design is an integrated process, which contains a process from mean line, through-flow, airfoil design and analysis. This study developed a design process, which can be easily adapted by industry.

The aerodynamic design procedures for turbo machinery airfoil used in this study are shown in Fig. 3. The design system consists of the mean line analysis, through flow analysis, airfoil section design, airfoil stackup, 3D blade row and multistage flow analysis. For obtaining the highest design efficiency, the optimizer was used to do the section optimization. The three-dimensional CFD code was used for blade stack up optimization. The optimizer may be used for three-dimensional balding although authors do not encourage to use optimizer for 3D optimization. The discussion about issues regarding to 3D optimization will be given in section of optimization.

Mean line analysis determines the loading of the stage and annular area. It plays an important role in the turbo machinery design. The mean line design for the first-stage compressor and last stage turbine is critical. The enthalpy rise for compressor and drop for turbine are fixed by the mean line analysis. The overall machine character is determined by meanline analysis. The compressor and turbine efficiency is strongly influenced by mean line design.

Through flow analysis is one of the preliminary design modules. The streamline curvature calculations can be used to optimize the overall parameters of a multi-stage turbo machine. This module establishes the definitions of the flow path and work distributions in radio direction. The velocity diagrams at design point for different blade row and different streamlines are determined. The optimization code can be used to do the optimization for selecting the best design parameters, for example, stage loading, and stage enthalpy change.

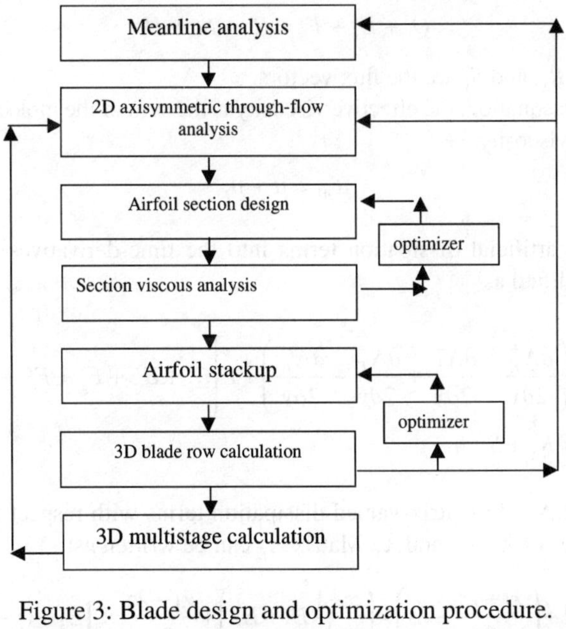


Figure 3: Blade design and optimization procedure.

The airfoil section design is a very important step for the aerodynamic design. All the airfoils are designed by the section design and reasonable stack up. In this design system, authors [2] developed a blade profile design system in which airfoil consist of four segments where tangent slopes as well as curvature are supposed to be continuous at the segment joints.

A three-dimensional blade method was developed in this study as shown in Fig. 3. It can be seen that CFD code for both section analysis and three-dimensional blade row analysis is very important. Development of the efficient and wide range application CFD codes is very important. In this study, an effective numerical method was developed for both two-dimensional and three-dimensional code. In this study, a brief introduction for the numerical method developed for two-dimensional incompressible and compressible flows was made. Details for the three-dimensional method and applications are given in reference [37].

## 2.3 Flow solver for section analysis

A time-marching algorithm [27, 36–41] was used in the present cascade flow computations. The computation starts with a rough perturbation, which develops under certain boundary conditions. In this approach, the governing equations are replaced by a time-difference approximation with which steady or time-dependent flows of interest can be solved at each time-level.

Assume that passage-averaging plane, where the through flow solution is calculated, is located at the radius r with the constant angular velocity Ω. The quasi-three -dimensional time-dependent Navier–Stokes equations on the passage-averaging plan can be written as:

$$Q_t + E_x + F_y = R_x + S_y + H \tag{1}$$

where $E_x$, $F_y$, $R_x$, and $S_y$ are the flux vectors.

In the above equation, the effective viscosity is the sum of the molecular viscosity and turbulent viscosity, i.e.

$$\mu_{\text{eff}} = \mu + \mu_t \tag{2}$$

Incorporating artificial dissipation terms into the time-derivatives [27, 41, 42], eqn (1) is modified as:

$$\left[\Delta t\left(\frac{\partial\Lambda_E^+}{2\partial x}-\frac{\partial\Lambda_E^-}{2\partial x}+\frac{\partial\Lambda_F^+}{2\partial y}-\frac{\partial\Lambda_F^-}{2\partial y}\right)+I\right]Q_t+E_x^+ +E_x^- +F_y^+ +F_y^-$$
$$= R_x + S_y \tag{3}$$

where $\Lambda_E$ and $\Lambda_F$ are matrix-valued dissipation terms with respect to positive or negative values of $\lambda_1$, $\lambda_2$, and $\lambda_3$. Matrix $\Lambda_E$ can be written as:

$$\Lambda_E = \lambda_3 I + \left(\frac{\lambda_1+\lambda_2}{2}-\lambda_3\right)\times\left(\frac{\gamma-1}{a^2}E_1+E_2\right)+\frac{\lambda_1-\lambda_2}{2a}\times\left[E_3+(\gamma-1)E_4\right] \tag{4}$$

where

$$
\begin{aligned}
E_1 &= [1, u, v, C_P T]^T \left[(u^2+v^2)/2, -u, -v, 1\right] \\
E_2 &= [0,1,0,u]^T [-u,1,0,0] \\
E_3 &= [1, u, v, C_P T]^T [-u,1,0,0] \\
E_4 &= [0,1,0,u]^T \left[(u^2+v^2)/2, -u, -v, 1\right]
\end{aligned} \tag{5}
$$

and where

$$
\begin{aligned}
\lambda_1 &= u + a \\
\lambda_2 &= u - a \\
\lambda_3 &= u
\end{aligned} \tag{6}
$$

However, in the calculations, we cannot choose the values of $\lambda_1$, $\lambda_2$, $\lambda_3$ according to eqn (6) because near the stagnation points $\lambda_3$ approaches zero, whereas the flow in sonic condition, $\lambda_1$ and $\lambda_2$ both approach zero. To solve this problem, we limit these values in the following manner:

$$
\begin{aligned}
\lambda_1 &= \max(u + a, u/M_r^2) \\
\lambda_2 &= \max(u - a, u/M_r^2) \\
\lambda_3 &= \max(u, u/M_r^2)
\end{aligned} \tag{7}
$$

and

$$
M_r = \begin{cases} 0.001 & M \le 0.001 \\ M & 0.001 < M \le 1 \\ 1 & M > 1 \end{cases} \tag{8}
$$

where $M$ is the calculated Mach number.

By integrating eqn (1) over space to form the hybrid scheme, we have

$$
\iiint \left[ \Delta t \left( \frac{\partial \Lambda_E^+}{2\partial x} - \frac{\partial \Lambda_E^-}{2\partial x} + \frac{\partial \Lambda_F^+}{2\partial y} - \frac{\partial \Lambda_F^-}{2\partial y} \right) + I \right]^n \frac{Q^{n+1} - Q^n}{\Delta t} dV
$$
$$
+ \iiint \left( \frac{\partial E^+}{\partial x} + \frac{\partial E^-}{\partial x} + \frac{\partial F^+}{\partial y} + \frac{\partial F^-}{\partial y} \right)^{n+1} dV = \iiint \left[ \frac{\partial R}{\partial x} + \frac{\partial S}{\partial y} \right]^n dV + \iiint H dV \tag{9}
$$

The evaluation of viscous terms $R_x$ and $S_y$ of eqn (1) requires first derivatives of the velocities and the energy value at each cell face. These are achieved by evaluating the gradient of every required flow quantity at the cell center from the known primitive variables at each time step. The terms can be written as

$$
\iiint \left( \frac{\partial R}{\partial x} + \frac{\partial S}{\partial y} \right) dV = \int (R_x \mathbf{n}_x + S_y \mathbf{n}_y)\, dS \tag{10}
$$

In the present study, the second order approximation form was used as:

$$\delta Q_{i,j}^{n+1} = Q_{i,j}^{n+1} - Q_{i,j}^{n} \tag{11}$$

Defining $A = \partial E/\partial Q$ and $B = \partial F/\partial Q$, we have

$$(E^{\pm})^{n+1} = (E^{\pm})^{n} + (A^{\pm})^{n}(Q^{n+1} - Q^{n}) \tag{12}$$

$$(F^{\pm})^{n+1} = (F^{\pm})^{n} + (B^{\pm})^{n}(Q^{n+1} - Q^{n}) \tag{13}$$

and the implicit Jacobians, $A^{\pm}$, can be written as:

$$A^{+} = (A + \rho_A I)/2 \tag{14}$$

$$A^{-} = (A - \rho_A I)/2 \tag{15}$$

where $\rho_A$ is the spectral radius of the Jacobian matrix $A$.

Finally, we can obtain,

$$\begin{aligned} &C_{i-1,j}\delta Q_{i-1,j}^{n+1} + C_{i,j-1}\delta Q_{i,j-1}^{n+1} + C_{i,j}\delta Q_{i,j}^{n+1} + C_{i+1,j}\delta Q_{i,j+1}^{n+1} \\ &+ C_{i,j+1}\delta Q_{i+1,j}^{n+1} = \Delta Q_{i,j}^{n+1} \end{aligned} \tag{16}$$

where

$$C_{i-1,j} = \frac{\Delta t}{2}(\Lambda_E^{-})_{i-1,j}^{n} S_{y\,i-1/2,j} - \Delta t(A_{i-1,j}^{-})^{n} S_{y\,i-1/2,j}$$

$$C_{i,j-1} = \frac{\Delta t}{2}(\Lambda_F^{+})_{i,j-1}^{n} S_{x\,i,j-1/2} - \Delta t(B_{i,j-1}^{+})^{n} S_{xi,j-1/2}$$

$$\begin{aligned} C_{i,j} = \Delta V_{i,j} &- \frac{\Delta t}{2}(\Lambda_E^{+})_{i,j}^{n} S_{y\ i-1/2,j} \\ &- \frac{\Delta t}{2}(\Lambda_E^{-})_{i,j}^{n} S_{y\ i+1/2,j} \\ &- \frac{\Delta t}{2}(\Lambda_F^{+})_{i,j}^{n} S_{x\ i,j-1/2} - \frac{\Delta t}{2}(\Lambda_F^{-})_{i,j}^{n} S_{x\ i,j+1/2} \\ &+ \Delta t(A_{i,j}^{-})^{n} S_{y\ i-1/2,j} - \Delta t(A_{i,j}^{+})^{n} S_{y\ i+1/2,j} \\ &+ \Delta t(B_{i,j}^{-})^{n} S_{x\ i,j-1/2} - \Delta t(B_{i,j}^{+})^{n} S_{x\ i,j+1/2} \end{aligned}$$

$$C_{i+1,j} = \frac{\Delta t}{2}(\Lambda_E^{-})_{i+1,j}^{n} S_{y\,i+1/2,j} + \Delta t(A_{i+1,j}^{-})^{n} S_{y\,i+1/2,j}$$

$$C_{i,j+1} = \frac{\Delta t}{2}(\Lambda_F^{+})_{i,j+1}^{n} S_{x\,i,j+1/2} + \Delta t(B_{i,j+1}^{+})^{n} S_{xi,j+1/2}$$

Since eqn (16) is an implicit scheme, it can improve the numerical stability even with a large time step. In this numerical scheme, $C_{i-1,j}$, $C_{i+1,j}$, $C_{i,j}$, $C_{i,j-1}$, and $C_{i,j+1}$ are all scalars, which are calculated using integration combined with the flux vectors. In this way, the computational efforts are greatly reduced as compared with other coefficient matrix implicit schemes for all time-steps [27, 41]. The implicit equation (eqn (16)) can be solved by using two sweeps along $i$ and $j$ directions as follows:

$$C_{i,j-1}\delta Q_{i,j-1}^{n+1} + C_{i,j}\delta Q_{i,j}^{n+1} + C_{i,j+1}\delta Q_{i,j+1}^{n+1} = \Delta Q_{i,j}^{n+1} - C_{i-1,j}\delta Q_{i-1,j}^{n+1} - C_{i+1,j}\delta Q_{i+1,j}^{n} \tag{17}$$

Since the present scheme consists of an explicit part to calculate $\Delta Q_{i,j}$ and an implicit part to calculate $\delta Q_{i,j}$, which possesses the advantage of both schemes.

In this study, the time-step is selected according to the CFL number constraints as:

$$\Delta t = \min(\Delta t_x, \Delta t_y) \tag{18}$$

where

$$\Delta t_x \leq \frac{CFL \cdot \Delta x}{|u| + a + (2\omega/\Delta x \rho)}$$

$$\Delta t_y \leq \frac{CFL \cdot \Delta y}{|v| + a + (2\omega/\Delta y \rho)}$$

where $\omega = \max\,(\mu/Re, \mu_t + 2\mu/Re, \mu\gamma/P_r Re)$.

The choice and development of a turbulence model remain important points to reproduce the flow features. However, there is still no such kind of turbulence model, which can well represent a true flow situation. The Baldwin–Lomax [27, 41] model has several good features, such as its usefulness in separated flows with a small separation region, its relatively smooth and continuous length-scale from transition of the boundary layer into the far wake region, and its ability to accurately predict the wall effect near the trailing edge. For these reasons, the above-mentioned Baldwin–Lomax turbulence model was employed in this study to handle the turbulent flow computing cascade flows around the turbine. The boundary conditions play an important role in determining accurate solutions and rapid numerical convergence. Several types of boundary conditions have been used which are similar to the previous studies [27].

The discussion has been extended for many years for the discussion of best type mesh, which should be used, for turbine or compressor blade flow calculation [43, 44]. The more orthogonal the grid, the smaller will be the numerical errors due to the discretization of governing equations. However, no one type of grid is ideal for blade-to-blade flow calculation. In this study, the H-type mesh was used. Another problem for the blade-to-blade mesh is the trailing edge mesh and trailing edge Kutta condition [45]. The author has noticed that the number of the mesh point near the trailing edge points strongly influences the loss calculation. Here, a realistic method is proposed, i.e. when the mesh on the trailing edge

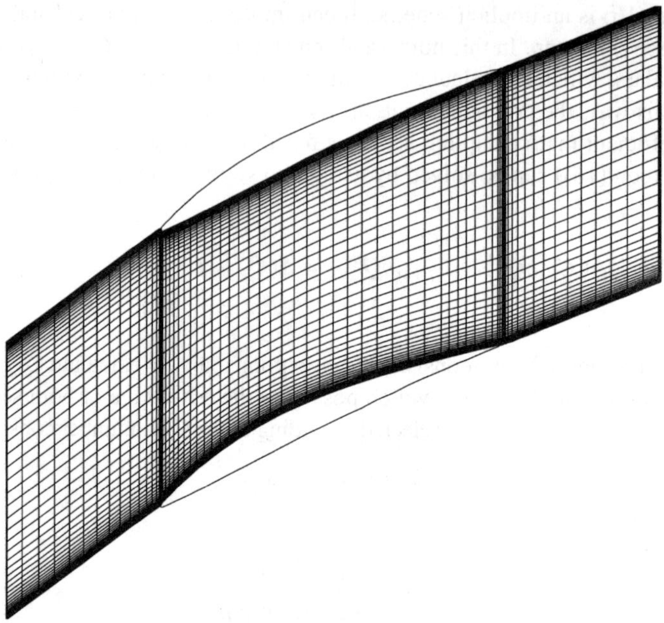

Figure 4: Computational mesh.

circle is more than 10 points, an explicit viscous Kutta condition [45] is applied on the trailing edge circle which flow can leave the blade smoothly.

The mesh refinement study was conducted prior to the calculations; as reported by Xu and Amano [27, 37, 41], the mesh size of $110 \times 45$ with 80 nodes on blade is sufficient. The H-mesh was used in all the computations. The computational mesh is shown in Fig. 4. A typical convergence history of the calculation is shown in Fig. 5. It is demonstrated that the present code has a good convergence. In this study, this code was used in the two dimensional airfoil section analysis and optimization.

## 2.4 Optimization

Turbine and compressor design faced the challenge of designing high efficiency and reliable blade design. Design techniques are typically based on the engineering experience and may involve trials and errors before an accepted design is found. The ability of the three-dimensional codes makes the detailed flow predictions possible; however, it is very time consuming to perform a three-dimensional flow calculation of the design in the design process. The three-dimensional approach is a way to help the designers do some parameters study and compare the numerical study with testing to improve the design. The three-dimensional code developed in this study [37] is used to do parameters studies, like lean, bow, and sweep effects. The two-dimensional analysis developed in the study is used to optimize blade section airfoil.

The blade design often started two-dimensional airfoil section. Some of the design parameters are obtained by through flow optimization and vibration, heat transfer

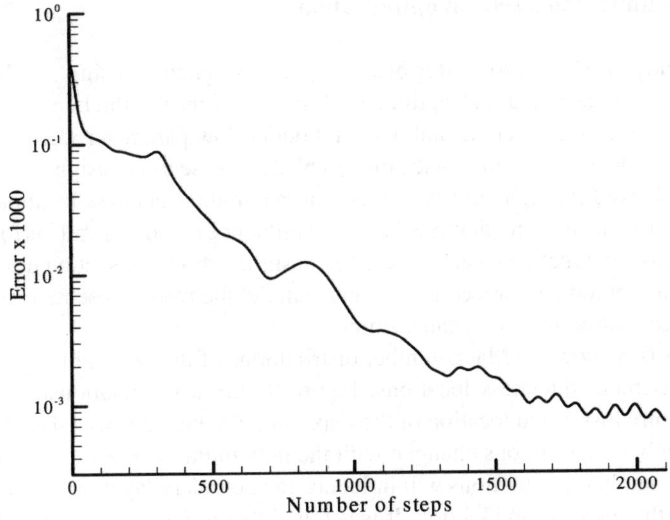


Figure 5: Typical convergence history.

analysis. For example, the solidity, the number of the blades, and most of the parameters were optimized through the sectional design. For example, leading- and trailing edge radii, stagger angle, and maximal thickness position. The final three-dimensional analyses are used to make lean, bowed and sweep modification of the blades.

## 2.5 Method of numerical optimization

The optimization process used in this study was based on the constrained optimization method [46, 47]. If the objective function to be minimized is $F(\mathbf{X})$, and constraint functions is $g_j(\mathbf{X})$, the optimization problem can be formulated by the objective and constraint functions as

$$\nabla F(\mathbf{X}) + \Sigma \lambda_j \nabla g_j(\mathbf{X}) = 0 \tag{19}$$

where $\lambda_j$ are the Lagrange multipliers and $\mathbf{X}$ is the vector of the objective. The finite differencing method can be used to obtain the gradients of objective and constraint functions with respect to design variables and constraints. Many commercial optimization packages are available as an optimizer to design problems [47].

The objective function of this study is adiabatic efficiency. The optimization objective is to obtain the highest efficiency under giving constraints. Foe more convenient, the total pressure loss was used as an objective variable to judge the design. The definition of the total pressure loss in this study is

$$\xi = \frac{(p^*{}_{in} - p^*_{out})}{\Delta H} \tag{20}$$

where $\Delta H$ is the outlet dynamic head of the exit plane.

## 2.6 Two-dimensional section optimization

In this study, a compressor stator blade original designed by using two NACA65 sections was redesigned and optimized. In this design, the thickness of the airfoil, chord, and stagger angle and inlet and outlet flow parameter were fixed. The designed blade was used to replace the old blade. The section design optimization mainly achieved through adjusts the section maximum thickness location. It was found that the maximum thickness locations influence the losses and performance. In this study, five sections were selected as design section. All section design used the similar method and procedure. As an example, the result presented here is the design information for 50% span section.

Figures 6–8 show the Mach number distributions of the design blades with different maximum thickness locations. Figure 9 shows the relationship between total pressure losses and location of the maximum thickness. It was shown that the Mach number distributions changed with the maximum thickness. The change of the Mach number distributions will influence the boundary layer development and influence the losses of the section. It was found that there is an optimum location

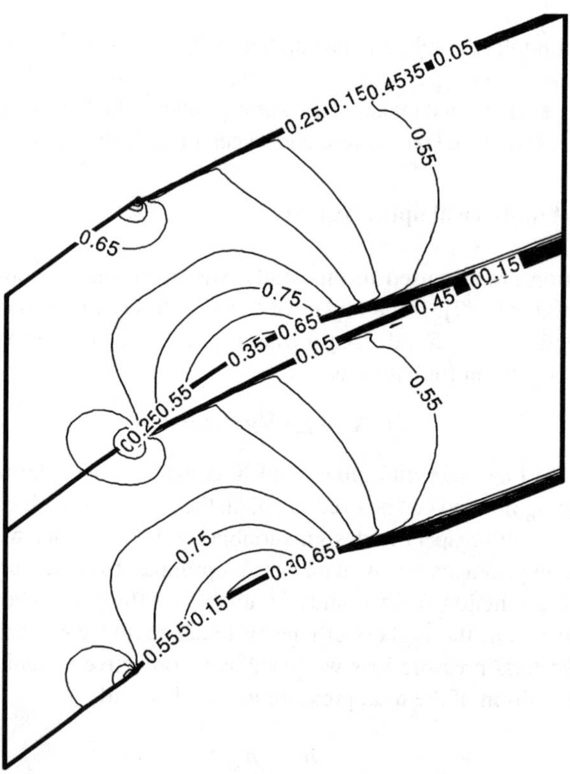

Figure 6: Mach numbers distribution at maximum thickness at 0.15c.

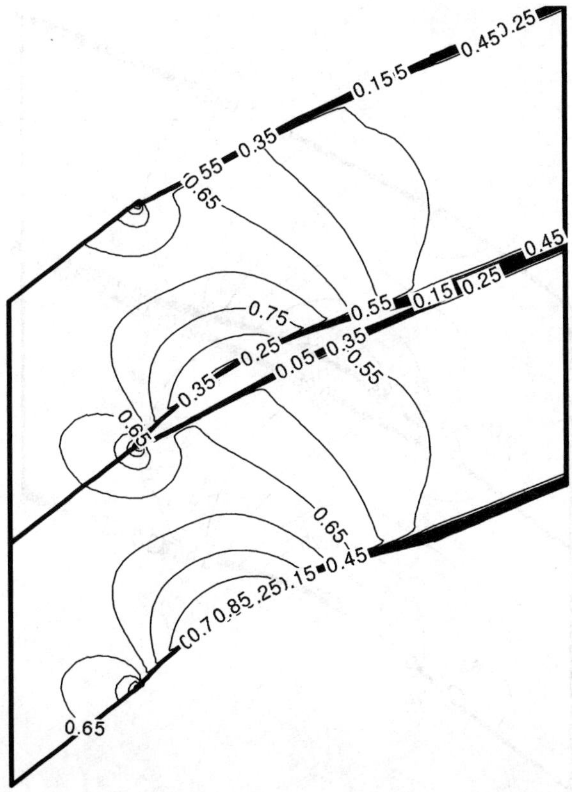

Figure 7: Mach numbers distribution at maximum thickness at 0.25c.

of the maximum thickness. Study also showed that this location is changed with the flow conditions and stagger of the airfoil. After optimization study, the airfoil section with highest efficiency was selected. The airfoil was stacked up using a smooth leading edge curve method. Before the airfoil was selected as a final blade, the three dimensional analysis was performed to investigate the benefits of the three-dimensional blade features. Three-dimensional analysis showed that after section optimization the airfoil performance was improved as shown in Fig. 10. However, the end wall region did not have a significant improvement. Some three-dimensional features were used to reduce the losses and increase the efficiency.

## 2.7 Three-dimensional CFD analysis and blading

In the present design system, three-dimensional CFD analysis codes were developed [37] to guide the three-dimensional final modification of the two-dimensional section stack up. Some parameters were selected to do the study based on design experience. In this redesign, the three-dimensional feature-bowed blade was used

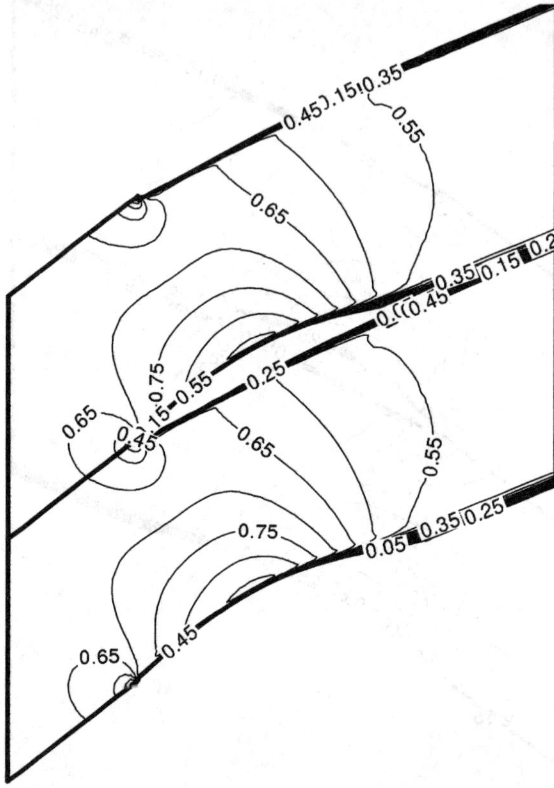


Figure 8: Mach numbers distribution at maximum thickness at 0.45c.

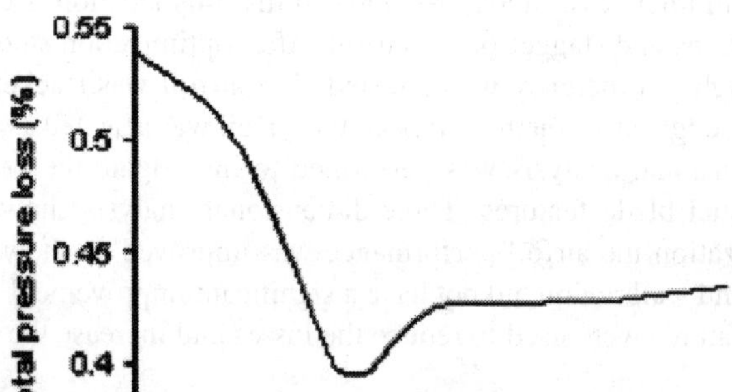


Figure 9: Total pressure losses and maximum thickness location.

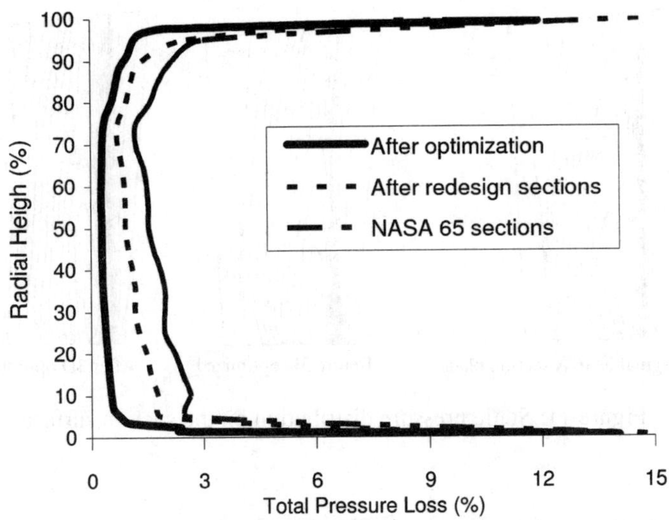


Figure10: Total pressure losses with span.

to reduce the secondary losses. In the three-dimensional blading, the optimizer will not be used in this study because performing the three-dimensional analyses is very time consuming. In the current study, three-dimensional optimization was obtained through parameter study.

In this study, only bow feature was applied. The bow location and degree of the bow were selected as the parameters to study. It was found that the 15° bow located at 30% of the span can eliminate separation and get better performance. Three-dimensional analysis showed that after two-dimensional section optimized design most of the span the flow remained attached and well behaved although it had a small region of end wall separation on the section side as shown in Figs. 11–14. After introducing the bows on the blade, the flow around the compressor blade on both tip and root end wall were attached. The total pressure losses were reduced as shown in Fig. 10. This study showed that total section efficiency was increased about 3% compared to the original blade. Both two-dimensional section design and three-dimension optimization increase the efficiency at similar percentage. It is shown that both two-dimensional optimization and three-dimensional study are important. However, three-dimension blading more depends on the experience of the designer.

It is interesting to point out, based on the design experience, that the three-dimensional optimization based on high efficiency normally does not produce final blade shape. The selection of the final configuration was performance through small adjustments based on experience to obtain the blade design with both high performance and good stability. After three-dimensional analysis, some small adjustments were made based on the possible manufacturing uncertainties and applications. For example, the compressor stator design in this study was recambered to increase exit angle about 1.5° to increase the compressor surge margin.

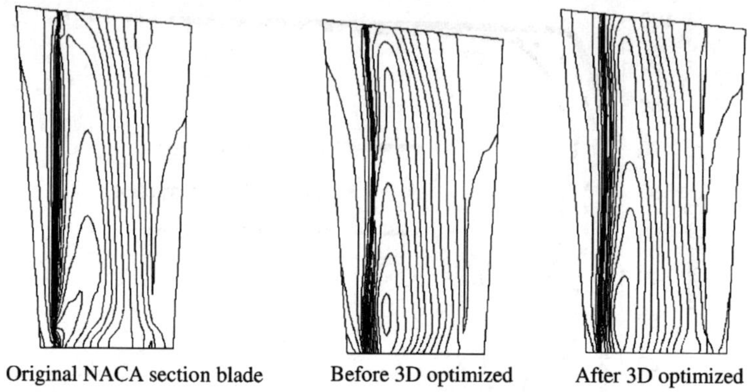


Figure 11: Static pressure distribution on the suction surface.

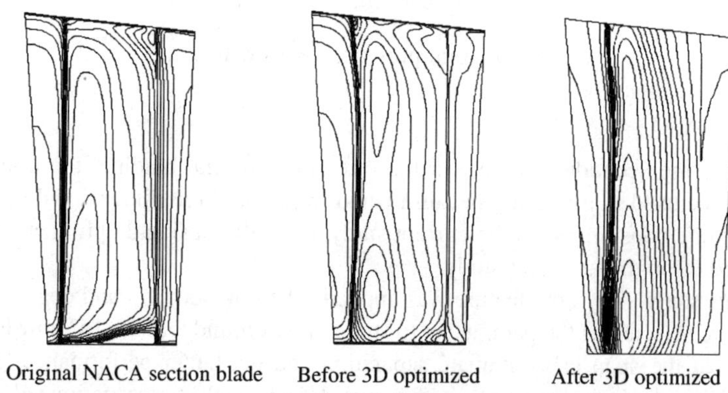


Figure 12: Axial velocity contour near blade suction surface.

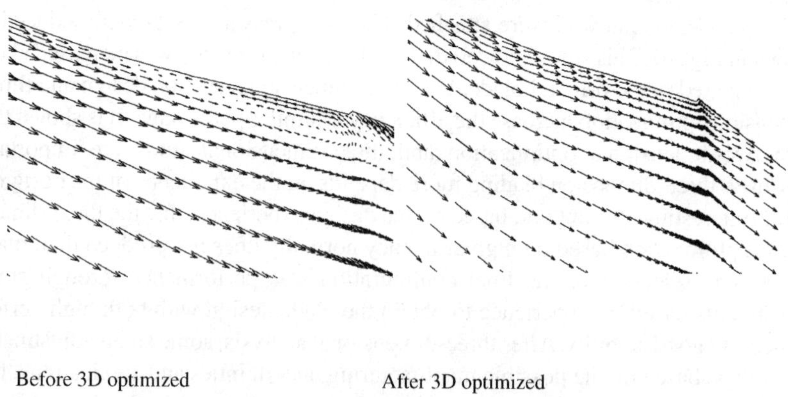


Figure 13: Velocity vector near root section near the trailing edge suction surface.

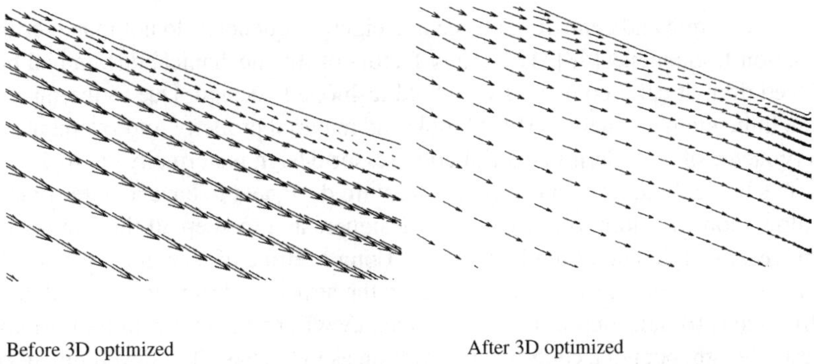


Figure14: Velocity vector near the tip section near the trailing edge suction surface.

### 2.8 Discussion

The flow field around turbine and compressor blades exhibits a very complex flow features. The flow field involves various types of loss phenomena and hence high-level flow physics is required to produce reliable flow predications. The blade design process is a very time consuming process and optimization process is very complicated. The method developed in this study for the airfoils section and blades design and optimization is one the possible processes and very easy for adapting in the aerodynamic design. In this study, a two-dimensional code was used for section design instead of three-dimensional code, which provides an economical way to design, and to optimize. The three-dimensional blades were optimized after stacked up the two-dimensional sections. The lean, bowed and swept features of the three-dimensional blade were created during the three-dimensional optimization. It was shown that both two- and three-dimensional design and analysis were backbones of the blade aerodynamic designs. Results show that the both airfoil shape optimization and three-dimensional optimization can be used to improve the performance of compressor and turbine blade.

## 3 Part III: Centrifugal compressor design experience

### 3.1 Compressor design

The new compressor design is focused on the customers' needs with shortest time to market, low cost and better performance. For pushing the design to the-state-of-art aerodynamic performance, the structural design needs to meet the suitable life of the compressors. The mechanical integrity is one of the important parts of the centrifugal compressor design. The mechanical constraints always are negative factors for aerodynamic design. The purposes of the mechanical analyses are to provide all compressor components at a reasonable time duration to sustain the aerodynamic and centrifugal force and eigen frequencies do not match critical excitation frequencies [48]. The safety factors of the mechanical design had been reduced dramatically compared with "old fashioned" design. Due to the maturity of the finite element analysis (FEA) tools and material property improvements, the safety factor of a modern industrial compressor design is normally set to 7–12%. The mechanical requirements require structure designers to have a better practice to allow more freedom to aerodynamic designers and to keep all the components at the least weight and at the lowest cost. Long lifetime of a single component of compressors is not a design goal. However, the structure design for defined period of lifetime is trivial. Emphasis on the efficiency will be the foundation of the compressor design, but not a central consideration as in the past. The development cost and development time are also the key factors that need to be considered for the modern compressor design. Industrial compressor design is required to design a state-of-the-art performance compressor without making second build for less cost and development time. This requires compressor design engineers to have extensive knowledge of centrifugal compressor design. The detail design considerations can reduce the time to perform the advance design studies and laboratory investigations. The wide variety of design subjects represents a very complex design world for all designers. The purpose of this paper is to provide information for new engineers and help them to understand the overall design before they start to design a centrifugal compressor.

The compressor market and business model have changed compared with a few decades ago. Industrial compressor design requires design for success in the marketplace, not just for scientific experiments. In the past, the compressor designers developed a new compressor in the development group, and passed the design to manufacturing. The manufacturing group would evaluate how to make it at the lowest cost. Some developments were rejected because they could not meet the market requirements. The new development model requires compressor designer's design for market, manufacturing and end users. Recently, some new business concept was proposed [49, 50]. New developments will consider an integrated system of manufactures and end users. The new compressor developments become a complex system task. Minimizing manufacturing cost of the compressor design is not enough. The compressor design must consider all aspects of the manufacturing and end users. If surplus is defined as total profit of manufacturing, end users and aftermarket, the new compressor development will focus on the design for maximum surplus. Therefore, in the compressor design stage, many choices of the design options need to be considered before final design and making decisions based on the surplus value. It is essential that design engineers begin to perform compressor design after fully understanding all aspects of the design considerations [51, 52].

### 3.2 Impeller designs

Impeller is one of the key components to influence centrifugal compressor overall performance [53–56]. Nowadays, the efficiency of centrifugal compressors has been increased dramatically especially low-pressure ratio centrifugal compressors. The big challenge for centrifugal compressor design is to keep efficiency level high at state-of-the-art and increase the compressor operating range [54, 55]. Increasing the compressor operating range without sacrificing compressor peak efficiency is difficult to achieve. Aerodynamic engineers not only need to understand the surge physics but also need to apply design experience to design. Surge is influenced by all components of the compressor. The physics of surge and stall are still not fully understood. We still cannot find any tool that can capture all features of surge and stall as shown in Fig. 15. Many theoretical studies [57] were focused on the better understanding of surge and stall, but none can be used as a design tool yet. More theoretical work and experimental studies need to be done in order to incorporate stall in the design system. Designs for wide operating range were mainly dependent on the engineers' experience and their understanding of the stall and surge. Another important objective for impeller design is to reduce the manufacturing cost. Manufacturing cost could be reduced when designs for manufacturability are effectively considered. The designed impellers should meet requirements to be easily withdrawn from a casting mold without destruction and disassembly of the mode. This requires the lean angle of blade change linearly with the impeller radius and axial direction as shown in Fig. 16. All these considerations for design will help the finial design to meet the compressor design target with less design iterations.

Different designers may have different design methodologies for impeller designs. What kind of area distributions, curvatures, velocity or pressure profile will lead to a good design is strongly dependent on the designers' practice and experience. Two totally different design philosophies could produce similar performance. For example, two impellers designed by Garret and Pratt Whitney [49] as shown in Fig. 17 had different shapes with similar performance at design point. The impeller designed by author also presented different features, which also provided good performance. It is shown that if design follows the basic design guideline, a wide range of solutions to the design can be used. The experience data presented in this paper will help designers to set up the design guideline for compressor initial design phase.

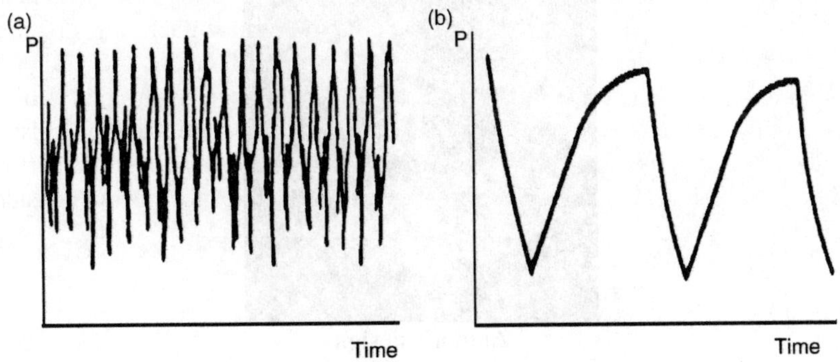


Figure15: (a) Stall and (b) surge pressure variation with time.

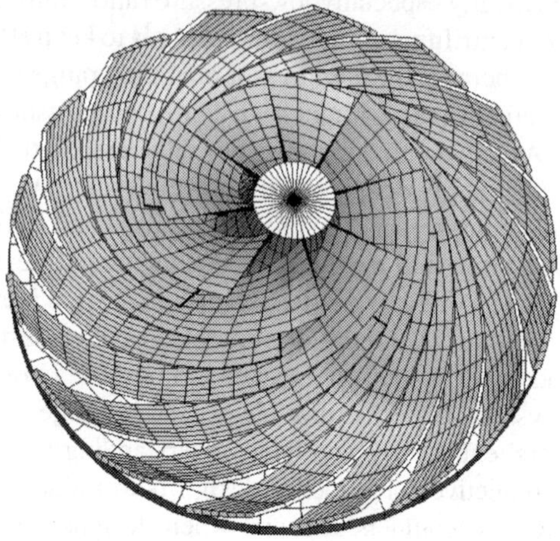

Figure16: Impeller for manufacturing.

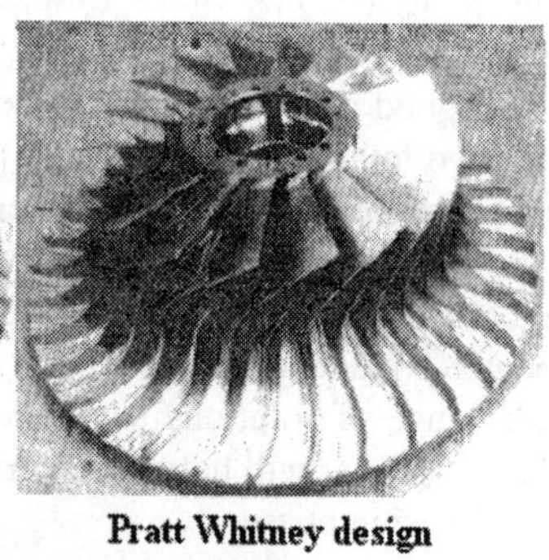

**Garret design** **Pratt Whitney design**

Author's design

Figure 17: Different impeller designs.

Authors' design shown in Fig. 17 is an example of recent developed single stage centrifugal compressors. At design point, the total to static stage pressure ratio is about 3.7 and the flow coefficient is about 0.12. The running clearance at impeller tip was 4.5% of the impeller exit blade width. Six builds were assembled and tested based on ASME PTC-10 test procedure [58]. The compressor performance obtained from average of six build tests is shown in Fig. 18. The differences of test results for different builds for adiabatic efficiency and head coefficient were within ±0.5% and ±0.75%, respectively. The test uncertainties for total pressure (in psi), static pressure (in psi), and temperature (in Fahrenheit) were ±0.25%, ±0.2% and ±0.5%, respectively, based on uncertainty analysis [59]. Test results showed that the compressor performance was encouraging at both design and off design point. The design met the low cost target and allows large manufacturing tolerances. The insensitivity of the impeller surface finish and large tip clearance make it easy for assembly.

### 3.3 Impeller geometry

The initial design of a centrifugal compressor is always started with requirements from customers or marketing analysis. Designers select basic configurations and provide basic performance to customers or marketing by using their experience data. Aerodynamic designers also need to provide estimation for the compressor basic geometry to engineers in other disciplines. For example, rotor dynamic engineers and bearing designers rely on the impeller geometry information to perform their work. Although the basic geometry design is not intended to yield optimization of the impeller, it can accelerate overall design process and reduce the development cost.

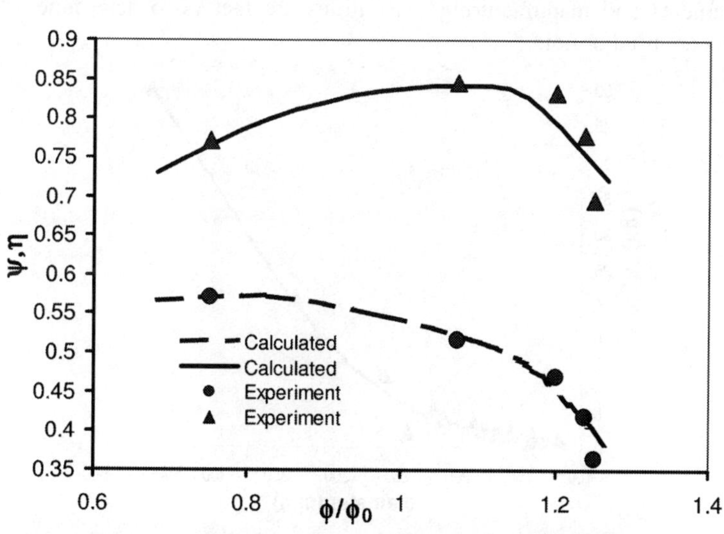


Figure18: Single stage compressor performance.

Before aerodynamic designers determine basic impeller geometry, the rotational speed of the impeller needs to be selected. If there are no special requirements for rotational speed, we normally optimize rotational speed based on the Balje's charts [60] by using optimal specific speed. Although Balje's charts are not very accurate tools, they are sufficient enough to provide the initial estimate for impeller geometries.

During the initial design, the important information needed for bearing designers and rotor dynamic engineers is impeller weight. Aerodynamic designers can estimate impeller sizes based on the required gas flow, pressure ratio and impeller rotational speed. Our design practices showed that weight of impeller is the function of impeller diameter. Fig. 19 summarizes the relationship between impeller diameter and weight for sixteen ASTM A564 stainless steel unshrouded impellers. The impeller weight mainly determined by impeller disk and blades only contribute a very small portion of the weight. Therefore, we plotted impeller weight and diameter relation in one figure for all designed unshourded impellers with different blade counts and with or without splitters.

In the initial stage of compressor design, selections of impeller inlet and outlet velocity vectors and choice of blade numbers are key initial design decisions. Velocity vectors may be obtained through a mean line program. The experience data showed that both inlet blade numbers and exit blade numbers were a function of stage pressure ratio. Relationships between numbers of blades and stage pressure ratio are shown in Figs. 20 and 21 for without and with splitter impellers, respectively. In general, high stage pressure causes blade-loading increase and impeller needs more blades to distribute loading. Variations in number of blades at similar pressure ratio were due to the size of the impellers. For small sizes impeller, manufacturing capabilities may limit the number of blades. Impeller sizes plotted in Figs. 20 and 21 were in the range 2–45 inches. The machine performance requirements and manufacturing feasibilities are factors to determine whether splitters are used or not.

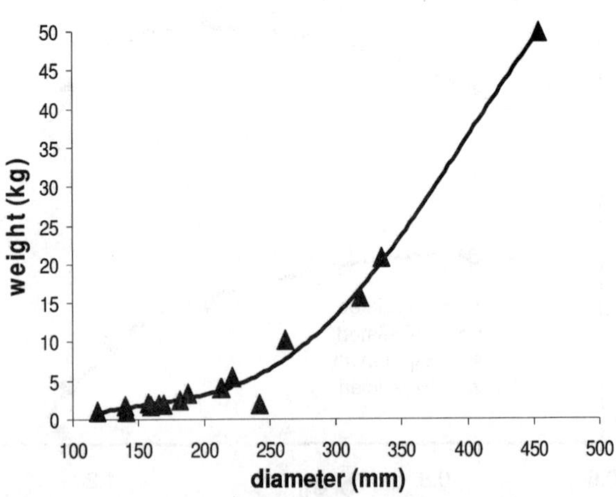


Figure19: Variation of impeller weight with diameter.

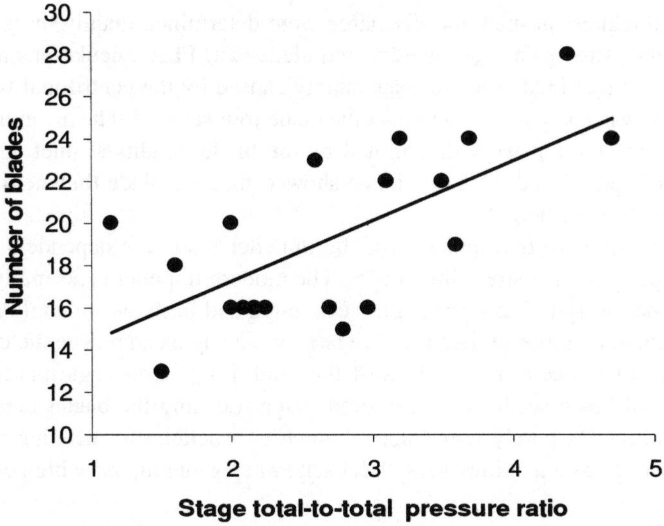


Figure 20: Number of blades at inlet versus stage pressure ratio.

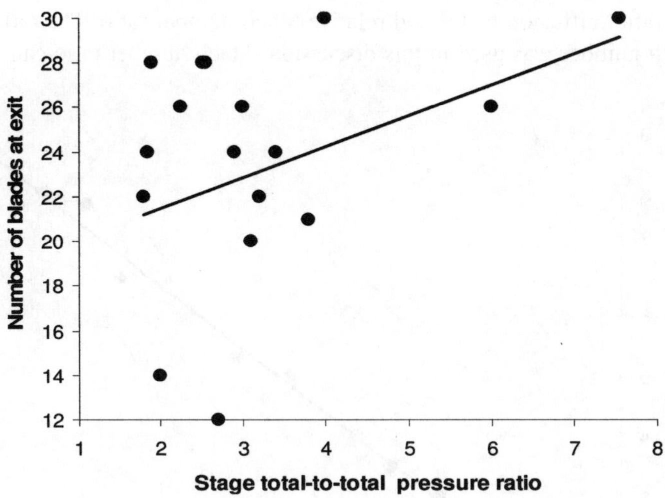


Figure 21: Number of blades at exit versus stage pressure ratio for split impellers.

The inlet blade height is determined by design inlet flow rate and impeller hub radius. Inlet hub radius is determined by the attachment of impeller. For overhung impellers, inlet hub radius is normally selected in the range between 10% and 20% impeller tip radius. For the shaft and bolt through impeller, selections of inlet hub radius are based on stress requirements.

Blade thickness at inlet and discharge were determined mainly by tensile and bending root stresses at leading edge and blade exit. FEA calculations and stress tests showed that blade root stresses mainly caused by the centrifugal force. The blade high was a key factor to impact the blade root stresses. The mean line thickness at inlet and exit were determined by the blade heights at inlet and exit as shown in Figs. 22 and 23. Experience showed that the blade thickness changed linearly with blade height.

The three-dimensional features of the impeller blade are dependent on engineers' experience and stress limitations. The modern impeller is normally a three-dimensional design. The wrap angle, lean angle and back sweep angle are using larger value than those in the past. The large wrap angle can reduce the camber of the blade but increase the frictions of the fluid. Large lean angle permits blade design at all blade sections with desired shape. Leaning the blades creates back sweep and retains purely radial fibers, which is beneficial for bending moments. Experience showed that impellers with back sweep generally have high efficiency.

### 3.4 Impeller aerodynamic design

One of important guidelines for impeller aerodynamic design is to set a reasonable diffusion ratio of some sort. The diffusion of the impeller can be represented by velocity ratio, diffusion factor and relative Mach number ratio. The ratio of relative Mach number was used in this discussion. Mach number ratio can avoid the

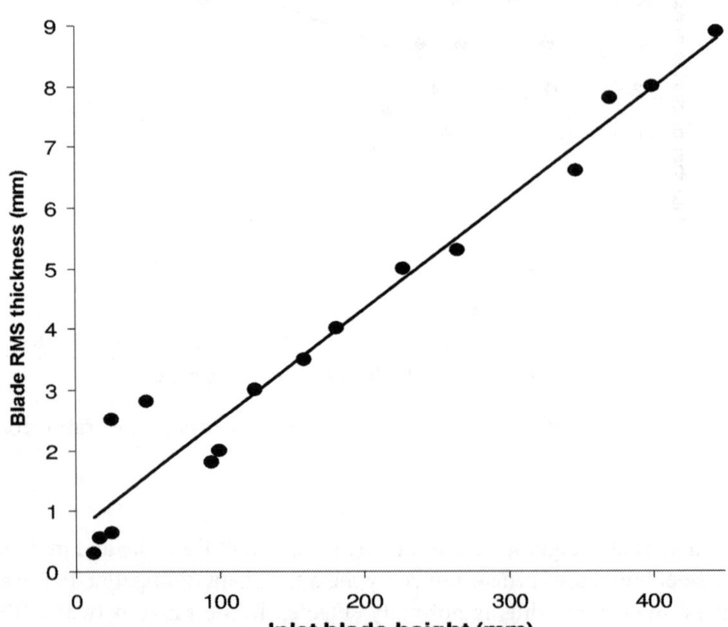


Figure 22: Relationship between impeller inlet RMS thickness and inlet width.

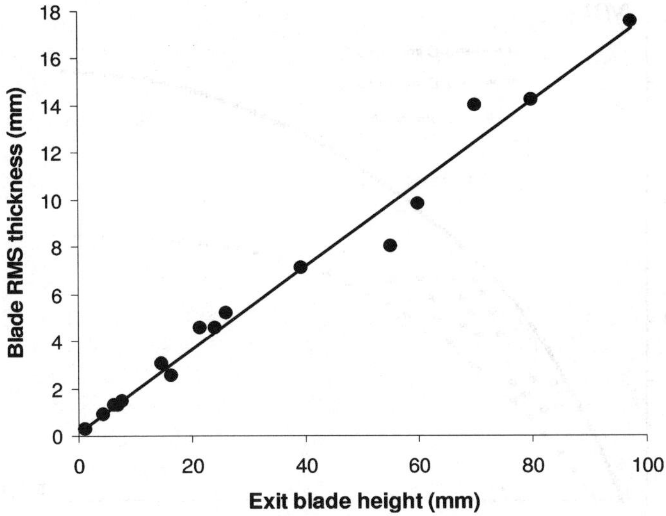


Figure 23: Relationship of impeller exit RMS thickness and tip width.

one-dimensional assumption at inlet. $MR_2$ is defined as the ratio of relative Mach number at impeller inlet to the average Mach number at impeller exit. Figure 24 also showed the up and low boundaries for maximum deceleration likely to be achieved for two-dimensional and three-dimensional impellers [54]. It is shown that the experience data fall inside theoretical boundaries. Experience showed that the Mach number ratio $MR_2$ fell between 1.15 and 1.4 gives a good overall performance. The upper boundary of $MR_2$ at 1.4 for industrial compressor and 1.7 for jet engine impellers are reasonable expectations. The experience indicated that large diffusion might cause a huge loss. The ratio of the Mach number can be selected within a large range. An important factor to impact the selection of diffusion level is inlet Mach number. Figure 25 is a relationship between relative velocity ratio, incidence and inlet relative Mach number for a typical industrial impeller. It can be seen that diffusion is not an absolute parameter, which influences the stall of the compressor. It is worth to point out that this test impeller was stall at inducer first. Inducer shroud velocity represented the rotational speed.

Traditionally, impeller inlet incidence is set to zero at design condition [61]. Modern impeller designs need not only to consider maximum efficiency at design point, but also to consider manufacturing cost and off-design performance for whole operating range [55]. Inlet blade angles are not necessarily the same as inlet relative flow angles. Experience data in Fig. 25 showed that changes of inlet flow incidence impact both efficiency and operating range of the impeller. Fig. 26 shows that little negative incidences could raise impeller operating-range. However, when negative incidence increased to a certain level, the operating range did not enlarge and efficiency dropped significantly. The impeller design should avoid this situation.

Estimations of the impeller exit width are critical for both primary performance estimation and basic dimension set up. The major impacts of impeller blade exit

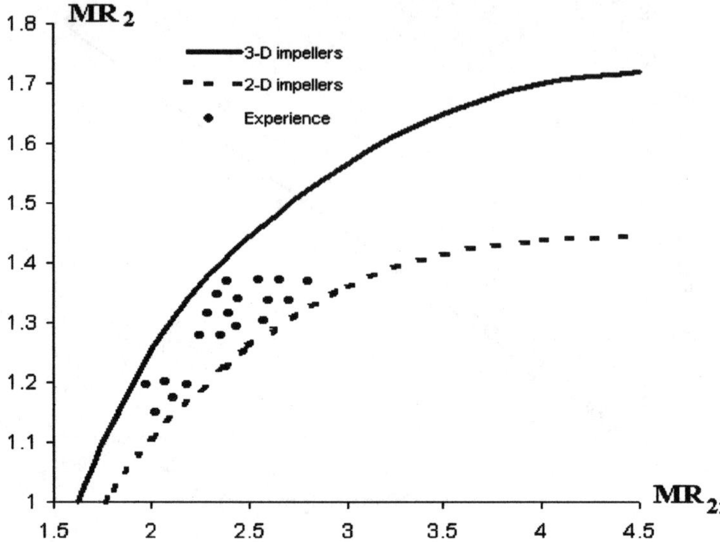


Figure 24: Relationship of $MR_2$ versus $MR_{2i}$.

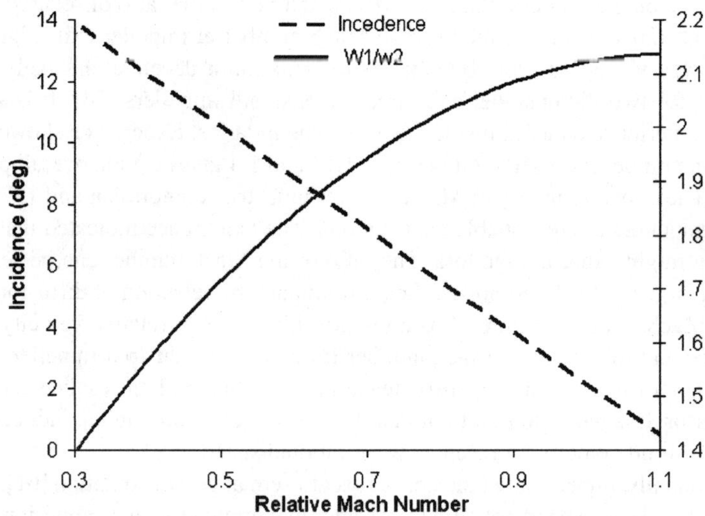


Figure 25: Stall incidence and relative velocity ratio versus Mach number.

width are flow capacity and pressure ratio of the stage. It is difficult to calculate the impeller exit width accurately in a simple way. Rodgers diffusion factor equation [62] provided a good estimated value for impeller exit width ($B_2$). If mean meridional blade length can be estimated as

$$L = \frac{2\pi(r_2 - r_1)}{4} \tag{21}$$

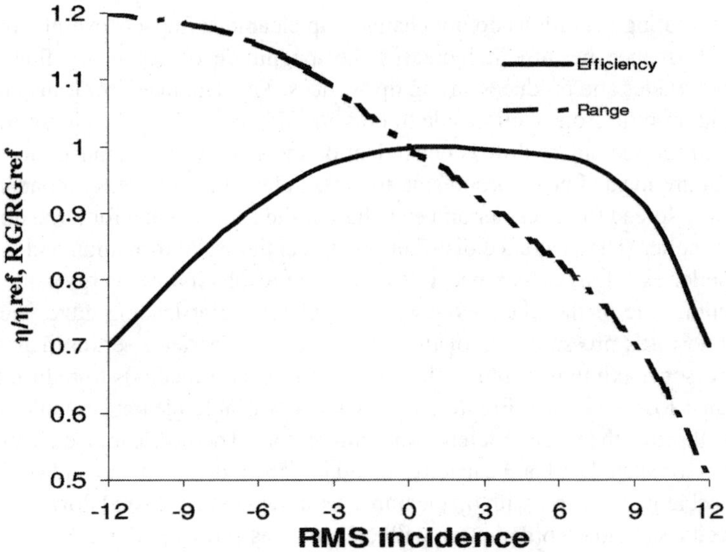


Figure 26: Incidence versus efficiency and range.

then impeller exit width can be estimated as

$$B_2 = \frac{10(D_2 - D_{s1})}{1+(W_2/W_1)}\left[DF - 1 + \frac{W_2}{W_1} - \frac{\pi D_2 C_{\tau 2}}{2LzW_1}\right] - \frac{(D_{s1} - D_{h1})}{2} \quad (22)$$

Secondary flows inside of the impeller caused by imbalance of static pressure and kinetic energy. One of the typical secondary flows, horseshoe vortex, has been well documented. It is shown that the strength of the secondary flows is governed by the vortex starting conditions. The further development of the vortex is determined by the conservation of angular momentum. Impeller meridional blade profiles influence secondary flow loss level. Laminar viscous dissipation function can estimate the secondary flow loss due to the blade profiles [62].

$$\Delta H_1 = N\int \mu\left[2\left(\frac{\partial u}{\partial x}\right)^2 + 2\left(\frac{\partial v}{\partial y}\right)^2 + \left(\frac{\partial u}{\partial y} + \frac{\partial v}{\partial x}\right)^2\right]dV \quad (23)$$

Tip clearance cannot be avoided for unshrouded impellers. Bearing clearances and manufacturing tolerances of impeller and intake ring controlled the minimum impeller tip clearance. The minimum tip clearance was normally defined at maximum rotational speed with hot weather conditions for most motor driven compressors. For compressors installed in the same shaft with gas turbine, minimum compressor tip clearance was estimated when compressor operating at maximum rotational speed with hot weather and machine overall net axial thrust load towards compressor. Tip clearance increases quadratically with impeller rotational speed if other operating conditions do not change. Tip clearance impacts overall compressor performance because it increases the magnitude of secondary flow inside impeller blades and produces strong tip vortices. Tip clearance flow transports low momentum fluid from suction side to pressure side of blades. The circumferential center of the secondary flow is dependent on the size of tip clearance. Secondary vortices are located near shroud side for small clearance, whereas secondary vortices may spread to the center and even hub of the flow channel for large tip clearance impeller. The clearance distribution affects the wake formation and location at impeller exit. Large clearance at leading edge results in low energy center close to suction side of the blade. Reducing the clearance at leading edge, the wake moves towards pressure side of the blade. The tip clearance setting depends on compressor maximum surplus value. There are several methods to reduce the tip clearance losses [2, 56]. Figure 27 shows that variable clearance could significantly improve the stage efficiency of compressors. The tip clearance changes the compressor stage head and capacity. Test data [55] indicated that head coefficient almost changed linearly with tip clearance as shown in Fig. 28 and flow coefficient followed a secondary order curve with clearance as shown in Fig. 29.

### 3.5 Reynolds number and surface finish

Reynolds number or Rossby number has big impacts on impeller maximum surplus value. Reynolds number is an indicator of inertial force versus viscous force for a moving fluid. Rossby number is a measurement of inertial force versus Coriolis force. Fundamental fluid dynamics theory [63, 64] showed that the flow inside a pipe for different Reynolds number represent different flow patterns. This is also true for flow inside of impeller blades. The experience showed that if a flat velocity

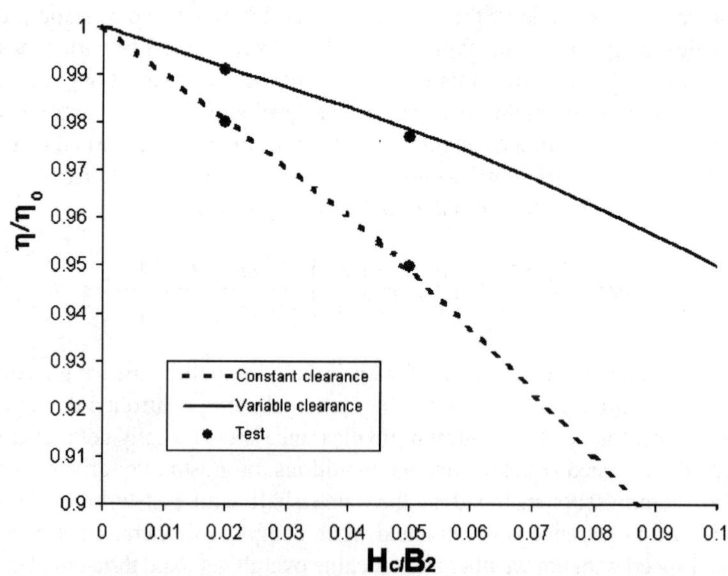


Figure 27: Compressor efficiency change versus clearance.

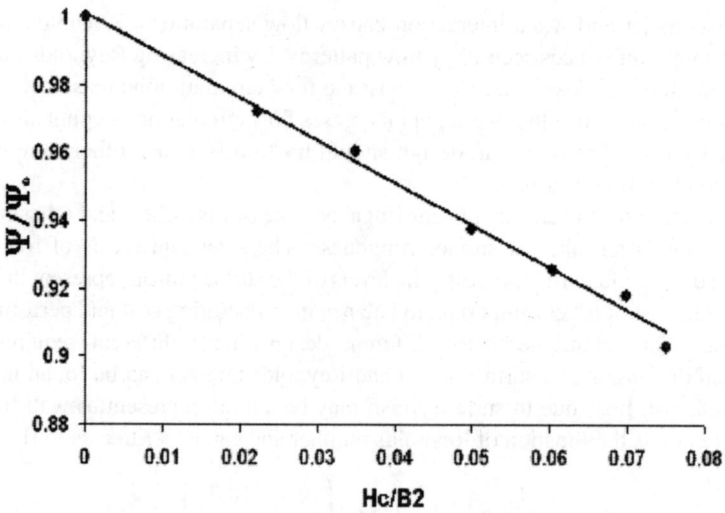


Figure 28: Head coefficient change with tip clearance.

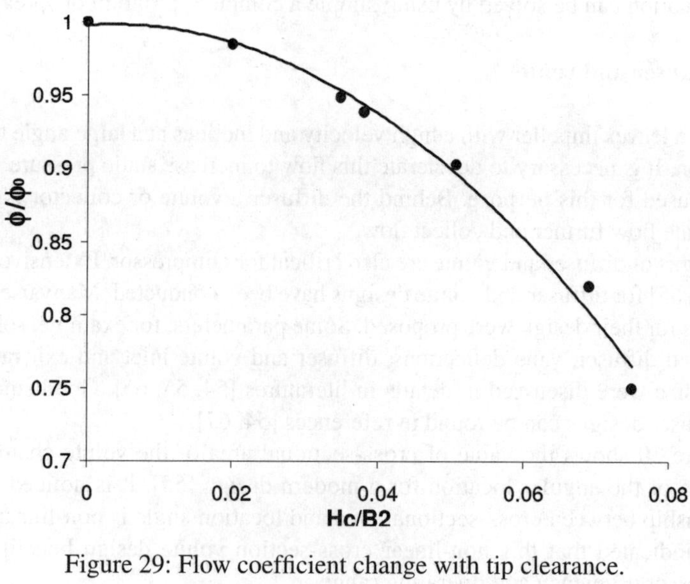


Figure 29: Flow coefficient change with tip clearance.

profile at inlet between two impeller blades or diffuser vanes, the flow development along the flow channel presents different profiles with different Reynolds numbers. For low Reynolds number flows, the exit velocity profiles are almost parabolic and only with small portion of flat profile. For high Reynolds number flows, the exit velocity profiles have large flat profiles.

The peak meridional velocities for high Reynolds number flows are normally located at the hub pressure sides of the blade due to potential flow effects. Low Reynolds number flow regions are located near suction side of the blades.

The viscous jet and wake interaction causes flow separations. Reynolds number also strongly influences secondary flow patterns. By increasing Reynolds number, the strength of clockwise secondary passage flow circulation increases. Reducing pressure and suction velocity gradient increases flow circulation of count clockwise secondary flow. The optimum design should try to offset each other to minimize the secondary flow losses.

The machining and casting of centrifugal compressor impellers and other components result in an inherent surface roughness. The sizes and forms of roughness depend on manufacturing process. The levels of the surface finish represent the manufacturing cost. It is very important to balance manufacturing cost and performance. Surface finish requirements for different designs have different requirements. Detailed discussions on surface finish and Reynolds number can be found in reference [65, 66]. Loss due to surface finish may be able to represent as wall friction. Wall friction is the function of Reynolds number and can be written as [65]:

$$\frac{1}{\sqrt{f}} = 1.74 - 2\log_{10}\left(\frac{k}{B_2} + \frac{18.7}{Re\sqrt{f}}\right) \tag{24}$$

This equation can be solved by using simple a computer program or spreadsheet.

### 3.6 Diffuser and volute

The fluid leaves impeller with a high velocity and inclines at a large angle to radial direction. It is necessary to decelerate this flow to increase static pressure. Diffusers are used for this purpose. Behind the diffuser a volute or collector is used to decelerate flow further and collect flow.

Designs of diffuser and volute are also critical for compressor. Extensive studies [54, 55, 65] for diffuser and volute designs have been conducted. Many useful suggestions for their design were proposed. Some parameters, for example, solidity of the vaned diffuser, vane deflections, diffuser and volute inlet and exit radii, and volute size were discussed in details in literatures [54, 55, 65]. Good summaries for diffuser designs can be found in references [54, 67].

Figure 30 shows the value of cross-sectional area of the volute channel as a function of the angular location for a modern design [55]. It is noticed that the relationship between cross-sectional area and location angle is non-linear. Experience indicated that this non-linear cross-section volute design benefited both compressor efficiency and operating range.

Recently, compressor designers started to pay attention to volute tongue studies [55]. The volute tongue is located at the intersection of scroll and volute discharge pipe. The volute tongue makes the flow circumferential discontinuity. One of the volute design objectives is to keep small incidences at the tongue area at design condition. However, due to widely operating range of the compressor operation and non-uniformity flow of the upstream, it is difficult to avoid tongue incidence. The non-uniformity of the static pressure produced by a volute tongue creates unbalanced aerodynamic force. This force may cause unacceptable levels of sub

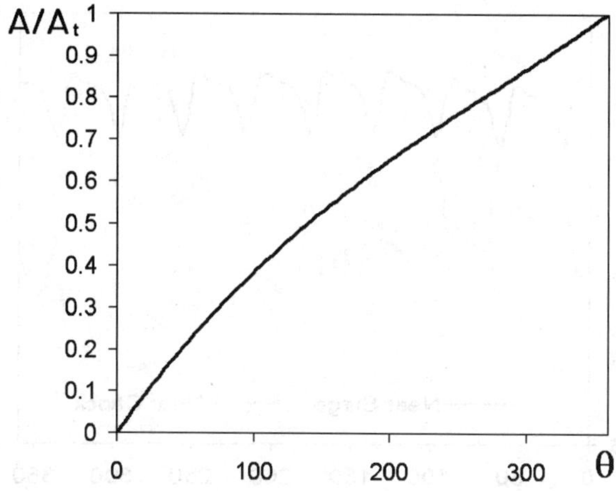


Figure 30: Volute section area distributions versus volute angle.

synchronous radial vibrations of rotor and excessive bearing loads. It is important to reduce or eliminate the static pressure distortions at the exit of the impeller.

Studies were conducted to discuss impeller–diffuser interactions [55]. It was found that rounded tongue could enhance off-design performance and eliminate unbalanced aerodynamic force. Figures 31 and 32 showed computed static pressure distributions at vaned diffuser exit and impeller exit at $\phi/\phi_0 = 0.7$ and 1.25, respectively. Results showed that only small unsymmetrical static pressure could be seen at vaned diffuser exit for both flow conditions. However, pressure distortions were eliminated for both low flow and high flow cases at impeller exit. Compressor tests [55] confirmed computational results. The pressure variations at impeller exit caused by blockages of diffuser vanes were observed. These pressure variations also could be seen at the down stream of diffuser vanes. Rounded tongue changed head coefficients of compressor stages. Different tongue radius has different impacts for compressor head coefficients as shown in Fig. 33.

### 3.7 Discussion

Developments of CFD methods and modern measurement technologies have drawn scientific societies to pay attention to CFD applications, CFD calibrations and optimization design by using CFD [2, 53, 54, 56, 68]. Although one-dimensional mean line and two-dimensional streamline calculations have achieved a certain point of maturity, test data and design experience still need to cooperate into design practices. Industrial compressor designs still rely on and will continue to rely on the one- and two-dimensional design procedures. Design experience data are one of important parts for performing advanced compressor development. It is important that the academic field and industrial side continue to study and share experience to improve centrifugal compressor design process.

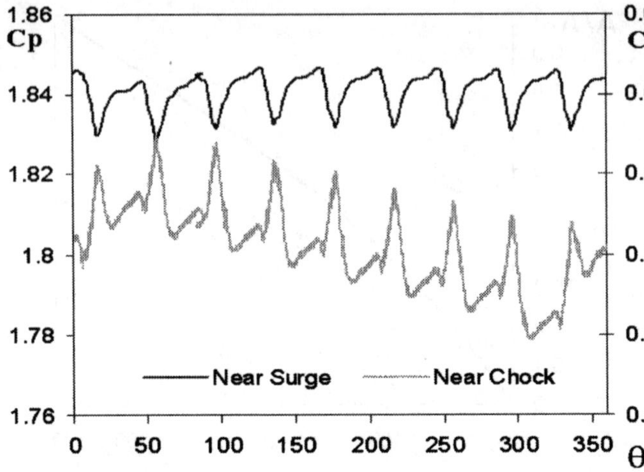


Figure 31: Static pressure coefficients at diffuser exit on the hub wall.

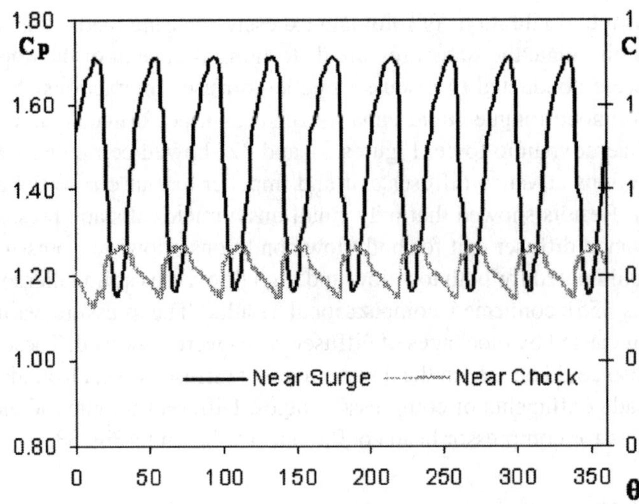


($\phi/\phi_0$ = 0.7 and 1.25)

Figure 32: Static pressure coefficients at impeller exit at the hub wall.

CFD and modern measurement technologies are very important in compressor designs. This paper presented some design experience to accelerate modern compressor design process. The detailed design data presented here are useful for designers to select design parameters. The experience data presented here also provide opportunities for improving design and manufacturing process to develop new compressors better, faster and cheaper.

Current design data are based on metals, steel or titanium components. The advanced materials will definitely play an important role in the future compressor

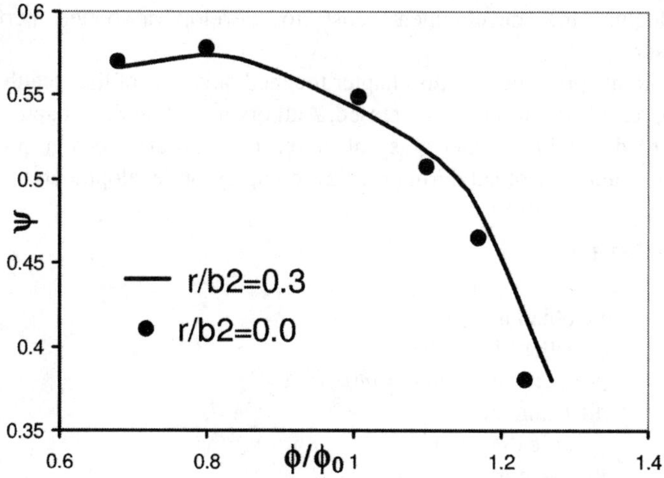


Figure 33: Volute tongue radius versus performance.

development. The possibilities for carbon composites, ceramic matrix composites and advanced matrix material will provide higher strength at low cost. Some design considerations will change with the progress of material science.

# 4 Summary

The design of the advanced compressors has been practiced with increasing degrees of sophistication while continuously advancing levels of efficiency and operating regions. It is very important to understand the compressor design process and optimization and learn from basic design experience. This section covered the basic compressor design process, compressor blade optimization and some design experience. The compressor development and design have not reached a level of full maturity. The studies of new design methods and new technologies will continue to draw the attention of scientists and research engineers.

The design optimization process discussed here is one of the many methods. With the progress of the numerical method and computer hardware, the current optimization method may become more mature or may replace by more sophistic methods. However, the method presented here can serve as reference for blade optimization. The present method was proved to have advantage than traditional method and easy to apply in the design process.

Due to the limitation of the computer CPU speed and memory limitation, full three-dimensional turbulent codes are still time consuming for using in the industrial compressor design and optimization. Major industrial compressor design systems are still reliable to two- and three-dimensional code with correlations from tests and experience. Basic experience and fundamental choice concerning the compressor design are very important for new compressor development. Basic parameter selections based on past design experience can accelerate the new compressor design. The design considerations also addressed the method for improving the manufacturability and reducing the development cost to develop a better performance compressor.

The contents presented in this chapter touched only part of the wealth of issues that compressor development addressed. Authors hope that this chapter will provide the readers with a better grasp of the basic compressor design process and design experience involved during the new compressor development.

## Nomenclature

| | |
|---|---|
| $A$ | Jacobian matrix |
| $B_2$, $b_2$ | impeller tip width |
| $C_p$ | pressure coefficient = $p/0.5\rho_{d2}U_2^2$ |
| DF | diffusion factor |
| $D_s$ | specific diameter |
| $E_x$ | flux vector |
| $F_y$ | flux vector |
| $F$ | function |
| $f$ | friction factor |
| $H$ | head |
| $K$ | sand grain roughness |
| $L$ | blade meridinial length |
| $r$ | RMS radius |
| $Re$ | Reynolds number |
| ICFM | compressor inlet flow in $ft^3$/min |
| $M$ | Mach number |
| $MR$ | Mach number ratio |
| $Q$ | volume metric flow rate |
| RG | operating range = $(Q_c - Q_s)/Q_c$ |
| RMS | root mean square |
| $R_x$ | flux vector |
| $S_y$ | flux vector |
| $\Delta t$ | Time step |
| $u$, $v$ | velocities |
| $V$ | velocity vector |
| $W$ | RMS relative velocity |

### Greek symbols

| | |
|---|---|
| $\Lambda_E$, $\Lambda_F$ | matrix-valued dissipation terms |
| $\Omega$ | angular velocity |
| $\mu$ | viscosity |
| $\rho_A$ | spectral radius of the Jacobian matrix $A$ |
| $\varphi$ | flow coefficient |
| $\eta$ | adiabatic efficiency (total to static) |
| $\psi$ | head coefficient |

**Subscripts**

| | |
|---|---|
| 1, 2 | impeller inlet and exit |
| C | chock or clearance |
| l | loss |
| O | zero clearance |
| ref | reference |
| S | surge |

## References

[1] Yiu, K.F.C. & Zangeneh, M., Three-dimensional automatic optimization method for turbo machinery blade design. *Journal of Propulsion and Power*, **16(6)**, pp. 1174–1181, 2000.

[2] Xu, C. & Amano, R.S., A turbo machinery blade design and optimization procedure, GT-2002-30541, 2002.

[3] Benini, E. & Toffolo, A., Development of high-performance airfoils for axial flow compressor using evolutionary computation. *Journal of Propulsion and Power*, **18(3)**, pp. 544–554, 2002.

[4] Oksuz, O., Akmandor, I. & Kasaoglu, M., Aerodynamic optimization of turbo machinery cascades using Euler/boundary-layer coupled genetic algorithms. *Journal of Propulsion and Power*, **18(3)**, pp. 652–657, 2002.

[5] De Meulenaere, A. & Van Den Braembussche, R.A., Three-dimensional inverse method for turbomachinery blading design. *Journal of Turbomachinery*, **120(2)**, pp. 247–255, 1998.

[6] Thygesen, R., Optimization of return channel blades for a radial compressor, VKI, June 2000.

[7] Xu, C. & Amano, R.S., Computational analysis of pitch-width effects on the secondary flows of turbine blades. *Computational Mechanics*, **34(2)**, pp. 111–120, 2004.

[8] Xu, C. & Amano, R.S., Development of a low flow coefficient centrifugal compressor, IGTI, GT2005-68006, 2005.

[9] Xu, C. & Amano, R.S., Static pressure distortion of a large cut back tongue volute, IGTI, GT2006-90001, 2006.

[10] Krain, H., Review of centrifugal compressor's application and development, IGTI, GT2003-38971, 2003.

[11] Griffith, A.A., *An Aerodynamic Theory of Turbine Design*, RAE, H. Report 11, July 7, 1926.

[12] Hamrick, J.T., Ginsburg, A., Osborn, W., Method of analysis for compressible flow through mixed flow centrifugal impellers of arbitrary design, Lewis Flight Laboratory, Cleveland, Ohio, 1950.

[13] Wu, C.H., A general theory of three-dimensional flow in subsonic and supersonic turbomachines of axial-, radial- and mixed flow types, NASA TN D-2546, 1964.

[14] Bosman, C. & El-Shaarawi, M.A.I., Quasi-three-dimensional numerical solution of flow in turbomachines. *Journal of Fluids Engineering*, pp. 132–140, 1977.

[15] Moor, J. & Moor, J.G., A calculation procedure for 3D viscous compressible duct flows. *Journal of Fluids Engineering*, **101**, pp. 415–428, 1979.

[16] Hah, C.A., Navier-Stokes analysis of 3D turbulent flows inside turbine blade rows at design and off-design condition, ASME paper 83-GT-34, 1983.

[17] Dawes, W.N., Development of a 3D Navier-Stokes solver for application to all types of turbomachinery, ASME paper 88-GT-70, 1988.

[18] Hah, C., Bryans, A.C., Moussa, Z. & Tomsho, M.E., Application of viscous flow computations for the aerodynamic performance of a back-swept impeller at various operating conditions, ASME paper 88-GT-39, 1988.

[19] Pinarbasi, A. & Johnson, M.W., Off-design flow measurements in a centrifugal compressor vaneless diffuser. *Journal of Turbomachinery*, **117**, pp. 602–608, 1995.

[20] Xu, C., Numerical and experimental investigation of separated flow in 3D bend duct. MS Thesis, Nanjing University of Aeronautics and Astronautics, 1992.

[21] Yeung, W.W.H., Xu, C. & Gu, W., Reduction of transient adverse effects of spoilers. *Journal of Aircraft*, **34(4)**, pp. 479–484, 1997.

[22] Yeung, W.W.H. & Xu, C., Separated flows around spoilers and forward-facing flaps. *Journal of Aircraft*, **35**, pp. 53–59, 1998.

[23] Xu, C. & Amano, R.S., Unsteady pressure field investigation of an axial fan-blade unsteady pressure field measurement. *International Journal of Rotating Machinery*, **8**, pp. 385–395, 2002.

[24] Xu, C. Amano, R.S. & Perez, R., Unsteady pressure field investigation of an axial fan-inlet and outlet unsteady pressure field measurement. *International Journal of Rotating Machinery*, **8**, pp. 375–383, 2002.

[25] Xu, C., Amano, R.S. & Lee, K.E., Investigation of an axial fan – blade stress and vibration due to aerodynamic pressure field and centrifugal effects. *JSME International Journal, Series B*, **47**, pp. 75–90, 2004.

[26] Wernet, M.P., Bright, M.M. & Skoch, G.J., An investigation of surge in a high-speed centrifugal compressor using digital PIV. *Journal of Turbomachinery*, **123(2)**, pp. 418–428, 2001.

[27] Xu, C. & Amano, R.S., A hybrid numerical procedure for cascade flow analysis. *Numerical Heat Transfer*, **37(2)**, pp. 141–164, 2000.

[28] Xu, C. & Amano, R.S., Flux-splitting finite volume method for turbine and heat transfer analysis. *Computational Mechanics*, **27(2)**, pp. 119–127, 2001.

[29] Benini, E. & Toffolo, A., Centrifugal compressor of a 100 KW microturbine: Part3 – optimization of diffuser apparatus, GT-2003-38154, 2003.

[30] Weigl, H.J., Paduano, J.D., Fréchette, L.G., Epstein, A.H., Greitzer, E.M., Bright, M.M. & Strazisar, A.J., Active stabilization of rotating stall and surge in a transonic single stage axial compressor *Journel of Turbomachinery*, **120(4)**, pp. 625–636, 1997.

[31] Weingold, H.D., Neubert, R.J., Behlke, R.F. & Potter, G.E., Bowed stators: an example of CFD applied to improve multistage compressor efficiency. *Journal of Turbomachinery*, **119(2)**, pp. 161–168, 1997.

[32] Sasaki, T. & Breugelmans, F., Comparison of sweep and dihedral effects on compressor cascade performance, ASME paper 97-GT-2, 1997.

[33] Kawagishi, H. & Kawasaki, S., The effect of nozzle lean on turbine efficiency. *Proceedings of ASME Int. Conf. on Joint Power Generation*, October, 1991.

[34] Wellborn, S.R. & Delaney, R.A., Redesign of a 12-stage axial-flow compressor using multistage CFD, ASME paper 2001-GT-345, 2001.

[35] Singh, G., Walker, P.J. & Haller, B.R., Development of 3D stage viscous time marching method for optimisation of short stage heights. *Proc. of European Conf. on Turbo Machinery*, Erlangen, 1995.

[36] Wallis, A.M. & Denton, J.D., Comparison of design intent and experimental measurements in a low aspect ratio axial flow turbine with three-dimensional blading, ASME paper 98-GT-516, 1998.

[37] Xu, C. & Amano, R.S., Computational analysis of pitch width effects on the secondary flows of turbine blades, ASME IGTI Turbo & Expo GT-2002-30200, 2002.

[38] Arnone, A. & Swanson, R.C., A Navier-Stokes solver for turbomachinery applications. *Journal of Turbomachinery*, **115(2)**, pp. 1556–1563, 1993.

[39] Chima, R.V., Explicit multigrid algorithm for quasi-three-dimensional viscous flows in turbomachinery. *Journal of Propulsion and Power*, **3(5)**, pp. 397–405, 1987.

[40] Inoue, M. & Furukawa, M., Artificial dissipative and upwind schemes for turbomachinery blade flow calculations. *Numerical Methods for Flow Calculation in Turbomachines*, VKI Lecture Series, 1994-06, 1994.

[41] Xu, C. & Amano, R.S., An implicit scheme for cascade flow and heat transfer analysis. *Journal of Turbomachinery*, **122(2)**, pp. 294–300, 2000.

[42] Turkel, E. & Vatsa, V.N., Effect of artificial viscosity on three-dimensional flow solutions. *AIAA Journal*, **32(1)**, pp. 39–45, 1994.

[43] Lin, H., Yang, D.Y. & Chieng, C.C., Variants of biconjugate-gradient method for compressible Navier-Stokes solver. *AIAA Journal*, **33(7)**, pp. 1177–1184, 1995.

[44] Li, H., Chen, S. & Martin, H.F., Two dimensional viscous blade flow code using shifted periodic grids, ASME paper 97-GT-516, 1997.

[45] Xu, C., Kutta condition for sharp edge flows. *Mechanics Research Communications*, **25(4)**, pp. 415–420, 1998.

[46] Vanderplaats, G.N., *Numerical Optimization Techniques for Engineering Design*, McGraw-Hill, 1984.

[47] Chung, J., Shim, J. & Lee, K.D., Shape optimization of high-speed axial compressor blades using 3D Navier-Stokes flow physics, ASME paper 2001-GT-0594, 2001.

[48] Singh, M. & Vargo, J.J., Reliability evaluation of shrouded blading using the SAFE interference diagram. *Journal of Engineering for Gas Turbines and Power*, **111(4)**, pp. 601–609, 1989.

[49] Kenny, D.P., The history and future of the centrifugal compressor in aviation gas turbine, SAE/SP-804/602, 1984.

[50] Collopy, P.D., 1997, Surplus value in propulsion system design optimization, AIAA-97-3159.

[51] Japikse, D., 2000, *Decisive Factors in Advanced Centrifugal Compressor Design and Development*. Presented at the International Mechanical Engineering Congress & Exposition (IMechE), November 2000.

[52] Rodgers, C. & Sapiro, L., Design considerations for high-pressure ratio centrifugal compressor, 72-GT-71, 1972.

[53] Krain, H., Hoffmann, B. & Pak, H., Aerodynamics of a centrifugal compressor impeller with transonic inlet conditions, ASME paper 95-GT-79, 1995.

[54] Japikse, D., Centrifugal compressor design and performance, Concepts ETI, Inc, VT, 1996.

[55] Xu, C. & Muller, M., Development and design of a centrifugal compressor volute. *International Journal of Rotating Machinery*, **2005(3)**, pp. 190–196, 2005.

[56] Krain, H., Flow in centrifugal compressor, von Karman Institute for Fluid Dynamics, Ls 1984–07, 1984.

[57] Van den Braembussche, R., Rotating stall in centrifugal compressors, VKI Preprint 1987–16, 1987.

[58] ASME, Performance Test Code on Compressors and Exhausters, PTC 10-97, 1997.

[59] Xu, C., A practical model for uncertainty evaluation in force measurements. *Measurement Science and Technology*, **9(11)**, pp. 1831–1836, 1998.

[60] Balje, O.E., *Turbomachines*, John Wiley and Sons: New York, NY, 1981.

[61] Schoonmaker, P.M., Preliminary experience with an expert system providing initial centrifugal compressor sizing for performance prediction and analysis, 91-GT-28, 1991.

[62] Rodgers, C., A diffusion factor correlation for centrifugal impeller stalling, 78-GT-61, 1978.

[63] Brun, K. & Kurz, R., Analysis of secondary flows in centrifugal impellers. *International Journal of Rotating Machinery*, **2005(1)**, pp. 45–52, 2005.

[64] Street, R.L., Watters, G.Z. & Vennard, J.K., *Elementary Fluid Mechanics*, John Willey & Sons, 1996.

[65] Wright, T., Comments on compressor efficiency scaling with Reynolds number and relative roughness, 89-GT-31, 1989.

[66] Childs, P.R.N. & Noronha, M.B., The impact of machining techniques on centrifugal compressor impeller performance, 97-GT-456, 1997.

[67] D'orsi, N.C. & Japikse, D., Advanced diffuser performance for centrifugal compressors and pumps, Tech. Mem, No. 247, Concepts ETI, Inc, Norwich, VT, 1994.

[68] Pazzi, S., Martelli, F., Michelassi, V. & Giachi, M., The use of artificial neural networks for performance predication of return channels for industrial centrifugal compressors, ASME paper GT-2002-30392, 2002.

# CHAPTER 10

# Advances in understanding the flow in a centrifugal compressor impeller and improved design

A. Engeda
*Turbomachinery Lab, Michigan State University, USA.*

## Abstract

The last 60 years have seen a very high number of experimental and theoretical studies of the centrifugal impeller flow physics at government, industry and university levels, which have been extensively documented. As Robert Dean, one of the well-known impeller aerodynamists stated, "The centrifugal impeller is probably the most complex fluid machine built by man". Despite this, it is still the widest used turbomachinery and continues to be a major research and development topic. Computational fluid dynamics has now matured to the point where it is widely accepted as a key tool for aerodynamic analysis. Today, with the power of modern computers, steady-state solutions are carried out on a routine basis, and can be considered as part of the design process. The complete design of the impeller requires a detailed understanding of the flow in the impeller and aerodynamic analysis of the flow path and structural analysis of the impeller including the blades and the hub. This chapter discusses the developments in the understanding of the flow in a centrifugal impeller and the contributions of this knowledge towards better and advanced impeller designs.

## 1 Introduction

Centrifugal compressors have the widest compressor application area. They are reliable, compact, and robust; they have better resistance to foreign object damage; and are less affected by performance degradation due to fouling. They are found in small gas turbine engines, turbochargers, and refrigeration chillers and are used extensively in the petrochemical and process industry. The centrifugal compressor finds a wide variety of applications, with each application placing its own demands on the design of the compressor. Given this spectrum of application, it is not surprising that the centrifugal compressor continues to command a great deal of attention, both from compressor designers and from those engaged in understanding the underlying engineering science.

Industrial applications require stage pressure ratios of 1.5:1 or even less, whereas helicopter gas turbines require the highest pressure ratios, with 6:1 being commonplace and the latest engines demanding 10:1 or more from a single compressor stage. Table 1 and Fig. 1 summarize the application areas of turbocompressors.

The wide range of demands on centrifugal compressors brings many design considerations. Most of the design requirements need solutions to two major problems: stress and aerodynamics. The stress problems are caused by the material strength limitations and the capability to accurately predict blade and impeller steady state and vibrational stress for complex impeller shapes and at high rotational speeds. The aerodynamic problem is to efficiently accomplish large air deflections and diffusion at high flow velocity, with the added difficulty of very small passage flow areas required to get good efficiency and high pressure ratio.

Table 1: Application areas of turbocompressors.

| Types of compressors | Pressure ratio | | | Efficiency | Operating range |
|---|---|---|---|---|---|
| | Industrial | Aerospace | Research | | |
| Positive displacement | Up to 30 | – | – | 78–82% | – |
| Centrifugal | 1.2–1.9 | 2.0–7.0 | 13 | 75–87% | Large 25% |
| Axial | 1.05–1.3 | 1.1–1.45 | 2.1 | 80–91% | Narrow 3–10% |

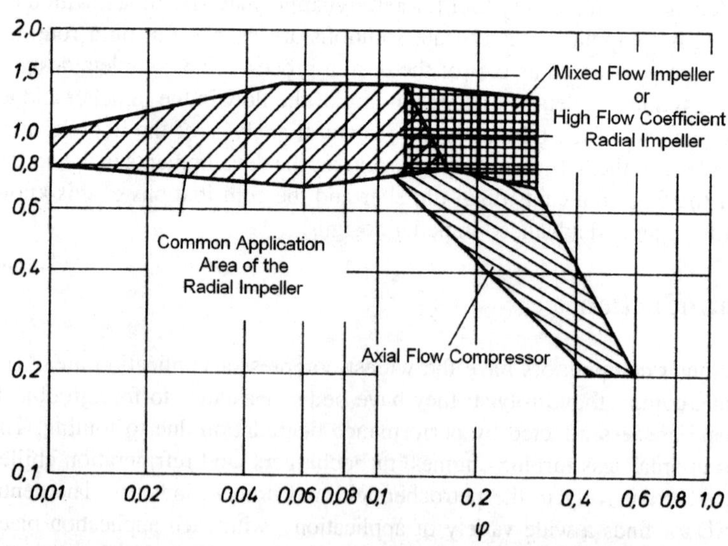


Figure 1: Application areas of turbocompressors.

Based on one-dimensional (1D) streamline theory, the aerodynamic design of the centrifugal impeller has significantly advanced in the last 60 years simply by using similitude, dimensional analysis, empirical knowledge, and control volume analysis.

Most early codes were limited in their applicability in that they were only two-dimensional (2D), or in three dimensions capable of analyzing a single bladed passage or one bladed element. Analyses could include stationary or rotating elements but not both. The loss and/or turbulence models built into the codes were inadequate to properly model the complex flows encountered in centrifugal compressors. The codes were limited to relatively simple geometric and topological (mesh) configurations. Finally, solutions provided were steady state or time averaged, and unsteady effects were not usually considered. As user demands grew for improved performance and more reliable equipment and as compressor vendors sought improved analytical and design methodologies, the application of computational fluid dynamics (CFD) in the industrial world became a necessity.

Figure 2 shows typical centrifugal compressor impellers. As already mentioned, CFD has been used extensively over the past three decades to predict the performance of compressors for both single and multiple blade row configurations. As CFD algorithms and software have continued to be developed and refined, it remains essential that validation studies be conducted to ensure that the results are both sufficiently accurate and can be obtained in a robust and predictable manner.

The complete design of the impeller requires aerodynamic analysis of the flow path and structural analysis of the impeller including the blades and hub. A typical impeller design procedure follows the pattern of specifying blade and hub geometry, performing aerodynamic and structural analysis, and iterating on geometry until acceptable aerodynamic and structural criteria are achieved. This requires the geometry generation to focus not only on blade shape but also on hub geometry. Thus, for interactive blade design to be an effective design tool, iterations in blade geometry, aerodynamic analysis, and structural analysis must be completed in a timely manner. This requires an integrated design system that is easy to use and capable of controlling the complete rotor geometry, and aerodynamic and structural analysis. To develop such a design system, it is critical that the impeller geometry generation procedure be clearly understood.

Modern solvers use robust schemes, which can accurately predict compressible and viscous flows around and in complex geometries. Current impeller design systems integrate geometry generation with aerodynamic analysis. Such design systems allow for graphical geometry manipulation of the impeller cross-section shapes, area diffusion and distributions, and outline connections. Aerodynamic analysis is performed using CFD. The CFD mesh node spacing and boundary conditions are automatically generated from within such design system, thus eliminating the need to manually setup third party mesh generators. Program output includes information for CAD and 3D viewing of the impeller. Currently two design approaches are practiced using CAD and CFD interactions:

- CAD system approach
- Design system approach

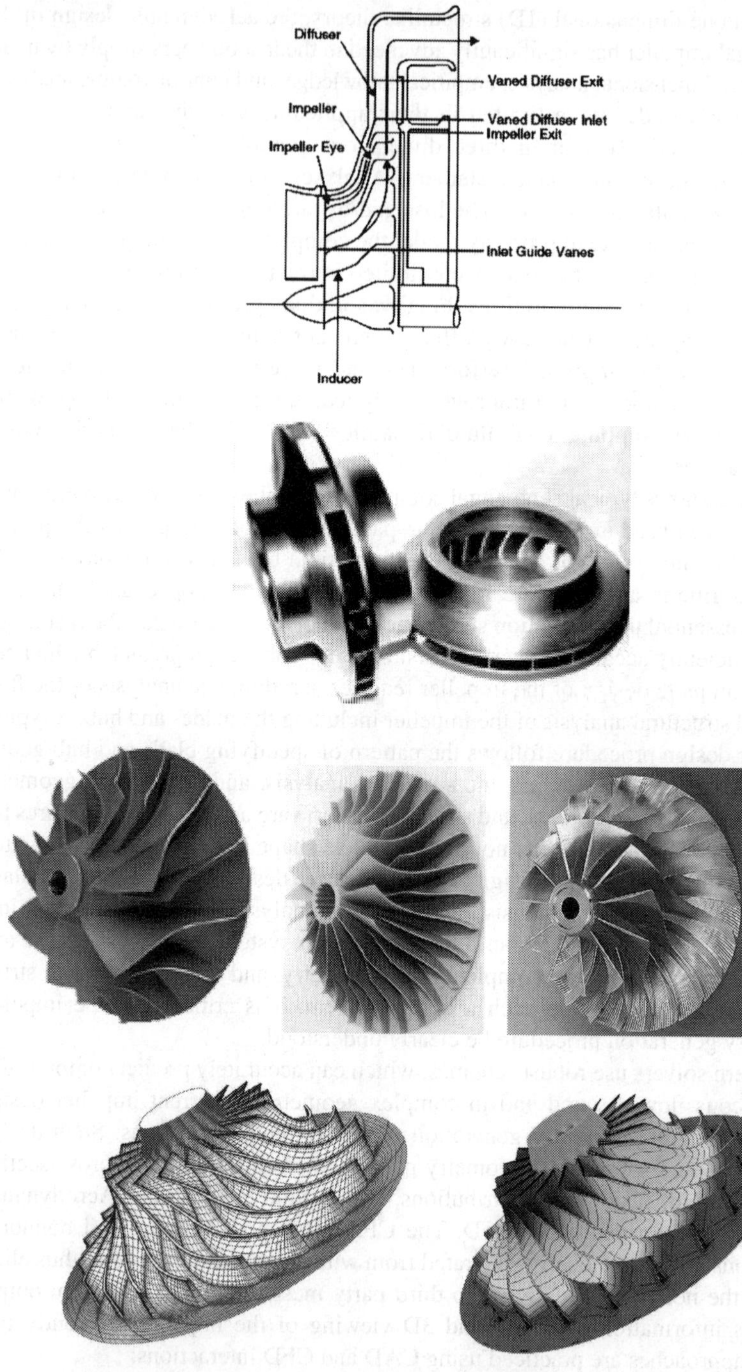


Figure 2: Examples of centrifugal impellers and modeling.

## 2 The historical development of the centrifugal impeller

Prior to 1800, the centrifugal compressor was able to deliver only 1 m head of water. It was Rateau [1], in 1899, who first showed that the centrifugal compressor can develop higher pressure and efficiency than ever thought. He was able to show that a centrifugal compressor impeller can achieve a tip speed of 250 m/s or higher. He first designed and tested a centrifugal compressor with an impeller of 250 mm diameter. This compressor developed 5.8 m head of water and 2000 $m^3$/h inlet flow while operating at 264 m/s tip speed and 20,200 rpm rotational speed. Following the success of this compressor, Rateau [1] designed two more centrifugal compressors. The first one was in 1903 for the steel industry; it developed 3.4 m head of water and 10,000 $m^3$/h inlet flow while operating at 14,500 rpm rotational speed. The second was a similar compressor in 1904 for a sugar refinery.

Just immediately before and after 1900, the major users of gas compressors were the steel and mining industries. As these industries grew, their demand for compressors with higher pressures increased. To respond to this need for higher pressure, Rateau [1] followed the experience from centrifugal pumps and designed the first multistage centrifugal compressor. The first known multistage centrifugal compressor was built in 1905 with five stages on the same shaft. It was designed for 2500 $m^3$/h inlet flow, 4 m head of water and 4500 rpm rotational speed.

About the same time as Rateau [1], Parsons [2] was investigating axial compressors for gas turbine applications but results showed that this compressor was poor in terms of efficiency and stability. Parsons [2] investigated some 30 different compressors and stopped the work with dissatisfaction. It took several decades afterwards to discover the superior qualities of the axial compressor. Following Rateau's [1] success in the period 1905–1910, many compressor manufacturing companies licensed Rateau's [2] patent. In the period 1910–1920, due to increased industrial application areas of the centrifugal compressor, the impeller designer focused more towards producing higher flow capacity, efficiency, and tip speed.

Since the beginning of the 20th century the theory of hydrodynamics and aerodynamics of centrifugal pumps and compressors has developed slowly. The major initial task was to establish the relationship between the performance parameters of the impeller, the form of the flow, and the form of the fluid passages. The most important of these relationships was that between the torque transmitted from the impeller to the fluid and the change of thermo-fluid properties of the fluid as it passes through the impeller, which is commonly known as the Euler equation. However, there were many intrinsic functions of the impeller, which went beyond the Euler equation and yet have equally fundamental significance.

Early in the 20th century, the centrifugal impeller designer developed a moderately efficient pump or compressor by considering only gross effects using 1D theory and then resorting to empirical development based on many years of experience in order to obtain acceptable designs, without paying much attention to the impeller flow details.

The period 1910–1930 saw the first intense research in centrifugal impellers both experimental and theoretical to establish the flow mechanism of the centrifugal impeller. Using the ideal assumption of a 2D irrotational flow of an inviscid and incompressible flow in the impeller, notably Busemann [3], Kucharski [4], König [5], Spannhake [6] and others attempted theoretical analysis of the flow. Stodola [7], Thoma, Kearton, Escher Wyss, Kucharski, Grunagel, and Stiess [1] experimented to gain an understanding of the flow mechanism in a centrifugal impeller. They were able to observe secondary flows but, owing to their lack of adequate measuring and analytical tools, they were unable to extend their observation with quantitative structure.

The first comprehensive study of impeller flow is that of Fischer and Thoma [8], where they showed that practically all flow conditions for a real fluid were actually different from those theoretically derived for an ideal frictionless fluid. The actual velocity distribution along circles concentric with the axis of rotation of the impeller was not the same as generally assumed. The relative velocity along the high pressure side of the blade was greater than the theoretical value, and reverse flow along the pressure side of the blade did not take place even at low discharges.

## 3 The centrifugal compressor stage

As shown in Fig. 3, a simple centrifugal compressor stage consists of four basic components:

- inlet,
- impeller,
- vaneless or vaned diffuser, and
- collector or a volute.

The fluid is drawn in through the inlet into the eye of the impeller parallel to the axis of rotation. To add angular momentum, the impeller whirls the fluid outward and turns it into a direction perpendicular to the rotation axis.

As a result, the energy level is increased, resulting in both higher pressure and velocity. The purpose of the following diffuser is to convert some of the kinetic energy of the fluid into static pressure. Outside the diffuser is a scroll or volute whose function is to collect the flow from the diffuser and deliver it to the discharge pipe. It is possible to gain a further deceleration and thereby an additional pressure rises within a volute.

An example of the contribution of each component of the compressor is shown in Fig. 4. The solid line represents the static pressure rise while the dotted line indicates the total pressure change for the individual component in a single stage of centrifugal compressor.

As a result, the energy level is increased, resulting in both higher pressure and velocity. The purpose of the following diffuser is to convert some of the kinetic energy of the fluid into static pressure. Outside the diffuser is a scroll or volute whose function is to collect the flow from the diffuser and deliver it to the discharge pipe. It is possible to gain a further deceleration and thereby an additional

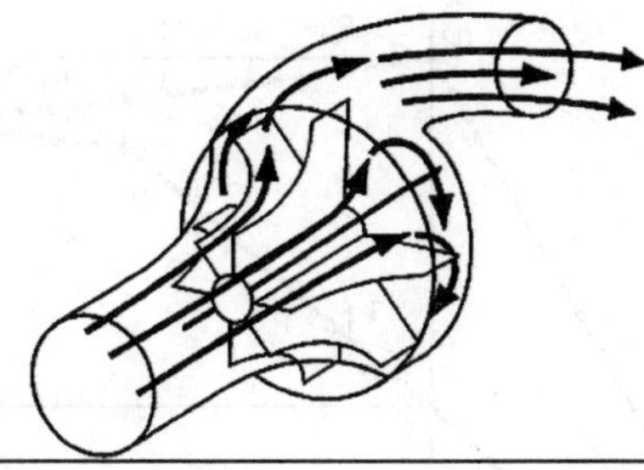

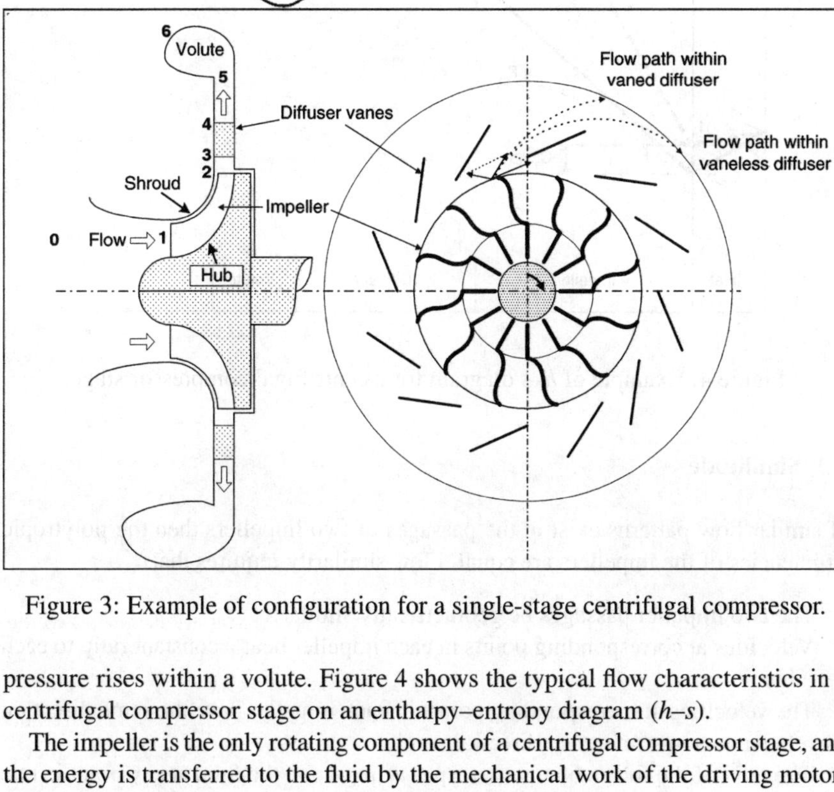


Figure 3: Example of configuration for a single-stage centrifugal compressor.

pressure rises within a volute. Figure 4 shows the typical flow characteristics in a centrifugal compressor stage on an enthalpy–entropy diagram (*h*–*s*).

The impeller is the only rotating component of a centrifugal compressor stage, and the energy is transferred to the fluid by the mechanical work of the driving motor.

## 4 Similitude, dimensional analysis and control volume analysis

Based on 1D streamline theory, the aerodynamic design of the centrifugal impeller has significantly advanced in the last 60 years without knowledge about the detailed internal flow in the impeller. This practice has led to very successful impeller designs simply by using:

- Similitude,
- Dimensional analysis,
- Empirical knowledge, and
- Control volume analysis.

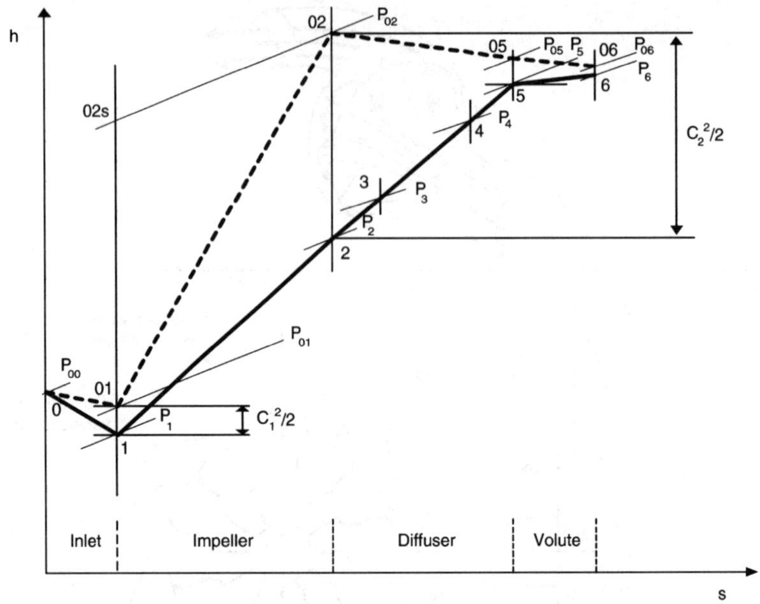


Figure 4: Example of *h–s* diagram for a centrifugal compressor stage.

## 4.1 Similitude

If similar flow patterns exist in the passages of two impellers then the polytropic efficiencies of the impellers are equal. Flow similarity requires that:

- The two impeller passages be geometrically similar;
- Velocities at corresponding points in each impeller bear a constant ratio to each other;
- The velocities at corresponding points in each impeller have the same direction relative to some reference direction; and
- The ratio of the forces of any two types acting upon the fluid at any point in one impeller equals the ratio of the same types of force acting on the fluid at the corresponding point in the second impeller.

## 4.2 Dimensional analysis

Using dimensional analysis, the characteristics of the impeller can be illustrated using dimensionless parameters involving all relevant variables. For impeller diameter *(D)*, inlet stagnation pressure ($P_{01}$), inlet stagnation temperature ($T_{01}$), mass flow rate ($\dot{m}$), rotational speed (*N*), fluid properties: the universal gas constant (*R*), kinematic viscosity ($v$), and the ratio of specific heats ($\gamma$), the performance of the impeller, may be represented by:

$$P_{02} = f\left(\dot{m}, P_{01}, R, T_{01}, \gamma, N, v, D\right)$$

$$\eta = f\left(\dot{m}, P_{01}, R, T_{01}, \gamma, N, \nu, D\right)$$

The mass flow rate (ṁ), the efficiency ($\eta$) and the stagnation temperature rise $\Delta T_{01} = T_{02} - T_{01}$ of an impeller can be expressed as functions of the independent variables:

$$\dot{m}\,;\eta : \Delta T_0 = f\left(P_{01}, P_{02}, R, T_{01}, \gamma, N, \nu, D\right)$$

These can be reduced to a set of non-dimensional groups:

$$\frac{\dot{m}\sqrt{RT_{01}}}{P_{01}D^2}; \eta; \frac{\Delta T_0}{T_{01}} = f\left(\frac{ND}{\sqrt{\gamma RT_{01}}}, \frac{P_{02}}{P_{01}}, \frac{\dot{m}}{\nu D}, \gamma\right)$$

Since the impeller operates on a specific gas, the values of $R$ and $\gamma$ are specified. Hence the non-dimensional groups become:

$$\frac{\dot{m}\sqrt{RT_{01}}}{P_{01}D^2}; \eta; \frac{\Delta T_0}{T_{01}} = f\left(\frac{ND}{\sqrt{\gamma RT_{01}}}, \frac{P_{02}}{P_{01}}, \frac{\dot{m}}{\nu D}\right)$$

The Reynolds number of the gas has little effect on the performance of the machine, and can usually be ignored. Hence:

$$\frac{\dot{m}\sqrt{RT_{01}}}{P_{01}D^2}; \eta; \frac{\Delta T_0}{T_{01}} = f\left(\frac{ND}{\sqrt{\gamma RT_{01}}}, \frac{P_{02}}{P_{01}}\right)$$

For the impeller efficiency a relationship between $\eta$, $\Delta T_0/T_{01}$, and $P_{02}/P_{01}$ can be established out of the definition of isentropic efficiency as:

$$\eta_{\text{imp-}TT} = \frac{\left(P_{02}/P_{01}\right)^{(\gamma-1/\gamma)} - 1}{\left(\Delta T_0 / T_{01}\right)}$$

For a particular machine, the diameter ($D$) is constant and therefore may be ignored. Thus, the complete performance of the compressor may be represented by the relationships (for a particular fluid, such as air)

$$\frac{\dot{m}\sqrt{T_{01}}}{P_{01}}; \eta = f\left(\frac{N}{\sqrt{T_{01}}}, \frac{P_{02}}{P_{01}}\right)$$

$$\frac{\dot{m}\sqrt{T_{01}}}{P_{01}}; \frac{\Delta T_0}{T_{01}} = f\left(\frac{N}{\sqrt{T_{01}}}, \frac{P_{02}}{P_{01}}\right)$$

By plotting $\eta$ and $P_{02}/P_{01}$ against the mass flow parameter $\dot{m}\sqrt{T_{01}}/P_{01}$ for a series of values of $N/\sqrt{T_{01}}$, the complete performance of the machine is determined.

### 4.3 Control volume analysis

Using a control volume analysis (Fig. 5), for any $N$ total amount of some property: mass, energy or momentum, the $n$ per unit mass amount of $N$ is expressed as:

$$N = n\rho dv$$

The time rate of change of N, can be expressed as:

$$\frac{dN}{dt} = \frac{\partial}{\partial t}\int_{cv} n\rho dv + \int_{cs} n\rho\bar{C}\cdot d\bar{A}$$

The equation states that the time rate of change $N$ is equal to the time rate change of the property within the control volume plus the net rate of efflux of $N$ across the boundary. For $N$ is momentum:

$$N = m\cdot\bar{C} = n\cdot\rho\cdot dv \qquad n = \frac{m\cdot\bar{C}}{\rho\cdot dv} = \bar{C}$$

$$\sum\bar{F} = \frac{d\left(m\bar{C}\right)}{dt} = \frac{\partial}{\partial t}\int_{cv}\rho\bar{C}dv + \int_{cs}\rho\bar{C}\bar{C}\cdot d\bar{A}$$

Applying this equation for the momentum component in the $x$ direction:

$$\sum F_x = \frac{\partial}{\partial t}\int_{cv}\rho C_x dv + \int_{cs}\rho C_x\bar{C}\cdot d\bar{A}$$

Applying this for the torque exerted by any force on the control volume, in this case an impeller and for steady state case:

$$T_{z-z} = \bar{r}x\bar{F} = \frac{\partial}{\partial t}\int_{cv}(\rho\bar{r})x\left(\bar{C}dv\right) + \int_{cs}\left(\rho\bar{r}x\bar{C}\right)\left(\bar{C}\cdot d\bar{A}\right)$$

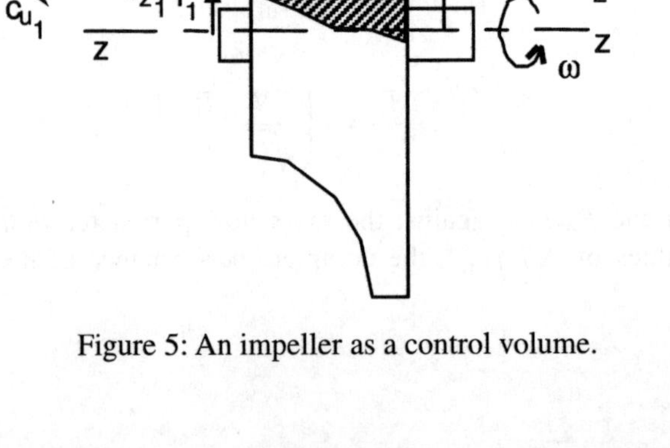


Figure 5: An impeller as a control volume.

$$T_{z-z} = \int_{A2} (r_2 C_{u2})(\rho_2 C_{r2} dA) - \int_{A1} (r_1 C_{u1})(\rho_1 C_{r1} dA)$$

$$T = \dot{m}\left(r_2 C_{u2} - r_1 C_{u1}\right)$$

The impeller is the rotating component of the centrifugal compressor stage, where energy transfer of the compressor stage occurs, stagnation enthalpy change across the impeller ($\Delta h_{0\text{-imp}}$).

$$\Delta h_{0\text{-imp}} = h_{02} - h_{01}$$

Thus, the overall energy transfer is given by:

$$\dot{m}\ \Delta h_{0\text{-imp}} = -\dot{W} = \omega T = \dot{m}\ \left(U_2 C_{u2} - U_1 C_{u1}\right)$$

From velocity triangles (Fig. 6) at impeller exit and inlet, it can be shown:

$$U_2 C_{u2} = \tfrac{1}{2}\left(U_2^2 + C_2^2 - W_2^2\right) \quad \text{and} \quad U_1 C_{u1} = \tfrac{1}{2}\left(U_1^2 + C_1^2 - W_1^2\right)$$

Thus,

$$\Delta h_{oc} = \tfrac{1}{2}\left[\left(U_2^2 - U_1^2\right) + \left(W_1^2 - W_2^2\right) + \left(C_2^2 - C_1^2\right)\right]$$

The sum of the first and the second terms on the right hand side represents the increase in static pressure and the kinetic energy increase is shown in the last term. In an axial compressor machine, there is no impeller tip speed variation ($U_2 = U_1$constant) explaining a higher enthalpy rise in a radial compressor. If the impeller inlet flow is purely axial (i.e. $C_{u1} = 0$) the theoretical total enthalpy reduces to:

$$\Delta h_{0c} = \frac{\dot{W}}{\dot{m}} = U_2 C_{u2}$$

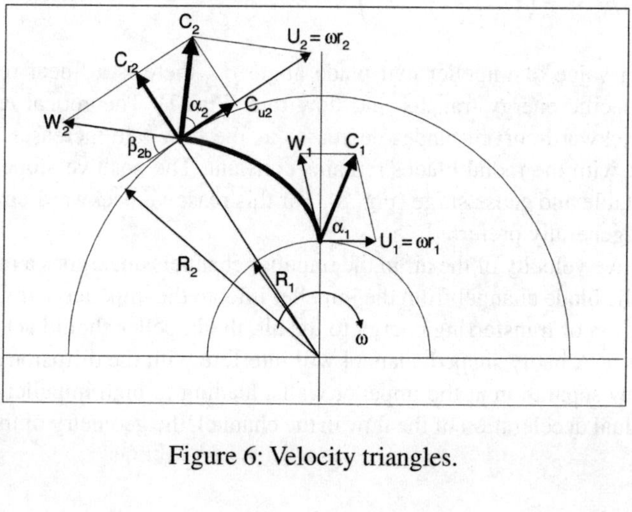


Figure 6: Velocity triangles.

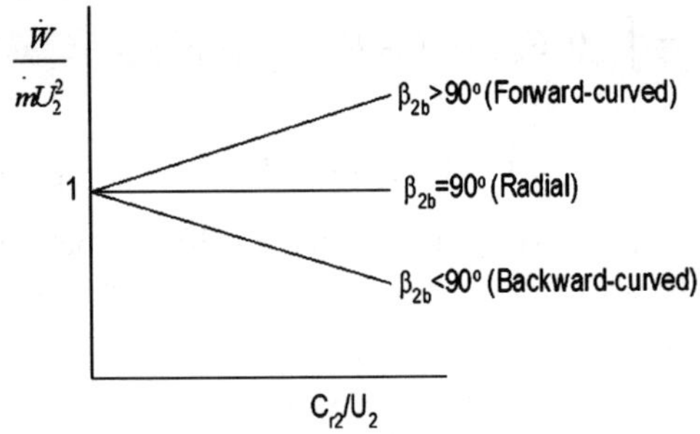


Figure 7: Linear relationship between specific energy transfer and flow rate for a given $\beta_{2b}$.

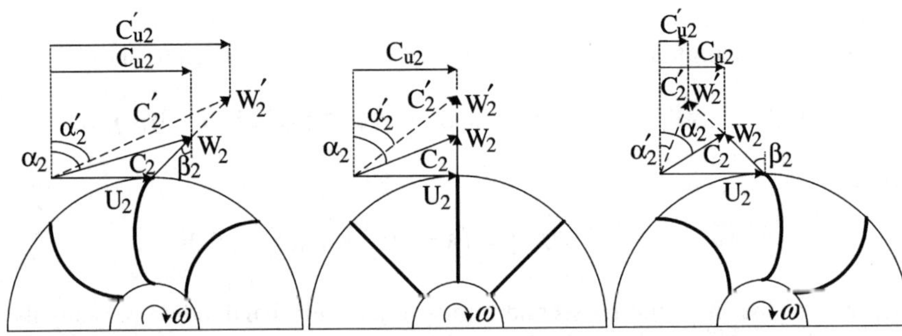

Figure 8: Exit velocity triangles for the three exit conditions.

Then the effect of impeller exit blade angle, $\beta_{2b}$ on the theoretical enthalpy rise becomes

$$\frac{\dot{W}}{\dot{m}} = U_2^2\left(1-\frac{C_{r2}}{U_2}\cot\beta_{2b}\right), \quad \text{since } C_{u2} = (U_2 - C_{r2}\cot\beta_{2b})$$

For a given value of impeller exit blade angle $\beta_{2b}$ there is a linear relationship between specific energy transfer and flow rate (Fig. 7). Theoretical enthalpy of impeller backward-curved blades decreases as the flow rate increases while that of impeller with the radial blades remains constant. The positive-slope condition can be unstable and cause surge (Fig. 8). For this reason a backward-curved blade impeller is generally preferred.

The relative velocity of the air in the impeller channels undergoes a rapid deceleration in the blade channel from the impeller inlet to the impeller exit. In addition to the function of transferring energy to the air, the impeller should act as an efficient diffuser. A badly shaped channel will interfere with the diffusion processes, causing flow separation at the impeller walls, leading to high impeller losses. To obtain gradual deceleration of the flow in the channel, the geometry of the impeller

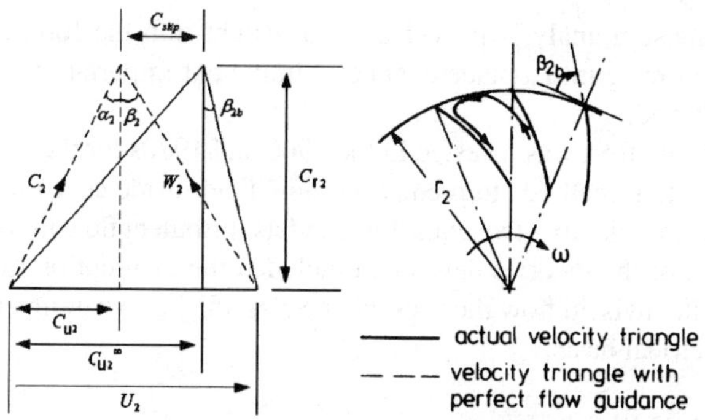


Figure 9: Definition of slip.

channels (defined by blade shape and hub and shroud profiles) should be chosen to give a smooth change in the relative Mach number along the mean flow path in the meridional plane. Consider the backswept impeller shown in (Fig. 9) rotating clockwise. The body of air passing through the channel between the two blades has inertia and will resist the turning effect of the impeller. The air therefore tends to rotate relative to the impeller channel, but in the opposite direction. Near the tip, this relative rotation causes the tangential component of absolute velocity ($c_{u2}$) to be less than the fluid leaving the impeller at the blade angle. Thus, the relative flow leaving an impeller is not perfectly guided by a finite number of blades, which is called *slip*, leading to the modified velocity triangle as shown in the figure below. The effect of slip is to reduce the magnitude of the tangential component of velocity, resulting in the reduction of the pressure ratio as well as compressor power consumption. Therefore, the slip effect requires a higher impeller rotating speed to obtain the same pressure ratio as the case of no slip, and this causes increased stress levels and relative velocities in the impeller, which finally leads to increased friction losses and reduced efficiency.

A slip factor in general form is defined by

$$\sigma = 1 - \frac{C_{\text{slip}}}{U_2} = 1 - \frac{C_{u2}^{\infty} - C_{u2}}{U_2}$$

From the velocity triangle (Fig. 9), the tangential component of velocity is reevaluated by

$$C_{u2} = \sigma U_2 + C_{r2} \tan \beta_{2b}$$

# 5 The development of 2D inviscid impeller flow theory

The flow in a centrifugal impeller is highly complex. It is three-dimensional, turbulent, viscous, and unsteady. The flow at the impeller exit is highly nonuniform and differs drastically from the 1D picture presented in 1D streamline theory. Although the viscous effect cannot be neglected when analyzing the flow in a centrifugal impeller, an inviscid analysis provides good insight into the forces applied to a fluid element moving into a blade channel. Simple but important physical knowledge can be gained.

A great deal of effort was invested in the 1960 and 1970s for the development of 2D flow calculation methods to predict the flow field inside centrifugal impellers. Because of the complexity of viscous flow models (turbulent flow in rotating curved passages) most of the theories developed included the assumption that the flow is not viscous. The inviscid flow theories can be classified into a number of groups:

On a geometrical basis,

- hub-shroud (H-S) theories,
- blade-to-blade (B-B),
- quasi 3D, and
- 3D.

The calculation method in each of these groups can be further classified on the basis of the computational scheme used:

- singularity methods,
- streamline curvature methods,
- finite difference methods, or
- finite element methods.

## 5.1 Impeller flow using 2D potential flow theory

The mathematically convenient simplification of a 2D, incompressible and inviscid flow field has been proven valuable in centrifugal impeller design, and has been employed by a number of notable authors since the early years of the last century. It was Kucharski [4] who pioneered this field of fluid dynamics by thoroughly examining the flow field of a simplified impeller fitted with straight radial blades with the inner tip placed at the center of the impeller. Spannhake [9, 10] presented improvements to this work by taking a more realistic inlet to outlet radius ratio for the impeller, and introduced the method of conformal mapping to solve the flow problem. Employing the method of conformal mapping Sorensen [11] and Busemann [3], treated the impeller, with logarithmic spiral blades. Uchimaru and Kito [12] applied the results of Schulz to compute slip coefficients, and Acosta [13] extended the work of Busemann by computing the pressure distribution along the blades and comparing the results with experiments. The above authors all contributed significantly to the solution of the potential flow problem for a 2D impeller flow. However, results were mostly obtained by numerical evaluation and computation; solutions in closed-form were limited to special cases only.

## 5.2 Impeller flow using 2D inviscid flow theory

Following the potential flow theories, H-S streamline curvature techniques were the next to be developed. Being based on first order ordinary differential equations they required relatively modest computer storage and simple numerical schemes. The price paid for this simplicity was the loss of ellipticity. Examples of these techniques were published by Katsanis [14], Novak [15, 16], Davis [17], Smith [18], and Adler and Iberg [19]. H-S finite difference methods were based in most cases on Wu's theory [20]. Typical solutions were developed by Wu [21], Marsh [22], Davis [23], Katianis and McNally [24, 25]. All the finite difference methods were physically more correct than streamline curvature methods because they are elliptic and therefore mathematically compatible with the subsonic elliptic flow field they ought to describe. The same is true for the H-S finite element methods, which were also based on Wu's formulation [20]. Here three methods can be mentioned: Adler and Krimerman [26], Oates [27], and Hirsch and Warzee [28].

Solutions on the B-B stream surface were essentially not much different from H-S methods because the differential equations of the B-B surface are basically analogous to the H-S equations. The major difference is that on B-B surfaces the domains in which the solutions have to be found lack well defined boundaries at the inlet to, and exit from the blade passages. These boundaries were given on the H-S surface by the hub and shroud extensions in up, and down-stream directions. In contrast, the B-B boundaries in form of the leading and trailing edge stagnation streamlines were determined in the course of the solution. Most frequently used in the B-B flow problem were, finite difference solutions. Stanitz was a pioneer in this approach [29–31], followed by Katianis and McNally [32–34] and Smith and Frost [35].

## 6 Current impeller design using CAD and CFD interactions

The emergence of CFD in the last 25 years provided a major impetus to solve the Euler and Navier-Stokes equations, which also govern the flow fields in turbomachines. This has been possible mainly due to advances in, grid generation, turbulence modeling, boundary conditions, pre- and post-data processing, and computer architecture. Most of the techniques used for the solution of the Navier–Stokes equations can be classified as finite difference, finite area/volume finite element and spectral methods. Only the first two techniques are widely used in turbomachinery.

Compressor designers and analysts require CFD software to provide accurate solutions to their applications of interest in a cost-effective and timely manner. With the widespread availability of affordable workstation class computing, the main issues are therefore accuracy and productivity. The information provided by the CFD must be of sufficient accuracy to allow the designer to make appropriate decisions based on the available information, and it must be available rapidly enough to fit within the time scales of the design cycle. For the designer, "accuracy" means providing reliable information of the following type:

- Qualitative: correctly reproduces the important flow features, such as swirl, boundary layers, shocks, wakes, separation zones, stagnation points, mixing layers, etc.
- Quantitative: efficiency, work input, pressure rise, blade profiles, component loss coefficients, flow distortion parameters, incidence, deviation or slip, etc.

While it is desirable that such features are perfectly reproduced, this expectation is unrealistic given limitations due to mesh size, turbulence modeling and other modeling assumptions such as steady state flow. To be used confidently, it is therefore of importance that CFD provides:

- "Sufficient" level of accuracy.
- Repeatability and consistency.

The two factors that influence accuracy most for compressor applications are the discretization accuracy of the flow solver and the computational mesh. Only after these two are adequately treated do other issues such as the turbulence model become of primary significance. In particular, the turbulence model is the usual scapegoat for poor CFD predictions, while in reality other causes are often more significant. Hence, it is often the case that "numerical errors" exceed "model errors".

## 6.1 Current impeller design trends

Even though the design of a centrifugal impeller has now advanced significantly, the design starting point is still the classical 1D and streamline-curvature theories.

- The design of bladed components typically begins in simplified codes such as 1D and streamline-curvature codes.
- This 1D flow design calculations uses empirical data from the literature and test results from a previous design to give preliminary results about the impeller performance, as well as impeller and diffuser geometrical parameters. The calculation routine is sequential, iterative and quick with modern personal computers.
- For the 1D calculations, input data include the impeller pressure ratio, inlet temperature, inlet pressure, fluid properties, rotational speed and mass flow. Simple 1D codes includes variables that can be altered later: specific speed, different geometrical ratios, number of blades in the impeller and diffuser, loss constants, inlet flow angles, volute pressure recovery factor, etc.
- 1D calculation of impeller efficiency is mostly based on measured data such as Balje [36]. The radial, vaneless diffuser is calculated analytically, including viscous effects, and also the volute is calculated analytically. Between the diffuser and the volute there is a semi-empirical loss coefficient model, based measurements in the literature. Reynolds number correction for efficiency is done by referring to literature such as Casey [37].

1D calculation results include:

- Pressure, temperature, enthalpy, density at inlet and outlet of the impeller, inlet and outlet of the diffuser and volute.
- Pressure, temperature and enthalpy must be given both in static and in total/stagnation values.
- The main geometry parameters for the impeller, diffuser and volute, as well as impeller specific speed and power consumption.

- Isentropic efficiency and pressure ratio in each point, pressure recovery in the diffuser and volute.
- Velocity triangles, deviation angle and relative Mach number in the impeller wheel inlet and outlet.
- Advanced 1D codes also estimate off-design performance, surge point and produce the impeller performance map.

Next follows the generation of the impeller blade geometry. The main geometry data of the impeller, such as the main dimensions and the blade angles obtained from 1D flow design are used as input data for 3D modeling. The blade design is started with defining the camber lines for the hub and shroud of an impeller blade. A 3D modeling software provides a graphical blade design environment for modifying the camber line geometry. Such a program is exclusively used for geometrical design; no flow calculation is done at this stage. Geometry specification usually takes place in special-purpose codes, as opposed to full CAD packages.

The next stage is to define the thickness distributions for the hub and shroud of the blade, and they can be modified graphically, as well. Based on the camber lines and the thickness distributions, the suction and pressure side co-ordinates for the hub and shroud are finally calculated. The blade can be displayed using various projections and enlargements of blade details can be displayed, as well. When the blade designer accepts the blade geometry, the numerical geometry data can then be generated for impeller flow CFD analysis: finite element method (FEM) based stress and vibration analysis (FEM) and computer-aided manufacturing (CAM). If any non-conformity is found in the blade geometry, or the results from either the CFD or FEM analysis do not meet the requirements, the design procedure must be repeated partially or totally, even starting from 1D flow design.

Specialized grid generators are now available, which take data directly from the geometry generation codes and provide automated templates for rapidly building grids. Non-axisymmetric components such as inlets and volutes are often developed in CAD packages or specialized codes. The CAD packages can export the data in IGES or in native CAD format, which can now be read into some CFD preprocessors and grid generators.

On the post-processing side, customized automated tools are now available. Postprocessors contain post-processing "macros" (a sequence of post-processor commands) which can provide both quantitative information and graphical images. Performance predictions for individual components or for the entire model can be obtained.

Available information includes averaged total and static temperature and pressures, loss coefficients, efficiency, flow distortion etc. In addition, useful circumferentially averaged meridional profiles of a variety of quantities can be obtained. Qualitative plotting of fully 3D, blade-to-blade and circumferentially averaged quantities can be automated, providing the analyst a detailed view of the flow field, enabling visual assessment of its quality. The high complexity of the flow, especially in the rotating impeller, makes the CFD modeling very difficult. Much research has been done in analyzing the flow and the different viscous phenomena. The selection of the turbulence model has a great influence on simulation accuracy, particularly if there is instability in the real flow. When time-dependent calculation of the whole impeller becomes a routine practice (at the moment it requires a lot of computation time), the accuracy of modeling obviously increases. The performance of the final impeller geometry is tested in the test facility.

Industrial designers and analysts require CFD software to provide accurate solutions to their applications of interest in a cost-effective and timely manner. With the widespread availability of affordable workstation class computing, the main issues are therefore accuracy and productivity (productivity also being directly related to cost). The information provided by CFD must be of sufficient accuracy to allow the designer to make appropriate decisions based on the available information, and it must be available rapidly enough to fit within the time scales of the design cycle.

Productivity in the usage of CFD has improved significantly over the past five years, for several reasons, including:

- More powerful computing, including PCs.
- The development of parallel processing.
- Links between CFD codes and other design tools, and with CAD.
- Customized and semi-automated grid generation tools.
- Specialized pre- and post-processing tools and environments.

The more powerful computing now available enables the analysis of larger grids and more components together, in a shorter period of time. The recent availability of parallel computing further expands the size of model that can be analyzed, and reduces turnaround time. The results are not only of higher quality, but obtainable sufficiently quickly to be useful within the design cycle. Figure below shows a typical CFD.

## 6.2 CAD system approach

With current commercial CAD and CFD systems, impellers with complex geometry can be analyzed. A typical CAD system approach is outlined in Fig. 10. The design process generally flows from a CAD generated solid model that is exported via IGES or STEP to a mesh generation package. The mesh generation package can either be contained within a commercial CFD system or an independent program. Once inside the mesh generation system, the mesh spacing and topology must be defined. This work also involves fixing errors in the geometry that results from importing IGES or STEP files. Critical areas of the mesh are examined and refined before applying the boundary conditions and CFD processing commences. Upon successful convergence, the results are examined and appropriate modifications are made to the geometry and the entire process is repeated. However, this "CAD System Approach" has one major drawback, it has a high total cost of analysis. This results from the extensive time required to repeatedly iterate on geometry until an acceptable design is achieved. Typically, the time to generate a 3D solid model to CFD processing is on the order of 80–120 hours. This time is a function of part complexity and ease of use of the CAD system. To reduce the complete process time, often the aerodynamic designer must know how to use and manipulate a 3D CAD system. This can be difficult and time consuming considering the complex volute geometry. A designer fluent in 3D CAD systems may also

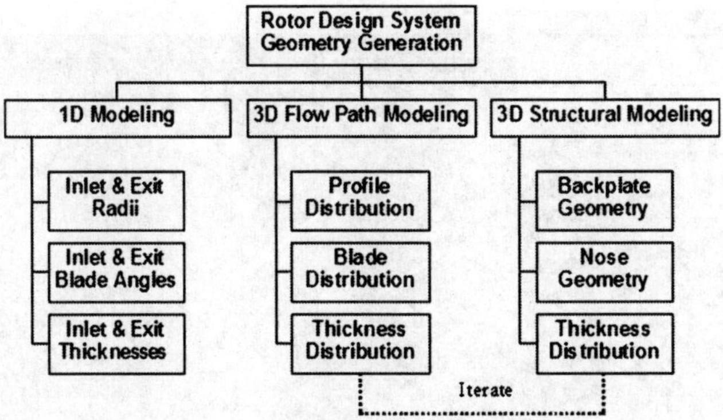


Figure 10: CAD system approach.

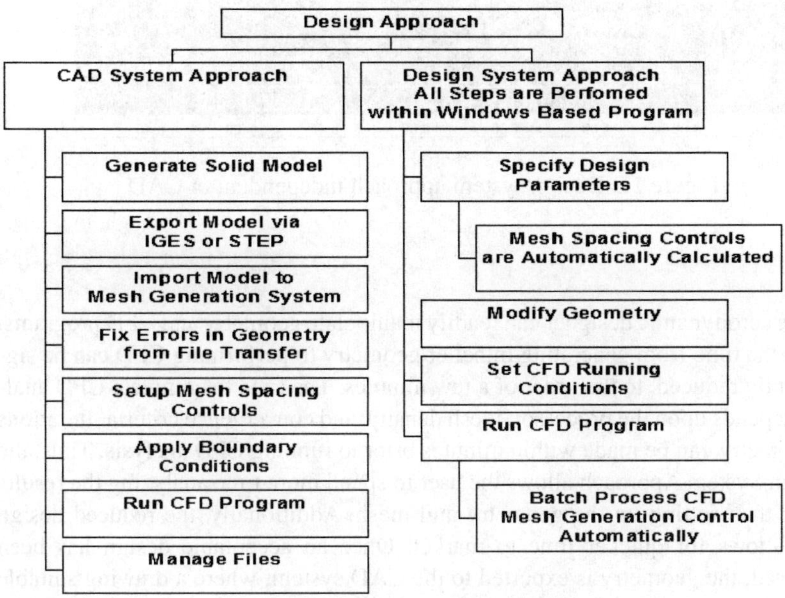


Figure 11: Design system approach.

be available, but at the expense of increases project resources by increased interaction time between the aerodynamicist and the designer.

### 6.3 Design system approach

In the design system approach shown in Fig. 11, geometry generation, mesh node spacing and boundary conditions are completely defined within an independent program that does not have a dependence upon CAD, shown in Fig. 12. Also, 3D geometry viewing is possible (Fig. 13).

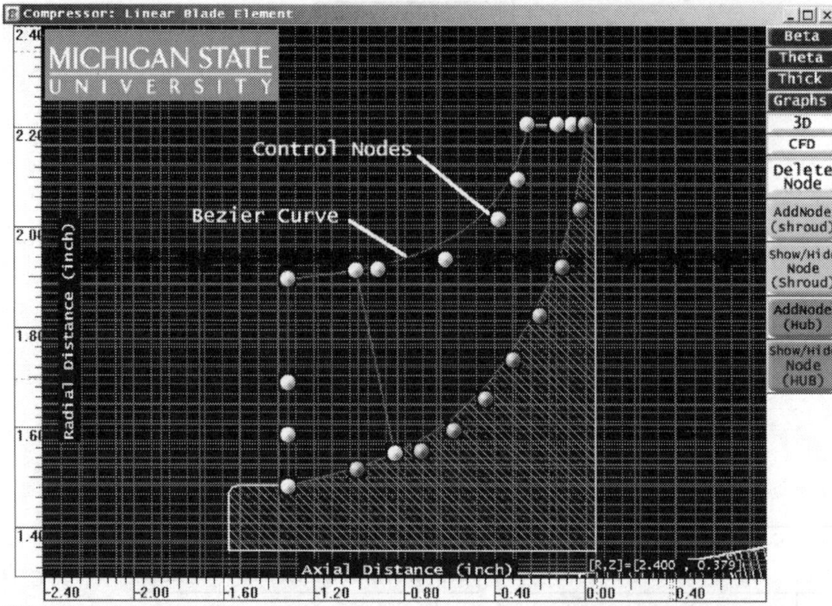


Figure 12: Design system approach independent of CAD.

The aerodynamic designer can readily manipulate geometry and CFD programs. Thus, the time from generating impeller geometry to performing CFD can be significantly reduced, to the order of a few minutes. The time to complete CFD analysis depends upon the processor, mesh density, and convergence criteria. Iterations in geometry can be made within minutes prior to running CFD analysis. Thus, the Design System Approach allows the user to spend more time analyzing the results rather than setting up the geometry and mesh. Additionally, the reduced design time allows for quicker time to market. Once an acceptable design has been achieved, the geometry is exported to the CAD system, where a drawing suitable for part procurement is produced. In this methodology, CAD is used as final documentation tool, not an interactive design system. While this approach may significantly reduce iteration design time, it requires the development of a geometry generator. The geometry generator development is heavily dependent upon Cartesian mathematics and convergence schemes.

The rotor blade geometry is comprised of a series of 3D streamlines. These 3D streamlines are determined from a series of mean-line distributions and are used to form the rotor blade surface. The design streamline distributions must be smooth for aerodynamic and structural analysis. One method of ensuring that design streamlines are smooth is through the use of Bezier polynomials.

Bezier polynomials are flexible smooth continuous curves that are controlled by the locations of node points as shown in Fig. 5. It should be noted that the end node

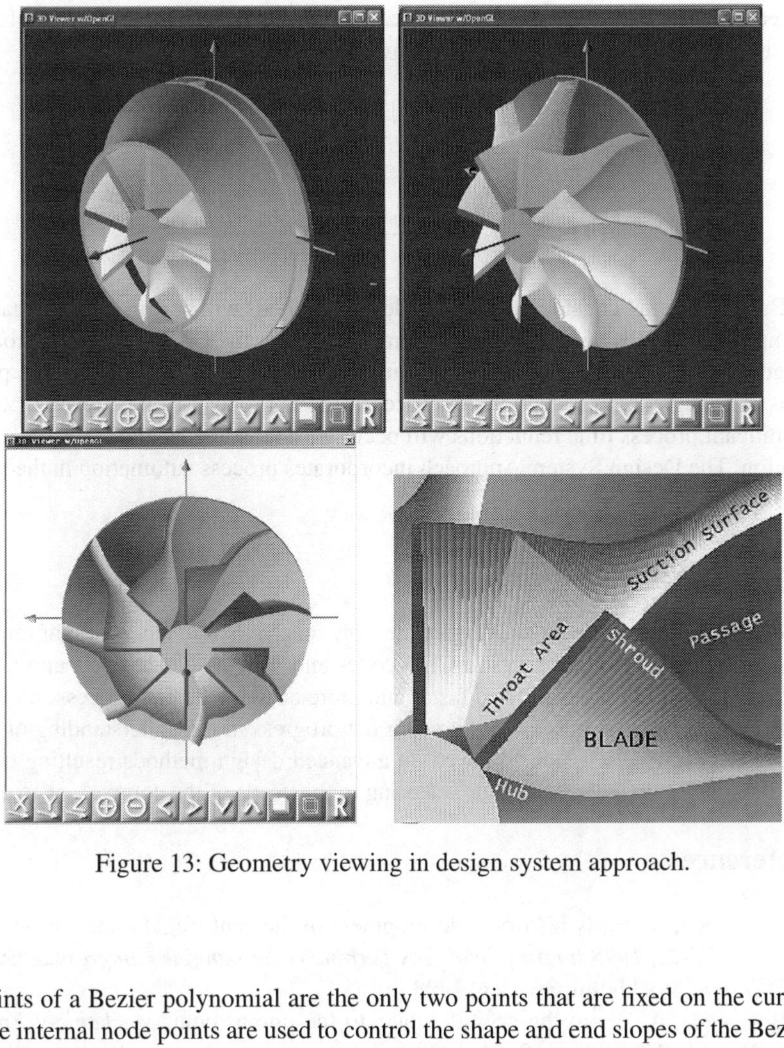


Figure 13: Geometry viewing in design system approach.

points of a Bezier polynomial are the only two points that are fixed on the curve. The internal node points are used to control the shape and end slopes of the Bezier curve. The general formulation for 2D Bezier polynomials is given by eqns. (1) and (2).

$$x(u)\Big|_{u=0}^{u=1} = \sum_{i=0}^{n_{\text{node}}} \frac{n_{\text{node}}!}{(n_{\text{node}} - i)!\,i!} u^i (1-u)^{n-i} X_{\text{node}\,i} \tag{1}$$

$$y(u)\Big|_{u=0}^{u=1} = \sum_{i=0}^{n_{\text{node}}} \frac{n_{\text{node}}!}{(n_{\text{node}} - i)!\,i!} u^i (1-u)^{n-i} Y_{\text{node}\,i} \tag{2}$$

Additional node points can be added to the Bezier polynomial without changing the original Bezier curve. This results in node locations that have less influence on other parts of the Bezier curve. Thus, certain regions of a distribution can be fined tuned without affecting the entire curve. The formula for degree elevation for a Bezier polynomial is given below in eqns. (3) and (4).

$$x(n)\Big|_{n=0}^{n_{\text{node}}+1} = \frac{nX_{\text{node }n-1} + (n_{\text{node}} + 1 - n)X_{\text{node }n}}{n_{\text{node}} + 1} \tag{3}$$

$$y(n)\Big|_{n=0}^{n_{\text{node}}+1} = \frac{nY_{\text{node }n-1} + (n_{\text{node}} + 1 - n)Y_{\text{node }n}}{n_{\text{node}} + 1} \tag{4}$$

Based on future considerations, both design methods will benefit from increased computer power by reducing analysis time. However, the CAD System Approach to generating a solid model and exporting to CFD will still follow the same process. Although slight reductions in process time will occur, it is believed that no significant process time reductions will occur without some level of process automation. The Design System Approach incorporates process automation in the system development.

## 7 Concluding remarks

The advances in the understanding of the flow mechanism in a centrifugal impeller and together with advances in CFD codes and their associated pre- and post-processing utilities have allowed faster and more accurate design processes.

A chronology was presented in tracing the progress in the understanding of the impeller flow. A discussion followed on advanced design methods resulting from the impeller knowledge and in turn catering to the needs of the compressor world.

## References


[1] Engeda, A., Early historical development of the centrifugal impeller, ASME 98-GT-22, *1998 International Gas Turbine & Aeroengine Congress & Exhibition*, Stockholm, Sweden, 1998.

[2] Engeda, A., From the crystal palace to the pump room. *Mechanical Engineering*, **121(2)**, pp. 50–54, 1999.

[3] Busemann, A., Das Förderhöhenverhältnis radialer Kreiselpumpen mit logarithmisch-spiraligen Schaufeln. *Z. angew. Math. u. Mech.*, **8**, p. 372, 1928.

[4] Kucharski, W., Stromungen einer Reibungsfreien Flussigkeit bei Rotation Fester Korper, R. Oldenbourg, 1918.

[5] König, E., Potentialströmung durch Gitter. *Z. angew. Math. u. Mech.*, **2**, p. 422, 1922.

[6] Spannhake, W., *Konforme Abbildung in Hydraulische Probleme*, VDI-Verlag: Berlin, 1926.

[7] Stodola, A., *Steam and Gas Turbines. Volumes I and II*, McGraw-Hill: New York, 1927.

[8] Fischer, K. & Thoma, D., Investigation of the flow conditions in a centrifugal pump. *Trans. ASME, Hydraulics*, **54**, pp. 141–155, 1932.

[9] Spannhake, W., *Kreiselraeder als Pumpen and Turbineno*. Julius Springer: Berlin, 1931.

[10] Spannhake, W., Problems of modern pump and turbine design. *Trans. ASME*, **56(4)**, April 1934.

[11] Sorensen, E., *Potential Flow through Centrifugal Pumps and Turbines*, NACA Technical Memorandum 973, April 1941.

[12] Uchimaru, S. & Kito, S., On potential flow of water through a centrifugal impeller. *Journal of the Faculty of Engineering*, **19**, pp. 191–223, 1931.

[13] Acosta, A.J. & Bowerman, R.D., An experimental study of centrifugal pump impellers. *Trans. ASME,* **79**, pp. 1821–1839, 1957.

[14] Katsanis, T., Use of arbitrary quasi-orthogonals for calculating flow distribution in the meridional plane of a turbomachine, NASA TN D-2546, 1964.

[15] Novak,. A., Strain line curvature computing procedures for fluid flow problems. *Trans. ASME*, 1967.

[16] Novak, A., *The Mean Stream-Sheet Momentum Continuity Solution Techniques for Turbomachinery*, Iowa State University, July 15–25, 1975.

[17] Davis, W.R., *A Comparison of Different Forms of the Quasi-Three-Dimensional Radial Equilibrium Equation of Turbomachines*, Carleton University Report No. ME/A 65-I, 1968.

[18] Smith,. L.H., The radial equilibrium equation of turbomachinery. *Journal of Engineering for Power*, **88**, 1966.

[19] Adler, D. & Ilberg, M., A simplified method for calculation of the flow field at the entrance of a radial or mixed flow-impeller, ASME Paper No. 70-FE-36, 1970.

[20] Wu, C.H., A general theory of three-dimensional flow in subsonic and supersonic turbomachines of axial- radial- and mixed flow types, NACA TN 2604, 1952.

[21] Wu, C.H. & Curtis, A., Analysis of flow in a subsonic mixed flow impeller, NACA 114-2749,1952.

[22] Marsh, H., A digital computer program for the through-flow fluid mechanics in an arbitrary turbomachine using a matrix method, RAM No. 3509, 1968.

[23] Davis, W.R., *Axial Flow Compressor Analysis Using a Matrix Method*, Carleton University, Report No. MW/A 73-1, 1973.

[24] Katsanis, T. & McNally, W.D., Fortran program for calculating velocity and streamlines on the hub-shroud mid channel flow surface of an axial- or mixed-flow turbomachine, NASA TN D-7343, 1973.

[25] Katsanis, T. & McNally, W.D., Revised Fortran program calculating velocity and streamlines on the hub-shroud mid channel stream surface of an axial-, radial-, or mixed-flow turbomachine or annular duct, NASA TN D-8430/1, 1977.

[26] Adler, D. & Krimerman, Y., The numerical calculation of the meridional flow field in turbomachines using the finite elements method. *Israel Journal of Technology*, **12**, 1974.

[27] Oates, G.C., Knight, C.J. & Carey, G.F., A variational formulation of the compressible through flow problem, ASME Paper No. 75-GT-32, 1975.

[28] Hirsch, C. & Warzee, G., A finite element method for through flow calculation in turbomachines. *Journal of Fluids Engineering*, **98**, p. 403, 1976.

[29] Stanitz, J.D., Two dimensional compressible flow in conical mixed flow compressors, NACN TN 1744, 1948.

[30] Stanitz, J.D., Two dimensional compressible flow in turbomachines with conical flow surfaces, NACA TI 935, 1949.

[31] Stanitz, J.D. & Ellis, G.O., Two dimensional compressible flow in centrifugal compressors with straight blades, NACA TR954, 1950.

[32] Katianis, T., Computer program for calculating velocities and streamlines on a blade-to-blade stream surface of a turbomachine, NASA TN-4525, 1968.

[33] Katsanis, T. & McNally, W.D., Fortran program for calculating velocities in a magnified region on a blade-to-blade stream surface of a turbomachine, NASA TN D-5091, 1969.

[34] Katsanis, T. & McNally, W.D., Fortran program for calculating velocities and streamlines on a blade-to-blade stream surface of a tandem turbomachine, NASA TN D-5044, 1969.

[35] Smith, D.J.L. & Frost, D.H., Calculation of the flow past turbomachine blades. *Proc. Institution of Mechanical Engineers*, Vol. 184, Part 3G(II), 1969.

[36] Balje, O.E., *Turbomachines: A Guide to Selection and Theory*, John Wiley and Sons: New York, 1981.

[37] Casey, M.V., The industrial use of CFD in the design of turbomachinery, AGARD Lecture Series 195, Turbomachinery Design Using CFD, 1994.

# CHAPTER 11

# Thermal engineering in hybrid car systems

K. Suga
*Department of Mechanical Engineering,*
*Osaka Prefecture University, Japan.*

## Abstract

To understand thermal engineering aspects in the recent hybrid cars, this chapter summarizes modern hybrid car systems and their thermal management issues. Since the recent commercial hybrid cars employ a gasoline/diesel–electric hybrid system, thermal issues related to internal combustion engines and electric systems are major factors for improving their tank to wheel efficiency. Although the combustion issues are important, this chapter focuses on the thermal management of the electric systems including batteries, traction/generation motors and inverters. Topics on heat sinks and combined thermal management of several electric parts are also discussed.

## 1 Introduction

The "hybrid car" system uses multiple power generation for providing propulsive force Although it has a history of over 100 years and although there are several hybrid car systems, the recent commercial hybrid cars employ the gasoline/diesel–electric hybrid system, which uses an internal combustion (IC) gasoline/diesel engine and electric motors. In order to enhance the system efficiency and to maintain its superior performance, thermal engineering is very important for developing the hybrid car systems. In fact, the modern hybrid car systems require highly sophisticated thermal management since their power electronic systems generate a large amount of heat and their performance is greatly affected by their surrounding temperature. In order to understand thermal engineering aspects in the hybrid car systems, this chapter introduces a brief history, system configurations and thermal engineering aspects of the modern hybrid car systems.

## 2 History of hybrid vehicles

Due to their environmental benefits and recent public concern over the global warming, hybrid cars are now highly demanded by the market. Its history started more than 100 years ago. Ferdinand Porsche, a German automotive engineer, designed the first successful hybrid car using an IC engine to generate and supply electricity to four wheel-mounted electric motors in 1898. Its driving speed was up to 56 km/h, which was the speed record in those days. He also designed Tiger II, a German heavy tank of World War II, with the same powertrain system where gasoline engines powered electric motors via electrical generators.

Victor Wouk, an American engineer, produced a working prototype by replacing the powertrain system of a Buick Skylark, which had a large volume under hood, with a Mazda rotary engine, electric motors, batteries and regenerative brakes in 1972. After Wouk's struggle, there have been no noticeable activities of building hybrid cars until the 1990s.

In 1993, Hino Motors Ltd., a Japanese truck manufacturer, put a diesel-electric hybrid truck on the market. Since then, many Japanese hybrid cars such as Toyota Prius [1] and Honda Insight [2] have been released. These automobiles have a direct linkage from the IC engine to the driven wheels, so the engine also provides acceleration power. Such power train mechanisms are the most practical ones for modern hybrid cars.

In contrast to the early years, the modern hybrid car systems require highly sophisticated thermal management since their power electronic systems generate a larger amount of heat but their performance is greatly affected by their surrounding temperature.

## 3 Configurations of hybrid car powertrains

One of the classifications of the hybrid car systems is "mild" or "full" hybrids. In the mild hybrid system, the IC engine provides the main propulsion force with the assist of the electric motors whenever extra power is required. In this system, the electric motor cannot operate independently of the IC engine. The electric motor also works as a generator to charge batteries in a recharging/regenerative brake system. The Honda Insight employs this system, so does the Honda Civic Hybrid of 2003–2005.

Full hybrid employs an integrated control system of the electric motors, IC engine and batteries, and thus the electric motors can operate on their own under a low speed condition. As shown in Fig. 1, the vehicle uses only the electric motors and thus energy from the batteries when it starts. When the vehicle reaches higher speeds, the IC engine starts up and takes over. Both the IC engine and the electric motors are used to provide the power for hard acceleration conditions. Unlike the mild hybrid system, the full hybrid system generates and consumes electricity at the same time. The regenerative braking system recharges the battery when the vehicle is braked. The Toyota Prius, Ford Escape Hybrid [3], etc., employ this system.

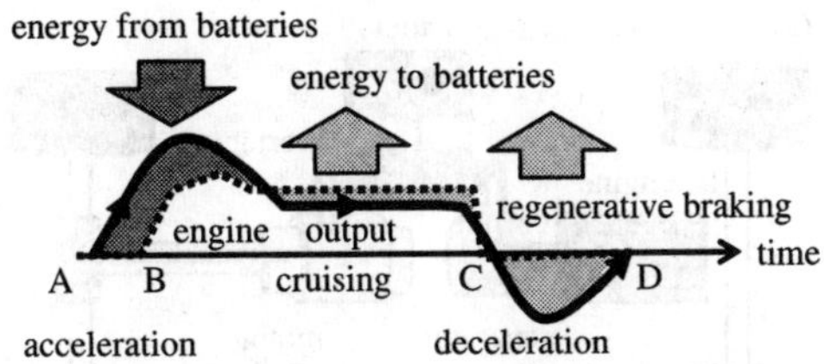


Figure 1: Energy management of the full hybrid cars: A – start (vehicle), B – start (engine), C – stop (engine), D – stop (vehicle).

Another classification can be made on the powertrain configurations. There are three major categories of the powertrain configurations of hybrid cars, which employ an IC engine and an electric powertrain system. They are called "series hybrid," "parallel hybrid" and "series-parallel hybrid."

### 3.1 Series hybrid

This is the simplest and oldest hybrid powertrain configuration originally employed by F. Porsche. (The system is similar to that of diesel–electric train locomotives.) In this configuration, as shown in Fig. 2a, the IC engine is not directly connected to the wheels and the electric motor is the only means of turning the wheels. The motor is supplied electricity from a computer controlled DC–AC inverter connected with both a battery pack and an electrical generator. The computer unit controls the power from the battery and/or the generator and excess electricity to charge the battery pack. A regenerative braking system is also used to charge the battery. When a large amount of power is required, electricity from both the battery pack and the engine-generator is supplied.

It is possible to tune the IC engine efficiently to operate at its most ideal speed whenever it is operating since the engine is only used to generate electricity. Consequently, the size of the engine can be typically smaller. However, the battery pack needs being more powerful than those in the other powertrain designs to provide driving power since the power delivered to the wheels is limited by the electric motors.

Although the omission of a mechanical connection between the IC engine-generator and the wheels provides flexible design, it requires separate motor and generator portions, which can be combined in the other design configurations.

This system is regarded to be particularly useful in driving cycles with many stops and starts, such as for delivery vehicles and city driving. It is also noticeable that some FC (fuel cell) cars use this series-style setup, replacing the engine-generator section with fuel cells. Such a car is called fuel-cell hybrid vehicle (FCHV).

### 3.2 Parallel hybrid

This powertrain system, employed in the Honda Insight, uses both an IC engine and an electric motor to generate the driving power of the wheels as shown in

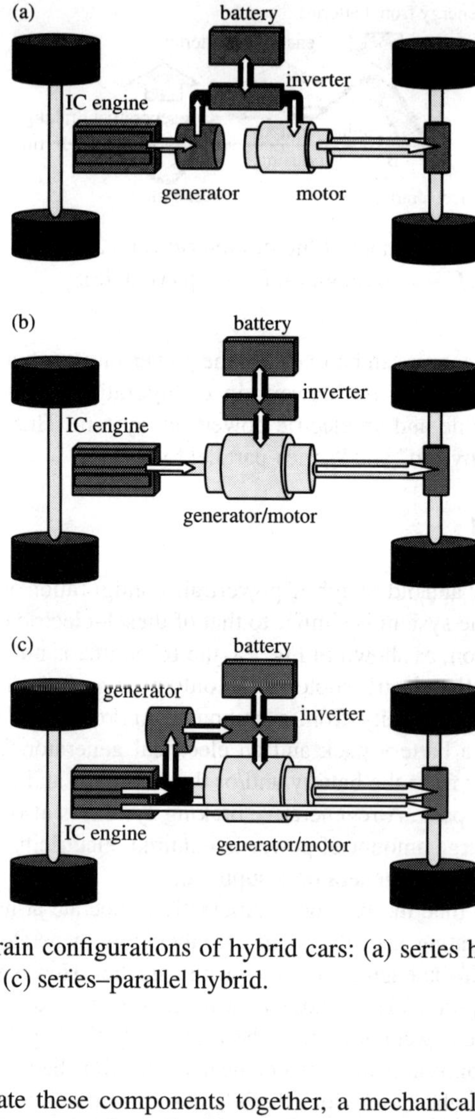


Figure 2: Powertrain configurations of hybrid cars: (a) series hybrid; (b) parallel hybrid; (c) series–parallel hybrid.

Fig. 2b. To operate these components together, a mechanical transmission and computer controls are required. In this configuration, it is possible to combine the electrical generator and the motor into one unit. The IC engine is the primary power source with motor assists in the cases of starts and acceleration. However, while the vehicle is cruising at a low speed, only the electric system operates and the IC engine is used to run the generator.

### 3.3 Series–parallel hybrid

This powertrain system combines the advantages of the parallel and series powertrain systems as in Fig. 2c. The engine can drive the wheels directly as in the

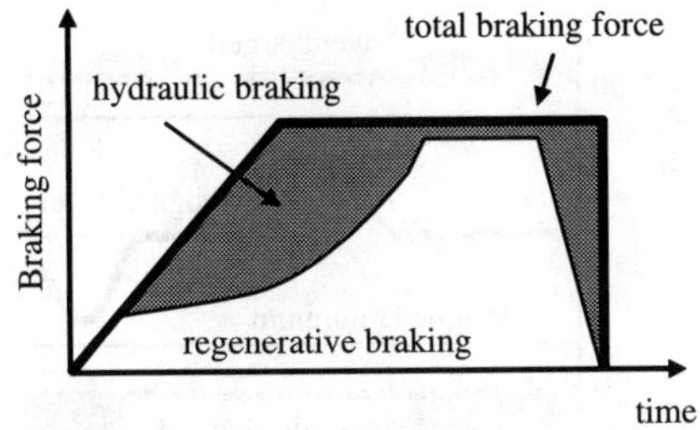


Figure 3: Cooperative regenerating brake system.

parallel powertrain and can also be effectively disconnected from the wheels so that only the electric motor powers the wheels as in the series powertrain in some occasions. The Toyota Prius employs this mechanism and so do some others. Consequently, the engine can almost always operate at near the optimum condition. At lower speeds, it operates more as a series hybrid, while at high speeds, where the series powertrain is less efficient, the engine takes over and energy loss is minimized. Although this system costs higher than the pure parallel hybrid due to its generator, larger battery pack, and more sophisticated controlling unit, the series-parallel powertrain has the highest potential to perform among the systems referred to.

It is clear that thermal management of the main components such as the battery, the electric motor, the DC–AC inverter, and the IC engine (which are common in any design configurations) is essential to achieve high performance of the hybrid car powertrain systems.

### 3.4 Regenerative braking system

In any power train configuration, a regenerative/recharging braking system is equipped in a modern hybrid car. When the vehicle is decelerated or braked, the system generates electricity by using the motors as generators. As illustrated in Fig. 3, the total braking force is obtained by both the regenerative braking and the conventional hydraulic braking. To obtain the maximum regenerating, the cooperative controlling system is optimized.

## 4 Technologies and challenges for thermal management of hybrid car system components

### 4.1 Battery pack

Although nickel-metal hydride (NiMH) batteries are normally used, Lithium ion batteries are expected to replace NiMH batteries in the near future. Irrespective of the battery types, the charging level is controlled within a designed range as in Fig. 4.

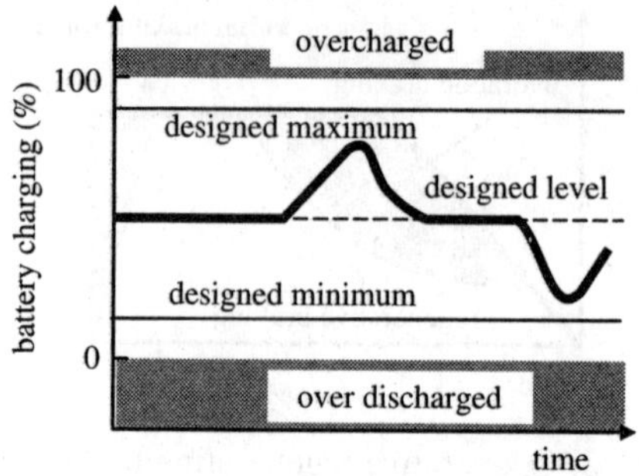


Figure 4: Example of a time history of battery charging in a hybrid car system.

Table 1: Permanent capacity loss ($\Delta_C$) of a Li-ion battery after 1 year usage.

| $T$ | $\Delta_C$ at 40% charging | $\Delta_C$ at 100% charging |
|---|---|---|
| 0°C | 2% | 6% |
| 25°C | 4% | 20% |
| 40°C | 15% | 35% |
| 60°C | 25% | 40% (after 3 months) |

Although a battery for a pure electric vehicle is fully charged, in a hybrid car system, the designed charging level is well under the full level (100%) to prolong the battery, which is frequently charged and discharged. The charging level reaches the designed maximum due to the recharging brake system when the car is braked whereas it goes to the minimum when the car is accelerated (the battery load is high).

In fact, the amount of permanent capacity loss, occurring after storage at a given charging level and temperature, varies [4] as shown in Table 1. It is clear, in Table 1, that the battery temperature needs controlling since a high temperature tends to lead to a large amount of capacity loss though the heat generation from the batteries is not very large (~100–150 W) [5]. In a hybrid car system is therefore equipped a battery cooling unit. It is normally by air cooling. An example of the Honda Insight is shown in Fig. 5. In this system, air from the passenger car compartment is used to cool the battery pack.

An additional air conditioning unit for the batteries is also possible. Fig. 6 shows an auxiliary air conditioning system [6] that provides cold air to a battery pack. It contains a separate evaporator, blower and air ducts, but is otherwise fully integrated into the passenger climate control system. The control scheme allows its automatic operation for any condition: on hot days, both the passenger air conditioner and the battery cooling module can operate simultaneously whilst on mild or cold days, either the passenger air conditioner or the battery cooling module can operate independently (the vehicle climate control system keeps passengers warm, while the battery cooling module keeps the battery cool).

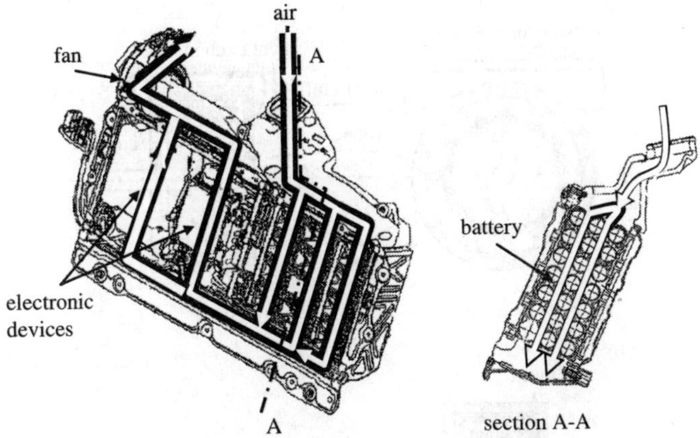


Figure 5: Battery cooling system of the Honda Insight.

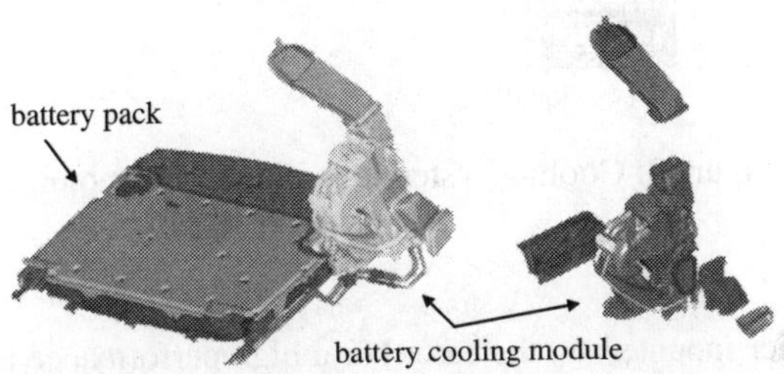


Figure 6: Auxiliary battery cooling unit [6].

## 4.2 Traction/generator motor

The thermal management of an electric traction/generator motor itself is for keeping its temperature under the limits, which are normally given by electric insulating materials used in the components. For example, the limit temperature of the coil is about 180°Cand that of the rotor magnet is about 120°C. A high-performance electric motor is estimated to produce heat of about 2500, 300 and 400 W due to its stator iron, copper and rotor losses under its design condition of 20 W/cm.[3]

One of the main issues of cooling motors is reducing the thermal resistance between the stator coil and the iron core to remove heat from the coil. For reducing the thermal resistance, the coil needs to be tightly packed after wiring to increase the contact surface area between the coated wire and the iron core. In fact, by doing so, the thermal resistance can be reduced to 15 K/kW from the standard value 50 K/kW.

Another more important issue is designing direct cooling system for the stator and the rotor by air/liquid. Cooling systems for a radial type motor are shown in Fig. 7. Figure 7a shows a system using air for cooling the stator and the rotor components. The air is recirculated inside the case to avoid getting outside dust and is cooled by

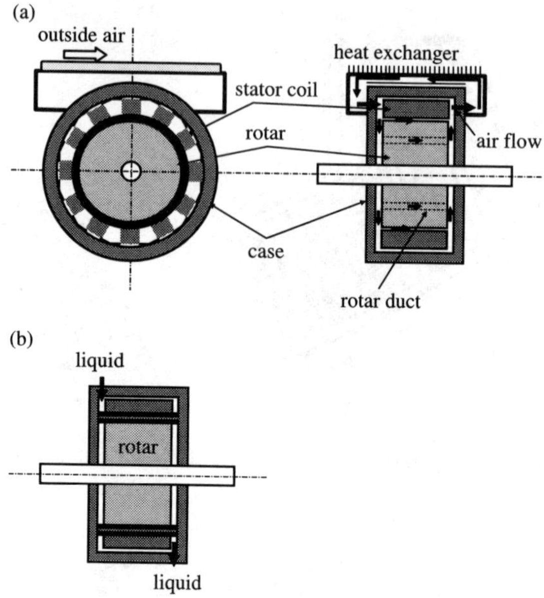


Figure 7: Cooling system of a radial type motor.

the heat exchanger mounted on the case. For a high performance motor/generator issuing much more heat, a liquid cooling system is coupled as in Fig. 7b.

An axial type motor has higher power density compared with that of the radial type and thus requires a liquid cooling system as shown in Fig. 8. The stator and the rotor are cooled by oil flows. Since heat generation from the rotor is <20% of the stator's, the required oil flow rate for the rotor is lower than that for the stator.

## 4.3 Inverter

Inverters and converters compose a power control unit (PCU) of hybrid cars. Currently, the volume of the PCU is about 9500 $cm^3$ due to the available space for it in engine rooms of sedans. Thus, one can see a linear relation between the motor output and the power density of the inverter [7] shown as the solid line in Fig. 9. However, the bigger and the more powerful motors are required, the smaller space remains for the PCU. Consequently, it is readily expected that a much higher power density (>10 $W/cm^3$) is required in the near future (shown as the broken line in Fig. 9) since required motor outputs are expected to be at least 100 kW. (In fact, the latest very powerful Toyota's hybrid sports utility vehicle (SUV) employs a motor whose maximum output is 123 kW [8] though there is a bigger PCU space in the SUV than those in sedans.)

Moreover, as shown in Fig. 10, for over 100 kW electric motor outputs, higher battery voltages and W/H (wire harness) currents are required. In order to

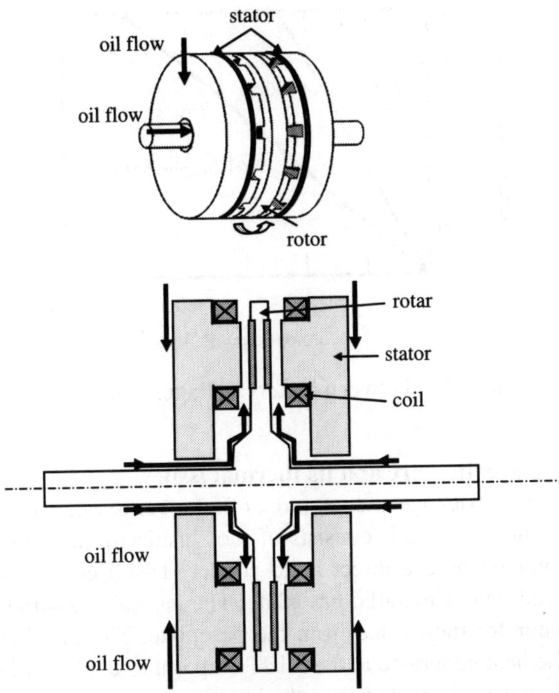


Figure 8: Cooling system of an axial type motor.

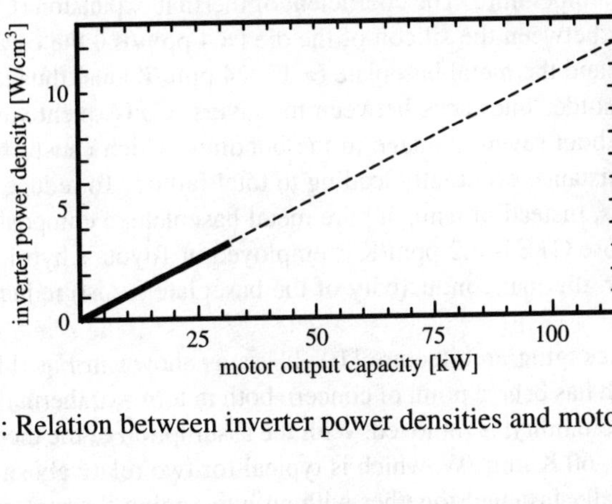


Figure 9: Relation between inverter power densities and motor outputs.

increase the W/H currents, it is necessary to minimize the distance between the motor and the inverter. This also requires smaller PCUs and thus their higher power densities.

Therefore, the thermal condition of an inverter is getting severer and its management may be the most important thermal engineering issue in hybrid car systems.

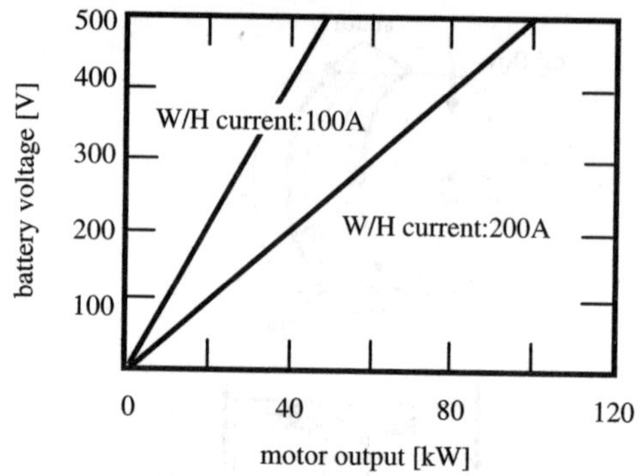


Figure 10: Relation between battery voltages and motor outputs.

### 4.3.1 Packaging architecture and its thermal issues

The currently applied package architecture of the power devices such as the inverters is illustrated in Fig. 11a. It consists of the insulated gate bipolar transistor (IGBT) chip soldered onto a direct bond copper (DBC) ceramic substrate that in tern is soldered onto a metallic baseplate. The ceramic substrate works as an electrical insulator for the device from the baseplate. The metallic baseplate is for spreading the heat generated at the IGBT chip out to the larger heat sink area. The heat sink is generally bolted onto the baseplate with an intervening thermal grease layer. Due to heating up and cooling down of the IGBT chip during its on/off cycle processes, each packaging layer expands and contracts at a rate proportional to its temperature. The coefficient of thermal expansion (CTE) is significantly different between the silicon of the die ($\simeq$ 4 ppm/K), the ceramic substrate ($\simeq$ 4–8 ppm/K) and the metal baseplate ($\simeq$ 17–24 ppm/K) and thus it leads to high stresses at the solder interfaces between the layers. Consequently, such repeated stresses bring about severe damage to the bonding, which causes higher electric and thermal resistance eventually leading to total failure. To reduce the thermally induced stresses, instead of using a pure metal baseplate, a composite material of Al and SiC whose CTE is 7.2 ppm/K is employed in Toyota's hybrids [9]. In such a case, however, thermal conductivity of the baseplate is also reduced to the half of copper's.

The latest packaging architecture [10, 11] is as shown in Fig. 11b. The metal baseplate, which has been a point of concern both in terms of thermal performance and long-term reliability, is removed. With the assumption of the thermal interface resistance of 20–60 K mm$^2$/W, which is typical for two relatively smooth flat surfaces mechanically fastened together with an intervening thermal grease layer, it can be estimated that reduction in operating temperature of 3–8 K is achieved per 1.0 kW heat dissipation, for a typical power module [10]. This corresponds to roughly 5–15% of the total temperature rise from the bottom of the baseplate to the die, which is significant improvement. This also leads to reduction of thermally induced stresses, which are the sources of reliability/durability issues.

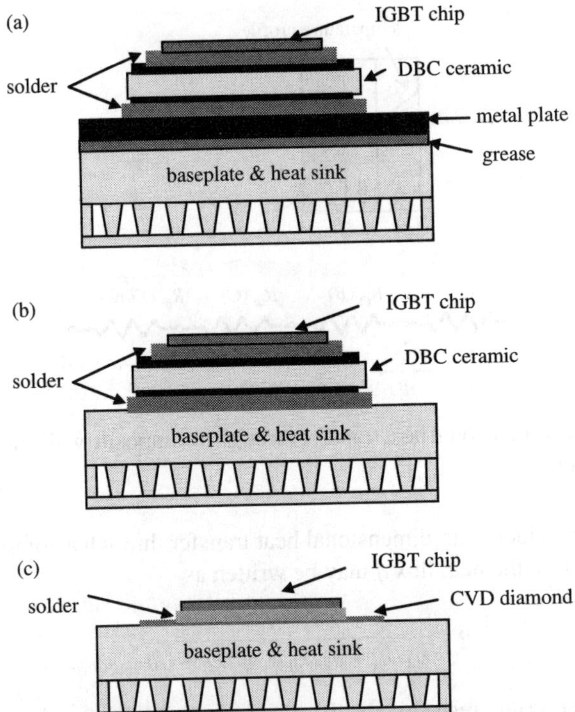


Figure 11: Chip-heat sink configurations: (a) standard architecture; (b) latest architecture; (c) future architecture.

Figure 11c is the ultimate packaging architecture currently considered [10]. It is an application of a thin diamond layer on the top of the baseplate surface by chemical vapor deposition (CVD). Diamond has the highest thermal conductivity (1–2 kW/(mK)) [12], which is 2.5–5 times as high as that of copper and 5–12 times as high as that of AlSiC. The use of CVD diamond would allow the heat from the silicon chip to spread out over the entire area of the baseplate with higher efficiency than is possible with either the first or second generation architectures shown in Figs11a and b. In addition, diamond is an electrical insulator and thus replaces the ceramic substrate, which exists exclusively to electrically insulate the power die from the rest of the package. This eliminates another layer from the packaging, providing additional enhancement of the thermal and mechanical performance.

#### 4.3.2 Thermal analysis of the power devices

In order to predict temperature distribution of a power device module, the finite element method (FEM) or the thermal resistance network analysis is usually applied. Since one can find many textbooks on the FEM, only the latter is reviewed here.

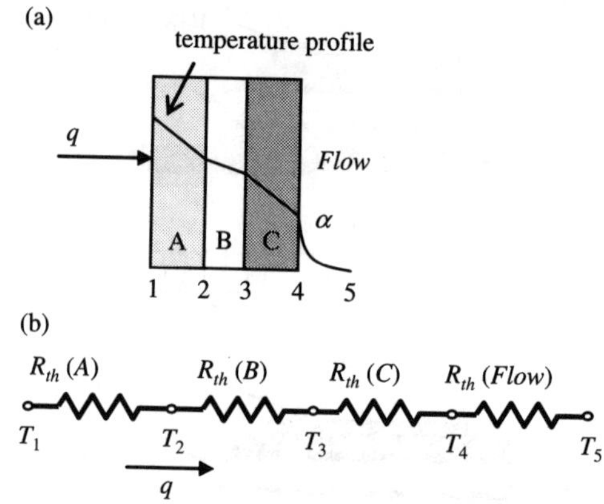


Figure 12: One dimensional heat transfer through a composite wall and its electrical analog.

When one considers one-dimensional heat transfer through a composite wall as shown in Fig. 12a, the heat flux $q$ may be written as

$$q = \frac{T_1 - T_5}{\delta_A/\lambda_A + \delta_B/\lambda_B + \delta_C/\lambda_C + 1/a}, \tag{1}$$

where $\delta_x$, $\lambda_x$ and $\alpha$ are, respectively, the thickness and the thermal conductivity of material $x$ and the heat transfer coefficient at the wall boundary. This form can be rewritten defining thermal resistance $R_{th}$ as

$$qS = \frac{\Delta T_{\text{overall}}}{R_{\text{th}}(A) + R_{\text{th}}(B) + R_{\text{th}}(C) + R_{\text{th}}(\text{Flow})}, \tag{2}$$

where

$$\begin{aligned} R_{\text{th}}(A) &= \delta_A/(\lambda_A S) \\ R_{\text{th}}(B) &= \delta_B/(\lambda_B S) \\ R_{\text{th}}(C) &= \delta_C/(\lambda_C S) \\ R_{\text{th}}(\text{Flow}) &= 1/(aS), \end{aligned} \tag{3}$$

for surface area $S$. Then it is readily see analogy between the above form and an electrical circuit as in Fig. 12b. In a more complicated composite wall case as in Fig. 13a, a parallel one-dimensional circuit shown in Fig. 13b can be applied to analyze heat transfer. (The thermal resistance in the parallel section is $1/[1/R_{th}(B) + 1/R_{th}(C)]$.) The thermal resistance network analysis is based on this analogy. For a power device module, a three dimensional network system can be applied as shown in Fig. 14. The top boundary may be connected with an electrical circuit which is a heat source nonlinearly depending on the thermal condition. Note that this analysis is based on simple analogy and thus more elaborate methods like the FEM should be applied for accurate predictions.

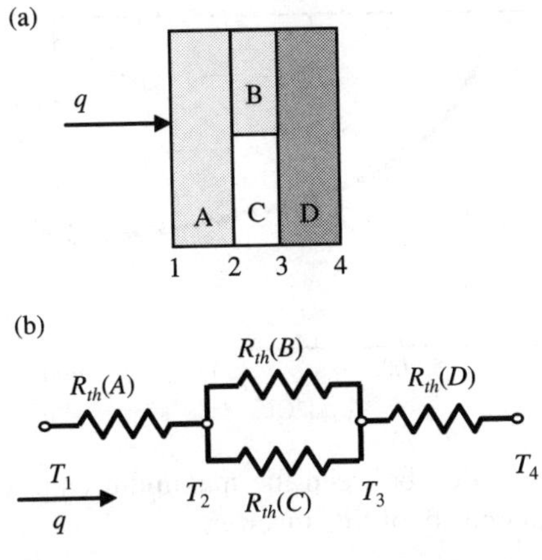


Figure 13: Parallel one dimensional heat transfer and its electrical analog.

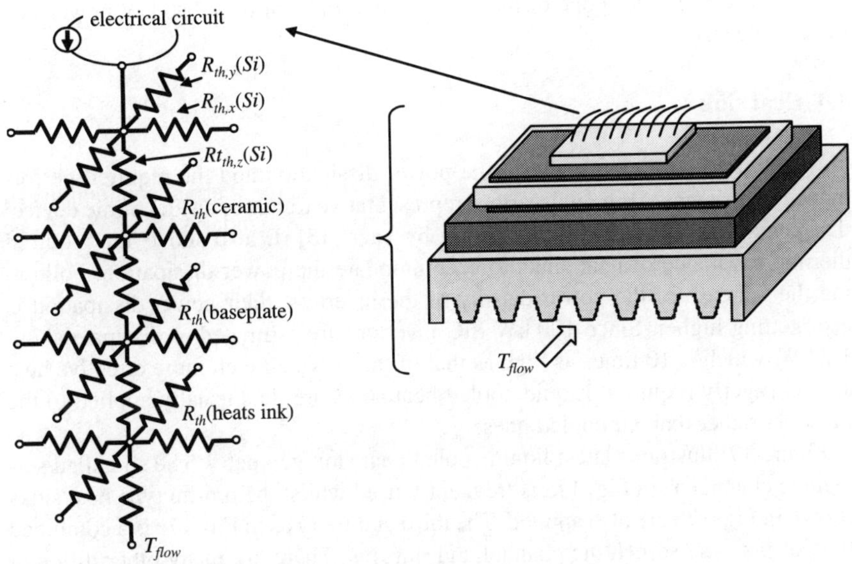


Figure 14: Thermal network analysis of a chip-heat sink configuration.

### 4.3.3 Thermal performance of SiC inverters

Since SiC power devices can work under more severe thermal conditions (>200°C) than those of the currently common Si devices, they are expected to be the next generation devices to meet the requirement for higher power densities. Energy Advancement Association of Japan [7] estimated the relation between the chip temperature and chip size/current density of 100 kW SiC inverters using the aforementioned simple thermal resistance model. The results are as in Fig. 15. At the maximum chip temperature of $T_{max}$ = 200°C, the chip size and the current

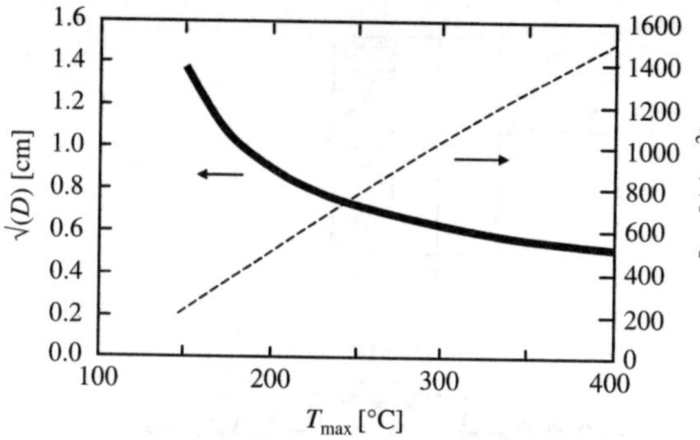


Figure 15: Estimated relation between the maximum chip temperature and chip size/current density of SiC inverter.

density would be $D = 0.9 \times 0.9$ cm$^2$ and $J_{max}$ = 500 A/cm$^2$, respectively. These values are well in the scope of manufacturing a PCU of over 100 kW power in a reasonable size.

### 4.4 Heat sink

To illustrate the relation between the power dissipation and the required performance of heat sinks, Fig.16 describes representative trends of an electronic device: the increase of CPU power dissipation by year [13]. It also shows the required thermal resistance of heat sinks to accommodate the power dissipation. Following the increase of the power density of the inverters, their power dissipation is also getting higher. Since 100 kW SiC inverters are estimated to dissipate about 1.2 kW, which is 10 times as high as that of the CPU, a much more effective heat sink is eagerly required. Liquid cooled heat sinks are thus usually applied to the inverters rather than air cooled ones.

Figure 17 illustrates latest liquid cooled heat sink geometry. The so-called serpentine channel as in Fig. 17a is frequently used whilst the pin-fin type heat sinks shown in Fig. 17b are also applied. The third geometry as in Fig. 17c is a combined type of both the serpentine channel and pin-fins. There are many other different pin–fin shapes and their arrangements depending on manufacturers and required performance.

The performance of pin-finned channels is, however, superior to that of the plate-finned ones. Figure 18 illustrates the geometry of the pin-finned and plate-finned channel heat sinks examined in [14]. The pin-finned channel has 73 pins (4 mm diameter) arranged in a staggered array whose streamwise and spanwise pitches are, respectively, 7.5 and 8.5 mm. The plate-finned channel has five plates whose thickness and spanwise pitch are 2 and 2.6 mm, respectively. Both are made of aluminum alloy whose thermal conductivity is 180 W/(mK) and the channel heights are 13 mm. Thus, the heat transfer surface areas are 12,836 mm$^2$ (pin-finned) and 30,380 mm$^2$

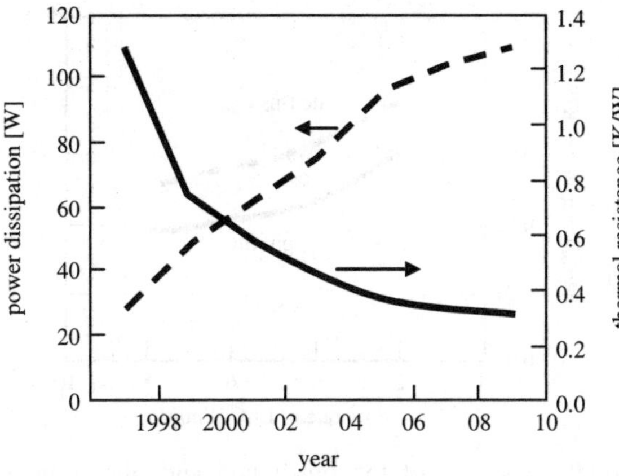


Figure 16: CPU power dissipation and required thermal resistance of heat sinks.

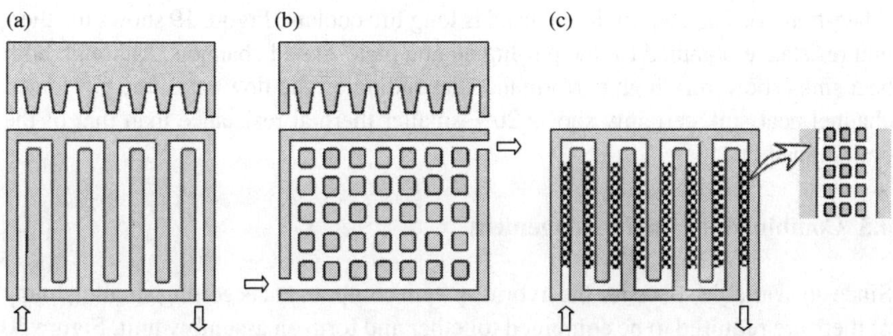


Figure 17: Liquid cooled heat sink geometry: (a) serpentine channel; (b) pin fin; (c) pin-finned serpentine channel.

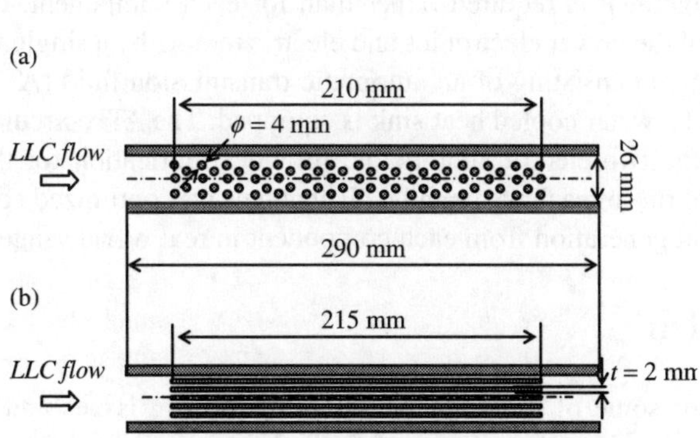


Figure 18: Finned channel heat sinks: (a) pin-finned channel heat sink; (b) plate-finned channel heat sink.

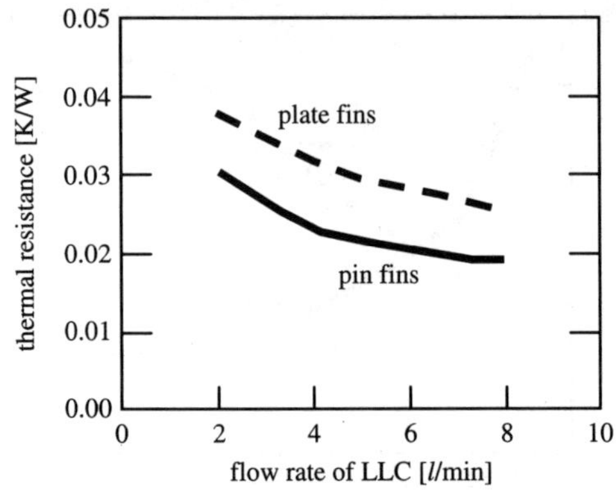


Figure 19: Thermal resistance of the pin-finned and plate-finned channel heat sinks.

(plate-finned). The cooling liquid used is long life coolant. Figure 19 shows the thermal resistance obtained for the pin-finned and plate-finned channels. Although both heat sinks show very high performance depending on the flow rates, the pin-finned channel heat sink certainly shows 20% smaller thermal resistance than that of the plate-finned one.

### 4.5 Combined thermal management

Since an available space for the hybrid system components is getting smaller, many of them are required to be combined together and form an assembly unit. Figure 20 shows the Aisin AW's HD-10 hybrid transmission [15, 16], which is employed in the Ford Escape hybrid. Traction and generator motors, a capacitor, a PCU and etc. are combined together to form a transmission unit there. In such a system, integral thermal management is required rather than for each component. In fact, in the HD-10, to cool the power electronics and electric motors by a single water cooling system, a system consisting of an automatic transmission fluid (ATF)/water heat exchanger and a water cooled heat sink is equipped. The ATF, circulated by an oil pump, cools the two electric motors and provides lubrication for the gear train. The number of fins of each water and ATF channels was optimized considering the balance of heat generation from each component in real world usage.

## 5 Conclusion

In this chapter, some of important thermal engineering issues and related technologies for designing components of the hybrid car systems are presented. As shown in Table 2, the current gasoline hybrid system enhances the tank to wheel efficiency of the conventional gasoline engine cars to 2.3 times higher. Therefore, the hybrid car systems are very effective technology to cope with the issues on

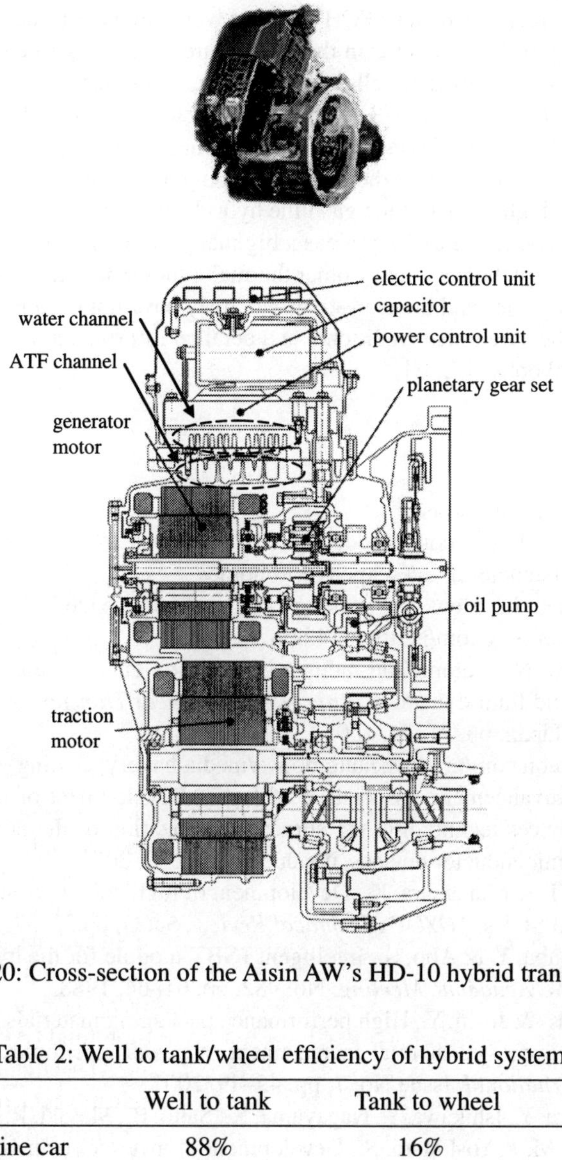


Figure 20: Cross-section of the Aisin AW's HD-10 hybrid transmission.

Table 2: Well to tank/wheel efficiency of hybrid systems.

| | Well to tank | Tank to wheel | Well to wheel |
|---|---|---|---|
| Gasoline engine car | 88% | 16% | 14% |
| Gasoline hybrid car | 88% | 37% | 32% |
| Hydrogen FCEV | 58% | 38% | 22% |
| FCHV | 58% | 50% | 29% |

the global environment and natural resources as well as economy. As discussed in this chapter, thermal engineering is certainly important to enhance the system efficiency and to maintain its performance for long term usage.

It must be supplemented that a FCHV system would replace the current gasoline/diesel engine hybrid car systems in the near future. The fuel-cell electric vehicle (FCEV) system employs fuel-cells instead of the IC engine in the series hybrid system as mentioned in Section 3.1. The hybrid system improves the efficiency of the FCEV. In fact, as in Table 2, the tank to wheel efficiency of the FCHV is enhanced to 50% from 38% of the FCEV. Although its current well to wheel efficiency is not as high as that of the gasoline hybrids, it has high potential since the well to tank efficiency of hydrogen has a big margin to be improved further.

Finally, although there are many other thermal issues in hybrid car systems such as those in IC engines and brake systems etc., they have not been included in this chapter since they are rather traditional and detailed information may be found in some other textbooks [17, 18].

## References

[1] www.toyota.com/prius/

[2] automobiles.honda.com/models/model_overview.asp? ModelName=Insight

[3] www.fordvehicles.com/escapehybrid/home/

[4] Buchmann, I., How to prolong lithium-based batteries, 2003, www.batteryuniversity.com/parttwo-34.htm

[5] Kobayashi, N., Automobile industry's action on environments and relevant thermal and fluid dynamics. *Proc. 43rd Nat. Heat Transfer Symp. of Japan*, Nagoya, Japan, pp. 71–74, 2006.

[6] www.visteon.com/products/automotive/media/battery_cooling_module.pdf

[7] Energy Advancement Association of Japan, Development of ultra low loss power devices technology: the survey on utilization of the next generation power semiconductor devices, conducted in 2002, 2003.

[8] Kikuchi, T. & Matsubara, K., Development of high output power control unit for hybrid SUVs. *TOYOTA Technical Review*, **54(1)**, pp.52–57, 2005.

[9] Nii, Y., Baba, Y. & Abo, S., Intelligent IGBT module for the hybrid vehicle. *Proc. JSAE Academic Meeting*, No. 982, pp. 61–64, 1988.

[10] Moores, K. & Joshi, Y., High performance packaging materials and architectures for improved thermal management of power electronics. *Future Circuits International*, Issue No. 7, pp. 45–49, 2001.

[11] Matsunaga, Y., Ishikawa, Y., Nagayama, K., Saito, R., Masaki, R., Hamada, H., Iwanaka, M. & Yoshihara, S., Development of inverter downsizing for HEV. *Proc. JSAE Academic Meeting*, No. 96-03, pp. 21–24, 2003.

[12] Lostetter, A., Barlow, F. & Elshabini, A., Thermal evaluation and comparison study of power base plate materials. *Advancing Microelectronics*, **25(1)**, pp. 25–27, 1998.

[13] Niekawa, J. & Kimura, Y., Introduction of micro-heat pipe and heat dissipation technology. *Nikkankogyoshinbun*, 1997.

[14] Matsuura, M. & Hirahara, T., Water-cooled heat sink, Unexamined publication number (patent) JP.2003-047258.A, 2003.

[15] www.aisin-aw.co.jp/en/02products/02at/05hev/index.html

[16] Hisada, H., Taniguchi, T., Tsukamoto, K., Yamaguchi, K., Suzuki, K., Iizuka, M., Mochizuki, M. & Hirano, Y., AISIN AW new full hybrid transmission for FWD vehicles, SAE Paper 2005-01-0277, 2005.

[17] Society of Automotive Engineers, *New Developments in Heat Exchangers for Automotive Design*, SAE International, 1997.

[18] Baumgarten, C., *Mixture Formation in Internal Combustion Engines (Heat and Mass Transfer)*, Springer: Berlin, 2006.